NOBEL LECTURES

PHYSIOLOGY
OR
MEDICINE
2006–2010

Nobel Lectures

Including Presentation Speeches and Laureates' Biographies

Physics

Chemistry

Physiology or Medicine

Literature

Economic Sciences

NOBEL LECTURES

INCLUDING PRESENTATION SPEECHES AND LAUREATES' BIOGRAPHIES

PHYSIOLOGY
OR
MEDICINE
2006–2010

EDITOR

Goran K. Hansson

Karolinska Institute, Sweden

W🌐 World Scientific

NEW JERSEY · LONDON · SINGAPORE · BEIJING · SHANGHAI · HONG KONG · TAIPEI · CHENNAI

Published by

World Scientific Publishing Co. Pte. Ltd.
5 Toh Tuck Link, Singapore 596224
USA office: 27 Warren Street, Suite 401-402, Hackensack, NJ 07601
UK office: 57 Shelton Street, Covent Garden, London WC2H 9HE

NOBEL LECTURES IN PHYSIOLOGY OR MEDICINE (2006–2010)

Published with permission from Nobel Media AB in 2014 by World Scientific Publishing Co. Pte. Ltd.
Nobel Prize® and the Nobel Prize® medal design mark are the registered trademarks of the Nobel Foundation.

ISBN 978-981-4630-20-7
ISBN 978-981-4630-21-4 (pbk)

Printed in Singapore by Mainland Press.

PREFACE

Nobel Week in December every year highlights the scientific and cultural accomplishments that are awarded with the Nobel Prize. It is a time of excitement not only for the Nobel Laureates who arrive in a wintry Stockholm, but also for the Committee members who have chosen them, for all those interested in science and literature, and for the general public that wishes to be part of the celebrations. Intense media coverage guarantees that everyone interested in science, literature, celebrities or festivities gets insights into the aspects of their particular taste.

The week from December 6 to 13 is loaded with activities ranging from symposia to concerts and from cocktail parties to visits to local schools. However, only two of these activities are mandatory for the Nobel Laureates according to the Statutes of the Nobel Foundation: to accept the Nobel Prize at the Nobel Prize Ceremony on December 10, and to give a Nobel Lecture that describes the discoveries awarded the Nobel Prize. The lectures are scheduled a couple of days before the Nobel Prize Ceremony, to allow the Nobel Laureates to enjoy the subsequent festivities in a more relaxed way.

The Nobel Lectures in Physiology or Medicine are given at Karolinska Institutet, the medical university that hosts the Nobel Assembly and the Nobel Committee for Physiology or Medicine. On the day of the Lectures, the main auditorium of the Institute is crowded with people — faculty members, postdoctoral fellows, students, and others — eager to hear the Nobel Laureates describe their work. Scientists all over the world are following the Lectures through webcast, adding to the excitement of the moment.

Records of the Nobel Lectures are kept as videotapes on www.nobelprize. org and in written form in the Nobel Lectures series (previously *Les Prix Nobel*). By reading the Nobel Lectures and autobiographic notes of the Nobel Laureates, the reader gets first-hand insights into the progress and breakthroughs of science over more than a century. Portraits of the Nobel Laureates, and the presentation speeches given by Nobel Committee members during the Nobel Prize Ceremony, add to the wealth of information available in this series.

The present volume contains Nobel Lectures, autobiographies, portraits, and presentation speeches by the Nobel Laureates in Physiology or Medicine for the years 2006–2010. The science contained herein presents a broad panorama of paradigm-shifting discoveries in medicine and life science. It includes fundamental principles of gene structure and regulation as represented by the discoveries of RNA interference and the telomerase machinery but also medical breakthroughs regarding the viral etiology of adult immune deficiency syndrome (AIDS) and cervical cancer, two major lethal diseases afflicting mankind. It ranges from breakthroughs in experimental genetics through the use of homologous gene recombination in embryonal stem cells, to the technology of in-vitro fertilization that offers new hope for infertile couples.

As a member of the Nobel Assembly and its Nobel Committee since more than 15 years and presently as its secretary-general, I have had the privilege of meeting the Nobel Laureates from the time period included in this volume and the excitement of hearing them present their Nobel Lectures. I hope that the reader of this book will feel the same excitement of learning first-hand how major discoveries were made and also get a glimpse of the individuals who made them.

Göran K. Hansson, M.D., Ph.D.
December 2014

CONTENTS

Physiology or Medicine 2006

Andrew Z. Fire and Craig C. Mello

"for their discovery of RNA interference – gene silencing by double-stranded RNA"

THE NOBEL PRIZE IN PHYSIOLOGY OR MEDICINE

Speech by Professor Göran K. Hansson of the Nobel Assembly at Karolinska Institutet.
Translation of the Swedish text.

Your Majesties, Your Royal Highnesses, Ladies and Gentlemen,

We live in an information society. A continuous flow of information reaches us through the news media and the internet. For modern man, the prioritisation and selection of information has become a necessary part of his survival strategy.

No information is more important to us than that which governs how we become humans. It instructs our stem cells to differentiate into nerve, blood and muscle cells, tells the organs how to develop and determines how we handle injuries and infections. That particular information is stored in our genome and it is used continuously in all our cells to make it possible for a nerve cell to function as a nerve cell and for a muscle cell to operate as a muscle cell.

It is remarkable that the entire instruction book on how to build a human being is present in each cell in our bodies. When reading that book, it is obviously of key importance that the muscle cell reads the muscle chapter only and does not end up as a frustrated nerve cell. Understanding how the genome is read has been a challenge for life science ever since Watson and Crick discovered, more than 50 years ago, that it is stored in the double helix of DNA.

It was soon established that genetic information is copied from DNA to a messenger RNA molecule, which in turn governs the production of proteins – the molecules that carry out the processes of life. And it was evident that if scientists could control the flow of information from DNA via RNA to protein, we would have marvellous tools at hand for use in medicine and biology.

Fifteen years ago, we thought we knew enough about the flow of genetic information to use it for practical purposes. But we did not achieve the expected results. Attempts to silence a gene in an experimental animal were sometimes fruitless, and attempts to use gene technology for improving the colours of flowers could even cause the plants to lose colour completely. These results perplexed the scientific community. Was there an unknown regulatory step on the way from DNA to protein?

This enigma was solved by the 2006 Nobel Laureates, Andrew Fire and Craig Mello. They suspected that RNA contained the solution to the problem and decided to test it in a simple model organism, the nematode worm *Caenorhabditis elegans*. Fire and Mello injected different types of RNA into the worms – and usually nothing happened. But they also made the ingenious decision to mix two RNA molecules in a test tube before injection. One RNA molecule was an exact copy of a messenger RNA and the other a mirror image of the messenger. In the test tube, the two RNA molecules bound to each other and formed a double strand. Injection of that double-stranded RNA led to the silencing of the gene. Fire and Mello had discovered a new mechanism for controlling the flow of genetic information.

In their brilliant paper from 1998, Andrew Fire and Craig Mello demonstrated that double-stranded RNA activates an enzymatic mechanism that leads to gene silencing, with the genetic code in the RNA molecule determining which gene to silence. Today, we call this mechanism RNA interference.

Continued research has shown that our cells use RNA interference to regulate thousands of genes. Through RNA interference, the pattern of gene expression is fine-tuned in such a way that each cell uses precisely those genes that are needed for building its proteins. Today we also know that RNA interference helps to protect us against viruses and jumping genes. Finally, RNA interference can be used to control gene expression in the laboratory – and hopefully soon also in clinical medicine.

Professor Fire and Professor Mello,

Your discovery of RNA interference has unravelled a new principle for regulating the flow of genetic information. It has added a new dimension to our understanding of life and provided new tools for medicine. On behalf of the Nobel Assembly at Karolinska Institutet, I wish to convey to you our warmest congratulations and I ask you to step forward to receive the Nobel Prize from the hands of His Majesty the King.

Andrew Z Fire

ANDREW Z. FIRE

Andrew Zachary Fire was born on April 27, 1959 at Stanford University Hospital in Santa Clara County California. Spending most of his early years (until age 16) in nearby Sunnyvale, he attended the local public schools: Hollenbeck Elementary School (1964–1970), Mango Junior High School (1970–1972), and Fremont High School (1972–1975).

Fire enrolled at University of California at Berkeley in the Fall of 1975, receiving an AB degree in Mathematics in 1978. Fire then entered the Ph.D program in Biology at Massachusetts Institute of Technology as a National Science Foudation Fellow in the Fall of 1978. Fire's Ph.D. thesis, titled "In Vitro Transcription Studies of Adenovirus", was submitted in 1983.

From 1983 to 1986, Fire recieved training in the *Caenorhabditis elegans* group at the Medical Research Council Laboratory of Molecular Biology in Cambridge, England as a Helen Hay Whitney Foundation Fellow. During this time, Fire initiated research directed toward improvement of microinjection technology and development of assays for expression of foreign DNA in *C. elegans* worms.

During his last year at the MRC lab, Fire applied for a research position at the Carnegie Institution of Washington's Department of Embryology in Baltimore Maryland, also applying for an independent research grant "Gene Regulation during early development of *C. elegans*" from the US National Institutes of Health. Both applications were successful and Fire moved to Baltimore in November of 1986. From his arrival at the Carnegie until 1989, Fire held the title of Staff Associate, an independent research position that was designed to facilitate the development of novel research programs in the absence of additional academic responsibilities. In 1989, Fire was appointed as a regular staff member at the Carnegie, with his group continuing to develop DNA transformation technology and collaborating on a number of studies to understand the molecular basis of gene activation in muscle cells. Along with the appointment as a full staff member at Carnegie Institution, Fire also acquired an adjunct appointment as a faculty member in the Department of Biology at Johns Hopkins, where he was involved in both graduate and undergraduate teaching and mentoring.

In 2003, Dr. Fire moved back to Santa Clara County, taking a position at the Stanford University School of Medicine, where he currently holds the title of Professor of Pathology and Genetics.

GENE SILENCING BY DOUBLE STRANDED RNA

Nobel Lecture, December 8, 2006

by

ANDREW Z. FIRE

Departments of Pathology and Genetics, Stanford University School of Medicine, 300 Pasteur Drive, Room L235, Stanford, CA 94305-5324, USA.

I would like to thank the Nobel Assembly of the Karolinska Institutet for the opportunity to describe some recent work on RNA-triggered gene silencing. First a few disclaimers, however. Telling the full story of gene silencing would be a mammoth enterprise that would take me many years to write and would take you well into the night to read. So we'll need to abbreviate the story more than a little. Second (and as you will see) we are only in the dawn of our knowledge; so consider the following to be primer... the best we could do as of December 8th, 2006. And third, please understand that the story that I am telling represents the work of several generations of biologists, chemists, and many shades in between. I'm pleased and proud that work from my laboratory has contributed to the field, and that this has led to my being chosen as one of the messengers to relay the story in this forum. At the same time, I hope that there will be no confusion of equating our modest contributions with those of the much grander RNAi enterprise.

DOUBLE STRANDED RNA AS A BIOLOGICAL ALARM SIGNAL

These disclaimers in hand, the story can now start with a biography of the first main character. Double stranded RNA is probably as old (or almost as old) as life on earth. Scientific recognition of this form of RNA is, however, a bit more recent, dating from the mid 1950s. The same kinds of base pairs of that can zip strands of DNA into a helix [1] were recognized just a few years later as being a feature of RNA structure [2–5]. When two RNA strands have extended regions of complementary sequence they can zip together to form a somewhat flexible rod-like structure similar in character (but distinct in detail [5,6]) from that of the DNA double helix.

The occurrence of double stranded RNAs in biological systems was uncovered in a number of experiments in the early 1960s [7–9]. Intriguingly all of the biological systems initially found to be sources for double stranded RNA involved virus infection. This data supported a proposal that many viruses might replicate from RNA to RNA through a double stranded RNA intermediate. At the time, the central dogma of molecular biology was being experimentally established, giving a clear indication that cells mainly used double stranded *DNA* and single stranded *RNA* for long and short term information storage respectively. This left no place in normal cellular information flow for double stranded RNA, while leaving a key role (at least transiently) for dsRNA in replication of RNA viruses.

Our story next jumps back almost thirty years to a set of experiments that were directed toward an understanding host cell responses to viral infection [10,11]. These experiments involve two different (essentially unrelated) viruses infecting a single host (Figure 1). One virus was quite virulent and would kill its unfortunate host animal, while the second virus was relatively benign, causing only minor symptoms. The surprising result was that a preliminary infection with the benign virus could provide resistance to a subsequent challenge by the more virulent, nasty virus. The conclusion from these results is that the host (a rabbit in this case) has a way of knowing that it has been challenged by a viral pathogen and somehow sends itself a signal allowing resistance to further challenge. Although the ability of viruses to induce immune responses had been known for a long time, these results were unexpected by virtue of the apparent lack of relatedness between the two viruses used in the experiment. The generalized response to infection was

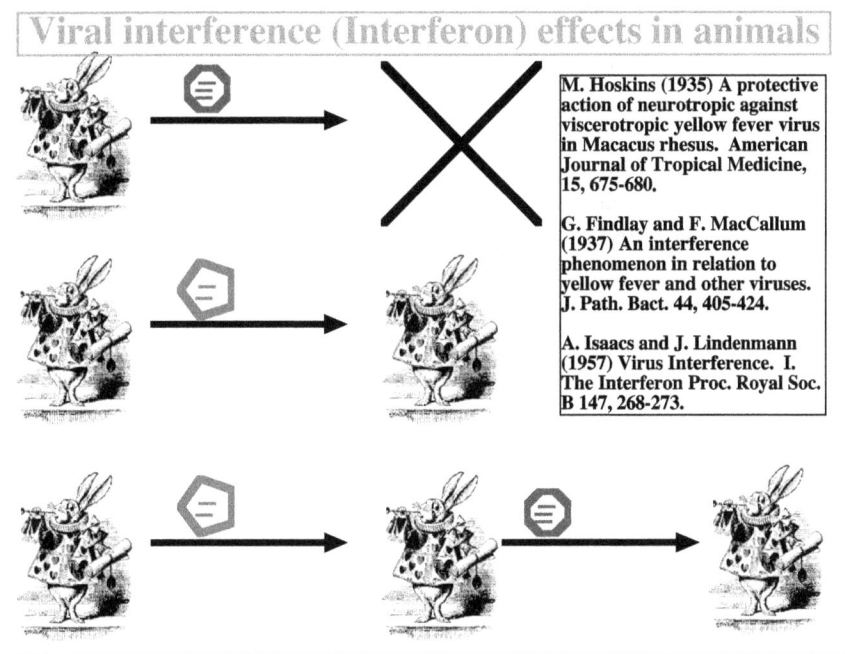

Figure 1. Diagram of viral interference effects ('innate immunity') in mammals. Top, A highly virulent virus (represented as a blue hexagon) will result in death if inoculated into a "naive" host animal. Middle, A less virulent and unrelated virus (represented as a red pentagon) infects cells but causes little or no systemic pathology, the animal remains alive. Bottom, a preliminary infection with the less virulent virus (red) leads to induction of an innate immune response which allows the animal to survive a subsequent challenge with the highly virulent (blue) virus.

a new phenomenon and led to an understanding of immune mechanisms that involve general alarm responses. A key follow-up to this observation was made about 20 years later when Isaacs and Lindeman [12] actually separated a protein component from the challenged animals that could transfer the general viral resistance when injected into naive animals. That protein component was called *interferon*.

In the course of this analysis, a physician/scientist named Richard Shope became interested in applying the innate immune response by finding treatments that would induce generalized immunity to provide viral resistance. Traveling the world at the end of the second world war, he collected biological materials looking for something that could be ground up and used as a starting material. His most notable success came from a fungus (*Penicillium funiculosum*) that he found in Guam growing on a picture of his wife Helen. Calling the extracts of the fungus "Helenine", Shope found that these could induce an interferon response in animals [13].

A next chapter in this early story was carried out by Maurice Hilleman's group at Merck, who used Shope's fungus as a starting point to purify the material that was actually responsible for the viral resistance. In a paper published in 1967 [14], they showed that double stranded RNA was present in the fungal extracts and was responsible for the induction of resistance. Given that there would have been little or no sequence similarity between the fungus-derived dsRNA and the viral target, they then tested additional very distantly related natural and synthetic double stranded RNAs and found that all could induce an interferon response [15–17]. There were (of course) many different questions raised by this study. Paramount perhaps was the question of why double stranded RNA was present in the fungus. Hilleman's publications suggested the intriguing hypothesis this was due to a fortuitous viral infection of the fungus. In fact, they had discovered an ancient system by which cells could sense a molecule that was a bellwether of viral infection (dsRNA) and respond by producing a signal that would tell the organism to dedicate its efforts and energies toward fighting viruses.

Early studies of systemic immunity were by no means limited to animal cells. Even as the first observations of an "interferon" response in animals were made in the 1930s, it had already been observed that plants could induce some remarkable immune responses. Applying a virus in one area of a plant could yield viral resistance (at least in some cases) that extended throughout the plant [e.g., 18,19]. Although these experiments indicated that plant had an immune system, it was known that they lacked the specific immune components (including antibodies and white blood cells) that had been studied for many years in animals.

This historical context of the gene silencing field thus includes the early recognition of an animal immune response (albeit a general one) dependent on double stranded RNA, and a plant immune response (albeit with trigger unknown) that could disseminate a specific signal over substantial biological distances.

GENE SILENCING ASSAYS IN A CONVENIENT NEMATODE

Now it's time to introduce another lead character into our story, one that is a close friend to Craig, myself, and to a few thousand other researchers world-wide. *Caenorhabditis elegans* is a nematode roundworm about 1 millimeter in length. In this lecture series, there were three talks on *C. elegans* in 2002 by Sydney Brenner, John Sulston and Bob Horvitz. Dr. Brenner credited "the worm" as deserving a significant portion of the scientific accolades (although he was reluctant to provide a monetary share to the worm) [20]. We should certainly credit this beast as well: *C. elegans* has turned out to be a very fortunate choice for studies of gene silencing. As you will see, the worm's vehement responses to foreign information have provided first great frustration and later some valuable insights.

One of the aspects of *C. elegans* that Craig and I have been very pleased with is the ability to microinject macromolecules (DNA, RNA, protein) into the animal [21–25]. Figure 2 is a picture that Craig took of this process, showing a fine glass needle injecting solution into an animal. After the needle pierces the cuticle, pressure is applied and some of the fluid comes into the cell that is being filled. The cell being injected in this photo is the germline or gonad of the worm, a large cell with hundreds of individual nuclei surrounding a common core of cytoplasm. Each gonad will generate hundreds of oocytes, making this is a remarkable technique for being able to influence a large population of animals with just a single microinjection. The micro-injection needle can be filled with almost any liquid including the great variety DNAs, RNAs, and proteins that we can now design and synthesize in the lab. The simplicity of microinjection for *C. elegans* provided an enticing experimental tool to manipulate the genome of the organism and observe the consequences to developmental events and physiology. At the same time, this technology has allowed a number of us in the field to study the diverse responses this system has to foreign information.

Among the goals pursued in early applications of *C. elegans* microinjection was to turn down or turn up gene expression for specific genes. In the mid 1980s, as a Helen Hay Whitney Fellow working at the Medical Research Council Lab of Molecular Biology, in Cambridge UK, I had begun doing experiments toward this goal, using among other tools the *unc-22* gene that provided some of the first characterized DNA clones for the worm. It was already known through some very nice classical genetics that reducing expression of *unc-22* led to a movement defect, a twitching behavior that is very characteristic of alterations in the activity of this gene [26,27]. Don Moerman, Guy Benian, and Bob Waterston prepared fragments of *unc-22* [28] that I then injected with the hope that the injected fragments might recombine with the normal *unc-22* allele and produce a loss-of-function character that could then be studied. The results of these experiments were a puzzle: although twitching worms appeared in populations derived from the injected animals, there was no direct alteration in the original *unc-22* gene. Instead of the sought-after recombination event, it appeared that the

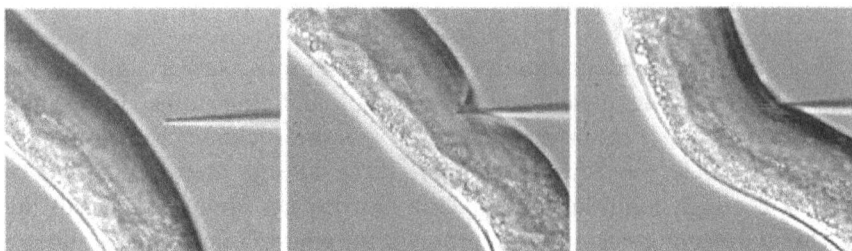

Figure 2. Micrograph of microinjection needle delivering a solution of DNA to the gonad of a *Caenorhabditis* adult hermaphrodite. Left, Microinjection needle poised at the side of the worm. The needle is filled with a solution for injection and is kept under a slight positive pressure until it is inserted into an animal (middle) whereupon an increase in pressure leads to microinjection of a volume of the material from the needle. After this, the needle is removed and the cuticle of the animal quickly recovers. Photographs courtesy of Dr. Craig C. Mello and reprinted from Mello and Fire, 1995 [25].

presence of extra DNA from the *unc-22* locus could induce the worm to turn down expression of the endogenous *unc-22* gene [29]. Several explanations for this unusual suppression effect seemed reasonable at the time: perhaps the endogenous *unc-22* locus DNA somehow paired with the foreign copies of this DNA; perhaps the foreign DNA was a template for synthesis of some amount of antisense RNA, which would then neutralize the activity of the normal transcript by base pairing, perhaps the fragments of *unc-22* were producing an aberrant protein or binding an essential regulatory factor, and perhaps there were some other mechanisms that were yet to be recognized.

Regardless of the actual mechanism of the interference in these initial experiments, the antisense strategy for "targeted" disruption of gene expression seemed particularly worthy of an explicit test. Such strategies were by no means novel at the time, having been pioneered some years earlier by Zamecnik and Stephenson [30], and by Izant and Weintraub [31]. In 1987, just after moving to Carnegie Institution in Baltimore, my co-worker Susan White-Harrison began to build DNA constructs to perform such an explicit test. Susan's constructions relied on our ongoing elucidation of muscle promoters (DNA sequences that instruct RNA polymerase to begin RNA synthesis in muscle cell nuclei). We expected a promoter hooked up to an *unc-22* fragment in the "antisense" orientation to give antisense RNA and thus perhaps gene silencing while the corresponding "sense" construct would give at most an excess of the sense strand and thus no expected silencing. We were hardly surprised when the antisense constructs produced a targeted interference effect (knockdown of the corresponding endogenous gene). This was consistent with a substantial number of reports of successful antisense intervention already in the literature. We were very surprised, however, when the control 'sense' constructs produced a similar interference effect [32,33]. The assumption for the "experimental" construct was that the antisense RNAs were finding their sense equivalents by standard Watson-Crick base pairing and taking the sense RNAs out of circulation. So what was going on with the sense constructs (where if anything, we might expect the fragment inserted into the expression vector to be over-expressed)? Although this mystery was intriguing, it was hardly compelling at the time. The propensity for DNA transgenes to produce unwanted RNA transcription was certainly a good starting point for potential models, and a reasonable explanation (that somewhat dampened any immediate research on our part) would have been that the transgenes for some reason produced sufficient antisense RNA to yield an interference effect.

A significant milestone in the study of silencing in *C. elegans* was the demonstration that direct RNA injection could induce an interference effect [34]. This observation came from work of Su Guo, who at the time was a graduate student in Ken Kemphues' lab at Cornell. Sue's insight that injection of RNA might provoke silencing turned out to be correct. Moreover, she was able to demonstrate effects with either sense or antisense preparations of RNA. This set of experiments had two lessons. First, the experiments established a remarkably efficient means of disrupting gene activity (particularly in embryos), thus facilitating a wealth of experiments in what we now call functional genomics (efforts to assign function to genes that are discovered by large scale sequencing). Second, the mystery of the interfering sense preparations was accentuated since 'sense' RNA preparations could still trigger an interference response.

After Su's experiments established RNA-triggered silencing as both a mystery and a powerful technique for studying gene function in the embryo, several other groups started working with the technique and marveling at it's unusual character. Craig Mello, first as a postdoctoral fellow working with Jim Priess at the Fred Hutchinson Cancer Research Center, and then as a new faculty member at the University of Massachusetts, began in particular to apply the technique [35] and to study the phenomenon as a window on a fascinating fragment of the tapestry of biological regulation. As I will describe later, a significant advance in understanding the concerted nature of the response came when Sam Driver and Craig discovered that the silencing could be evoked by a diffusible and specific molecular signal. As the experience from Craig's group and others with this odd form of gene silencing accumulated, much of the information was shared with the *C. elegans* community. Although the name "antisense" had initially been used to describe this process, it was clear (from the 'sense' results) that the phenomenon was not a simple one of antisense occlusion. There was thus a need for a new designation for the process, and after putting a few potential names to a vote, Craig chose the term "RNAi" ('RNA interference') to refer to the observed silencing process(es) [35].

TOWARD A STRUCTURAL UNDERSTANDING OF THE RNAi TRIGGER

For my perspective at the time (at that point as an observer of work in other labs on the worm's response to injected RNA), much of the accumulating data came together at an informal discussion on RNA-triggered silencing organized by Craig at the 1997 *C. elegans* meeting in Madison, Wisconsin. The workshop was held in the theatre of the Student Union, with the normal capacity of the room overwhelmed (I was sitting on the floor). At the time, there were several very clear but also very unexplained features of the response. In addition to the diffusible signaling data (reported by Driver at the previous year's *C. elegans* meeting), and the ability of both sense and antisense strands to produce the interference effect, there was a remarkable persistence to the effect. From work of Craig, Rueyling Lin, Morgan Park and Mike Krause, and from Patty Kuwabara [36], it was clear that injected RNAs could have effects for several days after the injection occurred (and in some cases generations after the initial injections). This contrasted with observations that Geraldine Seydoux had made several years earlier [37], showing that many native RNAs were comparatively unstable during the same time period in the same cells. The confluence of these two results suggested perhaps that the active interfering material had some kind of a privilege in its stability. Perhaps the injected material contained a fraction of particularly stable molecules that were responsible for the persistent interference.

Double stranded RNA was known to be relatively stable both chemically and enzymatically [e.g., 38]. In addition, dsRNA was a known low level contaminant in synthetic RNA preparations [e.g., 39]. From my graduate work with RNA polymerases, I was certainly also very familiar with the sometimes annoying ability of RNA polymerases to start in vitro at ends and other fortuitous sites. Thus the concept that double stranded RNA might be a component of the injected material was hardly a leap of logic. Arguing strongly against dsRNA as a potential effector was the fact that native dsRNA would have no free base pairs to interact with matching molecules in the cell. Thus a rational first guess would have been that injected dsRNA would have been unable to interact specifically with cognate sequences and thus rather useless for triggering genetic interference. A critical review of my research plan coming out of the 1997 worm meeting would certainly have brought this up as a major concern. One could imagine (in retrospect as well as currently) many different models and explanations for the phenomena. Some scenarios would have spawned interesting experimental investigations while others would have been of only limited interest; I was certainly fortunate that our research grant was not up for renewal for at least a few months.

The strength of the experimental system with *C. elegans* was that virtually any biochemical sludge could be concocted and injected into a worm, with a very rapid (and in most cases quite specific) assay at the end for targeted genetic modulation. This made it possible to test somewhat far-fetched hypotheses (like the involvement of dsRNA) without spending years or "breaking the bank". A second ingredient in testing the double stranded RNA was someone to make the experiments happen. SiQun Xu, with extensive experience with both nucleic acid synthesis and isolation and with *C. elegans* microinjection, was certainly the ideal person for this for many reasons. The setup was particularly comfortable for me since SiQun could thus do the syn-

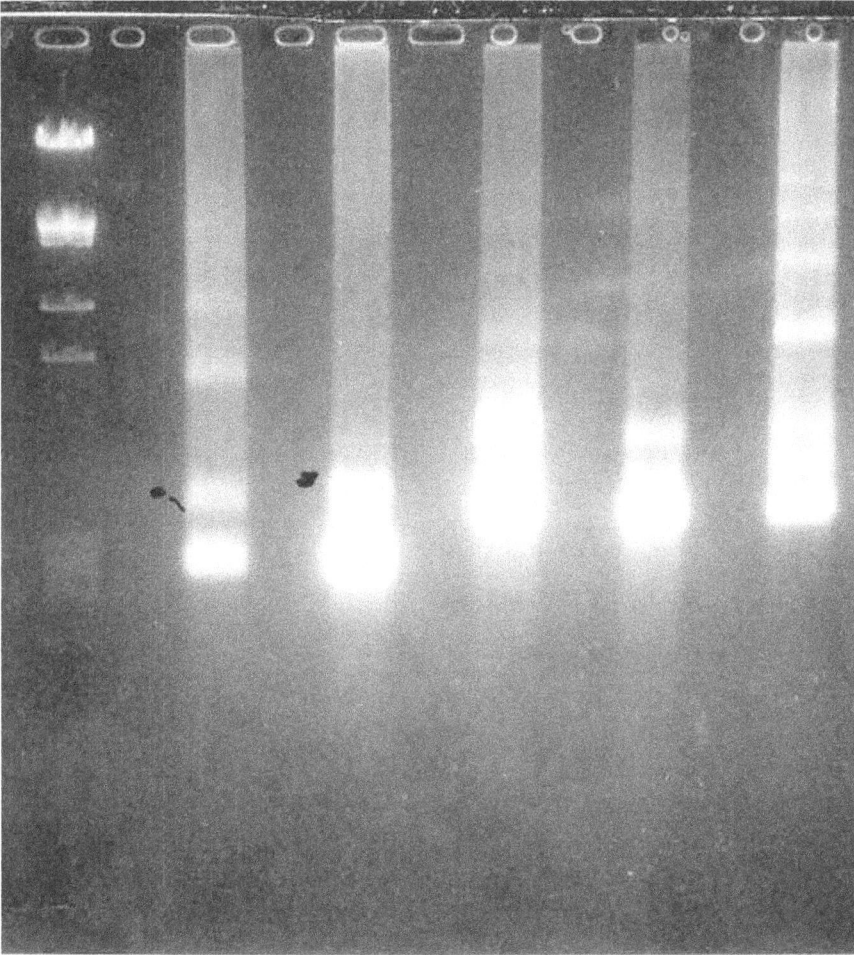

Figure 3. Electrophoretic separation of RNA prepared by *in vitro* synthesis. Left lane, marker DNAs. Remaining lanes show RNA populations with a strong band (bright signal) in the expected position for single stranded sense or antisense RNA (depending on the intended synthesis) and a number of unexpected bands (and a smear) in each lane that is visible due to overexposure of the photograph. RNA is resolved on agarose gels and visualized by fluorescence upon interaction with the included dye Ethidium Bromide. Source: Original Gel Photograph, SiQun Xu and Andrew Fire, 1997.

theses and injections and I just needed to visit my microscope in the lab for an hour or two every day to look at the injected animals and their progeny.

SiQun first repeated the kinds of RNA synthesis reactions and injections that others had done, using in this case our favorite gene, the *C. elegans unc-22* gene. This of course worked, generating a bunch or twitching worms as evidence for effective silencing of endogenous *unc-22* activity and setting the stage to use this assay in characterizing the relationship between structure and interference of the injected RNA. The picture shown in Figure 3 shows a series of the initial RNA preparations resolved using an electrophoretic

Quantitative assays for silencing: *unc-22*
- dsRNA is >100-fold more effective than sense or antisense
- dsRNA can produce interference at a few molecules per cell

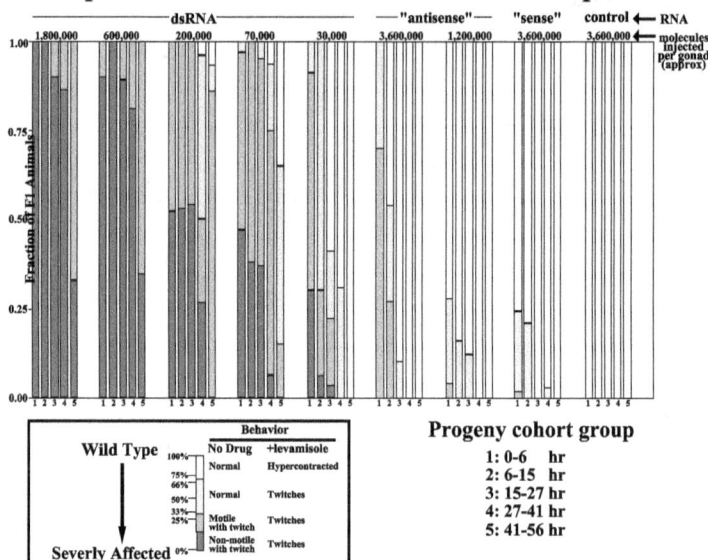

Figure 4. Quantitative assays for silencing of *unc-22*. Preparations of RNA similar to those in Figure 3 were enriched in the expected (sense or antisense) species by excising the major bands from agarose gels and extraction of RNA. Some unwanted dsRNA may persist in these samples but in general at a greatly reduced level when compared to samples not subject to purification. Individual sections of the graph show biological responses following injection of differing concentrations of single stranded and double stranded RNAs as diagrammed below (more highly affected animals are shown with a more intense red color). [Source: Reference 40 Supplement; see reference also for additional details].

field and an agarose gel. What you can see is a very prominent band, a bright spot, where the RNA that we expected was. This photo was deliberately over-exposed to reveal any other components that might be present, and one can certainly see additional (minor) bands and a general "smear" in addition to the major (expected) bands. After a few preliminary explorations of the dsRNA hypothesis using this assay with these impure RNA preparations, I was somewhat encouraged but still be no means convinced. It was clear that a cleaner preparation of starting material was needed. To achieve this, SiQun cut out the major bands from this gel, extracted the RNA and injected the purified sense or antisense RNAs into worms. This produced a result, albeit negative: almost all of the activity was lost by purification of single strands, suggesting that the sense and antisense weren't the material that was causing the interference.

SiQun's purified strands also provided a better starting point for testing the dsRNA hypothesis, since the two nearly-inactive strands could be mixed in a test tube to produce a well defined double stranded product. SiQun's injection of double stranded *unc-22* RNA formed in this way produced a remarkable result, with all of the resulting animals twitching strongly. To see how potent the effect was, SiQun injected smaller and smaller amounts of the double stranded material (Figure 4). The resulting animals showed an interference effect even after substantial dilution. When we finally did the calculation of how much material was being injected, we realized that we were seeing effects down to a few molecules of the double stranded RNA per cell. This was remarkable in that we knew from some previous work that we and Don Moerman and others had done that the target *unc-22* mRNA was much more abundant.

As with any uncharted phenomenon, the first job of the scientist is to look for explanations based on known processes. The summer of 1997 was a busy one for phone lines, email connections, and delivery services between Baltimore and Worcester, with numerous collaborative experiments with Craig and SiQun now joined by Steve Kostas and Mary Montgomery. In addition to the characterization of the specificity/generality/character of the effect on target genes, a major goal was to definitively ask whether double stranded RNA in the interfering sequence was directly responsible for the observed effects. An alternative explanation was still quite tenable: that double stranded RNA produced a non-specific response (either local or global) that potentiated the activity of small amounts of antisense. Settling this issue took a bit of molecular artistry to pursue. The most satisfying were a set of assays where we could look at gene-specific interference by complex RNA molecules that contained single stranded RNA matching one gene and double stranded RNA matching a second gene. All of these experiments pointed clearly to induction of specific interference by regions of double stranded RNA, and by the end of the summer we were all felt that a paper could be submitted definitively describing the ability of dsRNA to trigger a gene-specific and systemic silencing process [40].

dsRNA-TRIGGERED SILENCING PROCESSES AND THEIR ROLES: LESSONS FROM WORMS, PLANTS, FLIES, FUNGI, AND OTHER SUNDRY BEASTS

But of course we still did not know what was actually going on, in particular what was actually happening to the expression of the target gene. Mary Montgomery was certainly in an excellent position to pursue this question, having spent several years working around the apparent reluctance of *C. elegans* to translate injected RNA. The idea of an RNA injection experiment with a dramatic consequence (albeit strange and unexpected) was certainly enticing, so she took up the question of what happens to gene expression in the presence of injected double stranded RNA. At the time, one could imagine the interference affecting any step of gene expression or cellular homeostasis. Mary had observed that target genes lost their ability to accumulate mRNA in the cytoplasm [40]. Extending this analysis, she was able

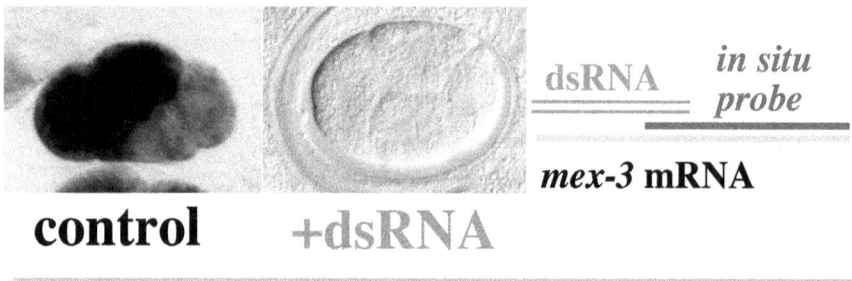

Figure 5. Injection of dsRNA results in disappearance of the targeted message. This experiment (from Mary Montgomery [40]) shows embryos of *C. elegans* with and without dsRNA injected corresponding to the mex-3 gene [142]. In control samples a strong signal is observed on *in situ* hybridization [AA] (intense blue stain, left panel), indicating a high level of mex-3 transcript throughout the four-cell embryo. Following dsRNA injection, the mex-3 transcripts are not detected (Center Panel).

to demonstrate that RNAi was accompanied by destabilization of the target mRNA in the nuclei and cytoplasms of infected cells [41]. In some ways we were lucky be working on one of the simpler dsRNA response systems; current knowledge of RNA-modulated gene expression has led to the realization that virtually every activity of genes can be affected by modulatory RNAs (replication, DNA structure and sequence, chromatin structure, transcription, processing, localization, ability to engage the translation machinery, and translational progression [e.g., 42–48]). Mary's experiments also provided a remarkably graphic description of the effectiveness of RNA interference in *C. elegans*. Figure 5 shows an example of this, with a test gene examined with and without interference at the level of messenger RNA abundance. In the case of a control sample, the messenger RNA for this gene is highly abundant and readily detected by the color reaction derived from a procedure called *in situ* hybridization [49]. After interfering with the test gene by injecting the corresponding dsRNA the messenger RNA was essentially undetectable.

The hypothesis that came from Mary's experiments was that the double stranded RNA produced a condition where the target transcript was produced but was very unstable. Restated, this postulates a sequence-specific RNA degradation system that could be triggered by dsRNA. An old TV show called "the twilight zone" was based on the idea that the universe contains many phenomena that go beyond our capacity to understand. As of early 1998, the data we'd accumulated was certainly consistent with the hypothesis that we were at least temporarily in the "twilight zone".

Accentuating this sense of unexplainable phenomena was a series of tests on the spatial requirements for dsRNA administration. These observations had a very rational starting point. When Su Guo did her original RNA injections at Cornell, she had intended to test for a biological effect of the injected material in the gonad. So she injected the gonad and indeed an effect

Levels of (im)precision in RNA delivery

S. Guo (Cornell): RNA into gonad --> gonadal affect
S. Driver (UMass): RNA into body cavity --> gonadal affect
L. Timmons (Carnegie): Feed [dsRNA+ bacteria] to worms

Eating control cells   Eating GFP dsRNA

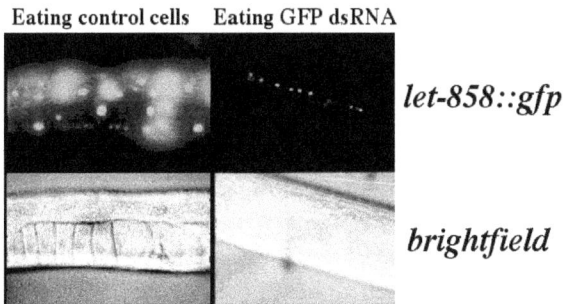

let-858::gfp

brightfield

Figure 6. RNA delivered outside of a cell can produce a potent interference effect. Above, schematic diagrams of RNA delivery experiments from Su Guo and Ken Kemphues [34], Sam Driver and Craig Mello [40], and Lisa Timmons [53]. Below, examples of feeding-based RNAi. Both animals are from a *C. elegans* strain where generalized somatic expression of a green fluorescent reporter is readily observed. The animal at the right is fed on bacteria expressing dsRNA corresponding to the gfp coding region. The animal on the left is fed on bacteria not expressing this construct. Note the dsRNA-dependent loss of gfp activity in this example in all visible cells except those of the nervous system.

was seen there [34]. The science/life-lesson that one can draw from this is "if you can do the experiment the way that seems most likely to be effective, do it just that way".

A subsequent observation from Sam Driver and Craig Mello, yields the lesson "if you can't do the experiment the way that seems most likely to be effective, still do it". In 1996 Sam was a beginning graduate student in Craig's lab at the University of Massachusetts. He was just starting out with injection and so putting the needle into the correct tissue was problematic. Sam and Craig realized that despite the improperly placed needles, the injections were still producing extremely efficient interference. When they then deliberately injected into the "wrong" place (the body cavity), they still observed a strong biological effect. Later, Craig and SiQun Xu each extended this set of observations to an extensive list of tissues where dsRNA injection produced a systemic effect.

Finally, we have a third lesson, this time derived from experiments initiated by Lisa Timmons, then a postdoc in my lab at Carnegie and now a faculty member at Kansas University. The lesson here, if you're a postdoc or perhaps a graduate student, is to do experiments that your advisor would never condone or suggest. Lisa engineered *E. coli*, which is a bacterium that is the food source for *C. elegans*, to produce double stranded RNA. When she fed this genetically modified food to the worm, she saw a gene-specific interference effect. Figure 6 shows a case where she had engineered the bacteria to make double stranded RNA corresponding to the fluorescent reporter GFP (a wonderful tool for following gene expression and cell patterns during development) [50–52]. Starting with a worm strain that produces GFP in essentially all somatic cells, Lisa found that the ingested RNA could silence gene expression throughout the animal [53]. (The picture tells another interesting story, which is that there is considerable resistance to RNA interference in nerves of the animal. Although we have yet to understand the basis or reason for this, the wholesale alteration in the efficacy of the pathway in different tissues provides additional evidence for a very deliberate biological process.) Hiroaki Tabara, a postdoctoral fellow working with Craig at the time, went even beyond the "feeding" experiment, showing that simply soaking worms in double stranded RNA could produce an interference effect [54]. These experiments were particularly surprising given our expectations that cell membranes would block all but the smallest diffusible molecules from moving between cells. We knew that there was little or no diffusion of DNA. A theme in macromolecular transport of large charged molecules has been that the cell transports only things that might be useful, with those transport mechanisms very specific and well controlled. I certainly had no idea of why the worm would be transporting dsRNA-derived signals in a facilitated manner.

So now we had every reason to think we were in "the twilight zone". Despite this, we were certainly pleased at our accomplishments in contributing to the development tools for manipulating gene expression in *C. elegans*.

We now step through a doorway from the limited world of our favorite model organism to the much richer real world inhabited by species too numerous to count. This transition is accompanied by the recognition that discoveries that we may initially view as our technical "accomplishments" are invariably a reflection of underlying processes that are a natural part of sustaining life.

Soon after the initial description of dsRNA-triggered silencing in *C. elegans*, several descriptions of similar processes appeared for other groups of organisms. These initially included observations from *Drosophila* (a fruit fly), Trypanosomes (single cell parasites), and plant systems [55–58], with many other organisms rapidly joining the list. Mammals were conspicuously absent from the initial list of organisms generally amenable to this type of manipulation. The exclusion of mammals from the list of easily manipulated species was not a surprise: the non-specific responses to dsRNA that were originally discovered by Hilleman and colleagues [14] were certainly sufficient to confound any analysis of specific genetic interference. Nonetheless, early efforts in this area provided both an indication of the potential existence of specific dsRNA responses in certain specialized mammalian cell systems (e.g. oocyte and ovary cells [59–61]) and of the predominance of the non-specific response in most others [e.g. 62].

In addition to establishing a broader biological occurrence of dsRNA-triggered genetic interference, the demonstration of dsRNA-triggered silencing in plants and fungi illuminated the process by connecting our rather fragmentary observations from *C. elegans* with a broad gene silencing literature. Indeed, papers starting a decade earlier from fungal and plant systems had been the first to describe sequence-specific effects of foreign DNA transgenes on the corresponding endogenous genes [63–67; also see ref. 68]. Intensively creative work had allowed workers in both plant and fungal fields to track down the sequence-specific foreign DNA reactions as a complex set of responses that could independently attack the target gene's chromatin or RNA [e.g., 46,69–71]. The distinctive spatial patterns of silencing for endogenous genes in plants [65,66] had been one of many features that had drawn a small cadre of highly innovative investigators to study this question for its own sake. Demonstrations of a systemic signal in the plant silencing [71,72] were particularly striking and certainly led to a clear recognition of potential similarities between the phenomena that had been observed in *C. elegans* and gene silencing in plants.

At this point, it is worth pointing out the substantial advantages of studying gene silencing (or any other important phenomenon) in more than one model system. The advantages of studying silencing in *C. elegans* turned out to be the flexibility of designing and making arbitrary RNA structures in a test tube and delivering them easily (by microinjection) into a rapid assay system (the nematode). This had circumvented many of the challenges faced by researchers working in plant systems, where such capabilities were not straightforward and complex issues of transgene structure and transcription confounded initial attempts to definitively assign a specific RNA structure as the trigger for the response. On the other side of the balance, plant systems offered a remarkable means to investigate the biological role of the interference response. Starting with the earliest recognitions of transgene-derived viral resistance [73] and observations that viral RNAs could be both triggers and targets for the silencing [46,74,75], it was rapidly clear that the silencing system might serve in the natural protection of plants from "unwanted information" in the form of viral pathogens. Definitive demonstration of this point came from a number of analyses of virus/host interaction.

To be a successful, one would expect a proposed antiviral system to effectively block pathogenesis of at least a subset of viruses that might otherwise menace the organism. Since it is well known that viruses still succeed in the world (much to our dismay), there must also be ways in which the virus can counteract any cellular defense mechanisms. A critical point in defining the role of RNA-triggered silencing process was the recognition that many successful plant RNA viruses produce protein components dedicated to the inactivation of the silencing mechanism [e.g., 76–79]. Deliberate suppression of host RNA-triggered silencing responses allows viral infectivity in at least a subset of plants for any given virus. The balance between the silencing mechanism and viral attempts to subvert it forms the basis for an ancient "arms race" between virus and plant. The character of this arms race was further evidenced in these studies by the ability to generate attenuated virus (by removing the anti-silencing function) and hyper-susceptible plants (by expressing a relevant viral anti-silencing protein or interfering with the endogenous RNAi machinery).

The emerging recognition that the transgene response mechanisms in plants were at least in part an antiviral response had raised the compelling question of how viral activity could be specifically recognized by a silencing apparatus. A rather remarkable proposal to explain this was put forth by Ratcliff, Harrison, and Baulcombe in mid 1997 [80], in a paper that arrived at Carnegie just as we had scored our first assays to test for the ability of dsRNA to trigger gene silencing in *C. elegans*. Baulcombe and colleagues had reasoned that unique features of viral replication intermediates might lead to improved transgene-based triggers for gene silencing, stating "It may be possible to increase the incidence of gene silencing by ensuring that trans-gene transcripts have features, such as double-strandedness, that resemble replicative forms of viral RNA" [80]. Combined with experiments suggesting an association between silencing effectiveness and certain secondary structures in the transgene and transcript [68,82], these proposals would almost certainly have inspired similar experiments to ours. The confluence of the two approaches, as always in science, proved to be the most powerful driver of further work, as the combination of chemical definition of the trigger in *C. elegans* and a biological explanation of its efficacy in plants led to a rapid explosion of scientific effort in the area.

TOWARDS A REACTION MECHANISM: EFFORTS TO PEER INSIDE THE BLACK BOX

Despite the great enthusiasm from those of working with plant, worm, and insect model systems, the mechanism by which dsRNA could silence gene expression was still an unknown. Seminar slides made at the time would show dsRNA and the mRNA target somehow entering a large and mysterious "black box", followed by degradation of the target RNA and some unknown fate for the effector dsRNA. This "black box" explanation limited our grasp of the RNAi system, both for understanding the underlying biology and for applying RNAi to organisms (like humans) where the response to dsRNA was more intricate than for "simple" invertebrates. The key questions (both in terms of molecular mechanism and in terms of potential roles of RNA-triggered gene silencing as an immune process) revolved around a need to understand the structure of the molecular assembly responsible for recognition of the target message by the effector RNA. Like antibody-antigen complexes in classic immunity, the identification of a "fundamental unit of recognition" seemed a key step in elucidating RNAi-based immunity in cells.

Some of this work could be done using *C. elegans*, and I will describe this in a bit of additional detail. Keep in mind (and I will describe at the end of this section) that much of the ongoing work was at this point being pursued in parallel in different systems by a plethora of research groups each with

Conclusions from Trigger Analysis

• Highly matched duplex in a region of target homology is required

• dsRNAs as short as ~25nt have can trigger specific RNAi responses

• '+' and '-' trigger strands contribute differentially to RNAi

The three strand problem

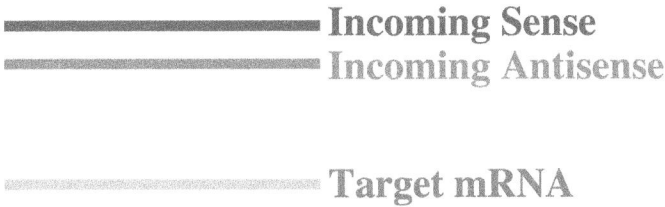

━━━━━━━━━━━━━━ Incoming Sense

━━━━━━━━━━━━━ Incoming Antisense

━━━━━━━━━━━ Target mRNA

Figure 7. Conclusions from experiments where RNA interference was assays after structural and chemical modifications had been made in injected RNAs. For details see text and Parrish *et al.*, 2000 [83].

their own angle on a specific model organism and interference assay. RNAi is a three strand process [Figure 7] involving a sense strand and an antisense strand in the trigger and a target transcript in the cell. We could manipulate the trigger strands extensively in an attempt to determine exactly what was required for the induction of specific interference. This analysis gave several specific results [83]. First, we found a different set of chemical requirements for the sense and the antisense strands in inducing interference. Second, there was a rather stringent requirement for sequence matching between the two trigger strands and with the target strand. Third, although there was a decrease in effectiveness as we used shorter and shorter triggers, we could obtain a response in *C. elegans* with triggers whose length was in the 20s of nucleotides. Combined with complementary structure-function experiments carried out at a similar time in other systems [e.g., 84] these data evidenced a very concerted chemical precision of effector RNA recognition and action in the (at that point still very unknown) black box.

A second area in which *C. elegans* could readily contribute to understanding of RNA-triggered silencing revolves around a genetic screen. The screen, originally executed by Hiroaki Tabara and Craig Mello [85], involved an important modification of Lisa Timmons' feeding experiment. Hiroaki engineered *E. coli* to produce a specific dsRNA, but in this case the dsRNA was targeted toward an essential gene in *C. elegans* (a gene called *pos-1* that

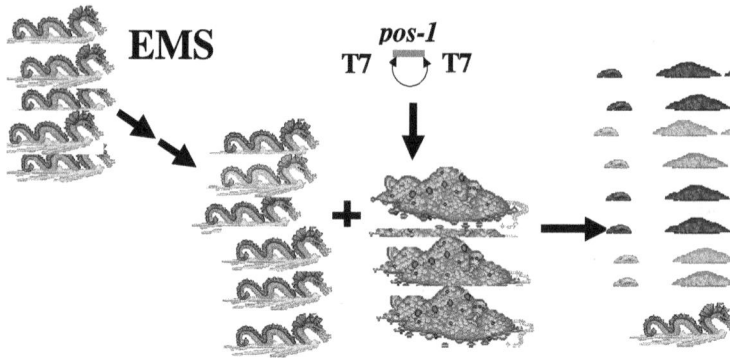

A mutational Screen for trans-acting factors involved in RNAi

See: Tabara, H., Sarkissian, M., Kelly, W., Fleenor, J., Grishok, A., Timmons, L., Fire, A., and Mello, C. (1999) "The rde-1 gene, RNA interference, and transposon silencing in *C. elegans*." Cell 99:123-132

Figure 8. Identification of mutations that eliminate responses to foreign RNA but are compatible with life for the worm. After mutagenesis with the chemical mutagen Ethyl Methane Sulfonate [26], animals were grown for several generations and then transferred to an *E. coli* food source expressing dsRNA corresponding to the *C. elegans pos-1* gene. Embryogenesis is arrested in the vast majority of the resulting population and only mutants such as those eliminating RNAi can continue growing as a population. For details see text and Tabara *et al.*, 1999 [85].

Hiroaki had characterized during his graduate work with Yuji Kohara [86]). Without the activity of this gene, worm populations could not survive, so that the engineered bacteria are an exceedingly poor food source for *C. elegans.* Selecting the extremely rare animals that can grow on that food source was then possible [Figure 8] and was facilitated by working with populations that had been chemically treated several generations earlier to produce mutations. Among the animals that grew on this food source were a subset that lacked the responses to all the kinds of foreign dsRNA that we had used for interference. For at least two genes, Hiroaki found that a complete loss of function resulted in a worm that looked normal (or nearly normal) in the laboratory, but which was unable to respond to our dsRNA challenges. The existence of these mutations provided further (and very compelling) evidence that RNAi was a concerted process. If the ability of dsRNA to silence genes had been a simple reflection, for instance, of the physical chemistry of dsRNA, then we would have been unlikely to find mutations that abrogated this activity. That *C. elegans* could survive without the process and grow normally (at least in the artificially pristine conditions of an isolated Petri plate) was a demonstration that the organism relied on a dedicated mechanism to facilitate dsRNA-triggered silencing. Through considerable effort, mostly from Hiroaki and Craig, it was possible in a relatively short time to identify the genes which had been mutated in the resistant strains. The identities of the corresponding gene were both illuminating and frustrating.

rde-4 encoded a protein with a structure clearly suggestive of an ability to bind to dsRNA [87]; although certainly reassuring, this identity by itself (and the expected ability of the protein to bind dsRNA non-specifically [88]) was not sufficient to illuminate the underlying mechanism.

rde-1 encoded a protein from a large family (now called the "Argonaute" family) for which there was at the time only a trace of biochemical data. Proteins from related families had been shown to play key developmental roles [89–91]. There was some indication of an RNA interaction [92], but there was little biochemical information beyond this. As it became clear that other genetic model organisms shared a dsRNA response mechanism, it likewise became clear (from genetics in plants, fungi, and flies [e.g., 93–95]) that at least a subset, like *C. elegans*, could survive without this mechanism. The ability of diverse organisms to encode proteins of similar character to those involved in *C. elegans* gene silencing, and the eventual identification of homologous genes as functionally required for RNAi in distinct model systems [e.g., 95–97] supported the argument that we were all looking at a similar and conserved biological process. Beyond the standard "model" organisms, the existence of homologous coding regions in mammals supported the argument that mammals might indeed also have similar responses if it were possible at some point to tease away the non-specific response.

Despite these hopeful suggestions, the RNA structure-function and genetic analysis had not put us in a position either to propose a unifying mechanism for RNAi or to design experiments to test for the efficacy of the system in mammals. Even in hindsight, going forward in either direction would have been complicated; in particular, the shortest RNAs that we had initially tested for interference in *C. elegans* [83] were too long to have fit into the what we now know as the RISC complex [see below], and were not of the proper structure to provide side-effect-free gene silencing response in mammalian cells.

Getting into the black box required a series of keen biochemical observations. I won't go into these observations in too much detail here, as the small RNAs that mediate exogenous and endogenous genetic control in diverse biological systems are certainly worthy of their own narrative. Still a summary of the small interfering RNA story serves to provide some context for how we now think about RNAi.

The first indication that a small RNA population might be key to the RNAi process came from experiments in plant systems that were carried out by Andrew Hamilton and David Baulcombe [98]. Studying plants undergoing experimental gene silencing, they found a population with a narrow size range of 21–25 whose presence was closely associated with the silencing. Critical to this analysis was the decision to look for RNAs in a small size range and the rather impressive chemical trick of actually detecting these RNAs.

With small RNAs identified as potential additional characters in the story, biochemical research gained considerable momentum. To know anything about what was happening in the black box required an ability to study the reaction not within the complex environment of living cells, but in some type of isolated system. Two groups initially took up this challenge: one at MIT (Phil Zamore, Tom Tuschl, Ruth Lehman, David Bartel, and Phil Sharp) and one at Cold Spring Harbor (Scott Hammond, Emily Bernstein, David Beach, and Greg Hannon). Each succeeded independently (using very different approaches) in recapitulating the RNAi reaction in soluble extracts of *Drosophila* cells [99,100]. As the analysis of the biochemical reaction proceeded from these groups and others, it became clear that the small RNAs that Hamilton and Baulcombe had observed in plants were indeed central to the interference reaction. The reaction was, at least conceptually, divided into three phases, the cleavage of a long dsRNA trigger into shorter dsRNA segments, the loading of chosen single stranded products of this cleavage into a tight ribonucleoprotein complex, and the scanning of potential target RNAs in the cell by this complex [99–102]. The Hannon lab, perhaps while watching late-night television, coined catchy (and now standard) names for the two enzyme complexes central to the reaction: *Dicer* (which cleaves the dsRNA into short segments) and *Slicer* (which assembles around a single strand of processed effector RNA and goes on to cleave target messages [somewhat equivalent to the term RISC]).

The pathway that resulted from the confluence of biochemical and genetic analysis is shown graphically in Figure 9. The reaction initiates with cleavage of the large dsRNA fragment into small double stranded fragments. Selected strands of single stranded RNA then get incorporated into the "slicer" complex, which then searches around the cell looking for target RNAs in a manner that is not yet understood. When those target RNAs are found, they are cleaved by an enzyme activity which is intrinsic in the RISC, leading eventually to target degradation. Although this mechanism certainly didn't explain all of the phenomenology, it has proven remarkably general as a working model on which to base further study of RNAi.

Among the consequences of this model were some predictions of how to achieve specific RNAi in human cells. A key step in this was the detailed chemical description by Elbashir, Lendeckel, and Tuschl of the first small RNA intermediate in the silencing process [103]. This mostly-double-stranded small RNA population formed by Dicer had a characteristic set of termini with two slightly overhanging bases on each strand and the negatively charged phosphate group on the non-overhanging end. Recall that long double stranded RNA induced a nonspecific effect which prevented us from looking for any specific effects. Determination of the intermediate structure sparked an informed guess that RNAs of this structure, known from earlier studies to be too short to induce a strong non-specific response [e.g., 104], might produce a much more specific response. This was indeed the case, with reports appearing first from Elbashir, Tuschl and colleagues [105], and then in rapid succession from other groups including our colleagues Natasha Caplen and Richard Morgan at NIH [106]. The relatively straightforward nature of these assays led quickly to adoption of siRNA-mediated interference as a preferred method for certain analyses of gene function in mammals.

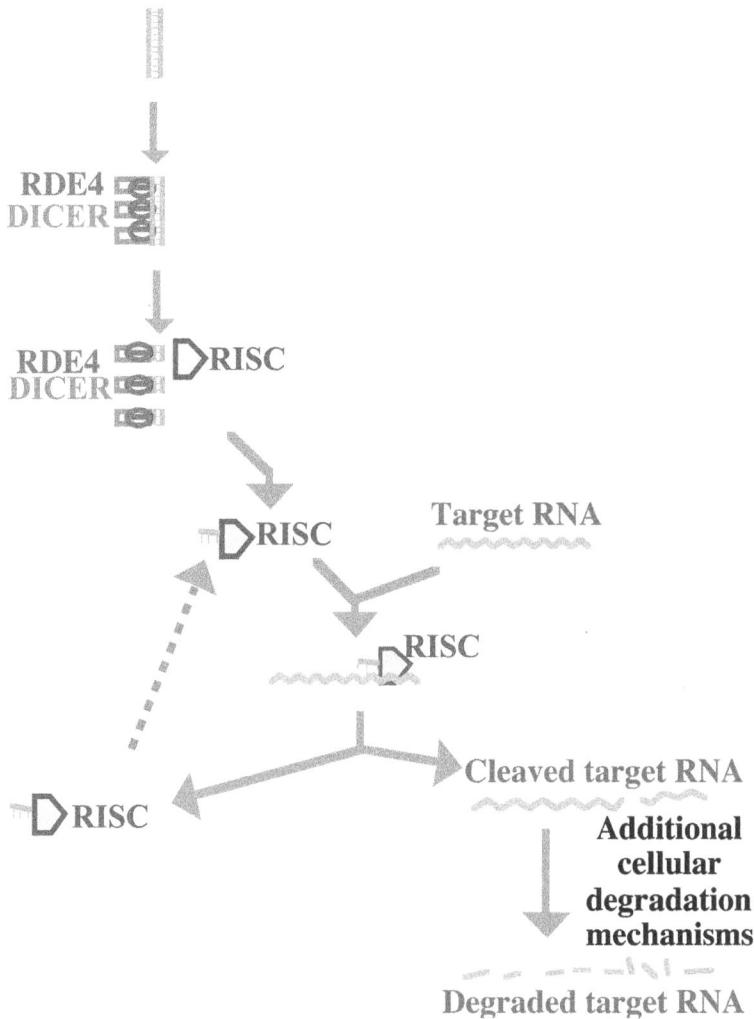

Figure 9. A basic model for the conserved central core mechanism in RNAi. Based on bio-chemical and genetic experiments as described in the text, double stranded RNA enters the cell, is set upon by a complex of a dsRNA-binding protein (RDE4 for *C. elegans*) and a dsRNA-specific nuclease (Dicer). Following dicing of the dsRNA into short double strand-ed segments, individual small RNAs are loaded into a second protein complex including a protein member of the *Argonaute* family (to assemble an RNA-Induced-Silencing Complex, also called a 'RISC' complex). These can then survey the existing RNA population in the cell for matched targets, which are then subject to degradation.

RNA INTERFERENCE AS IMMUNITY: SOME ANALOGIES AND QUESTIONS

The genetic and biochemical elucidation of RNAi also raised some interesting questions of analogy between the classic immune response (involving antibodies and lymphocytes) and RNAi (Figure 10). First will come the question of specificity. For the classic immune system, specificity is enforced by a series of interactions between recognition proteins (Antibodies and T-cell receptors) and their potential partners (including foreign proteins and other molecules). The flexibility of specific protein recognition repertoire thus serves as major basis for the classic immune response. For intracellular responses to foreign RNA, it appears that nucleic acid complementarity plays a similar role. Hybridization of short effector RNAs to a target message provides both rapid and specific recognition on which to base an immune response. The critical length of the duplex, in the 15–25 nt range, turns out from first principles to be optimal for achieving specific recognition without burdening the system by non-specific hybridization that would be more common with longer effector molecules (a point made many years ago by Tom Cech, in giving an introductory lecture in ~1991 to a group of scientists who hoped to use antisense technology for therapeutic goals).

A second challenge for the RNA interference pathway is how to ensure that no self-attack occurs that might harm the host cell; essentially there is a need to be sure that none of the cell's own essential genes are targeted by the RNAi mechanism. A part of this assurance relies on the use of dsRNA

RNAi versus Our "Traditional" Immunity

Specificity: How to find a "needle in a haystack"?

How to react to diverse pathogens without self-attack?
Pre-existing "innate" repertoire
Infection-specific "acquired" repertoire

How to focus on small pieces of each pathogen?

How to mount a systemwide response?

How to conserve resources for useful responses?
by Stabilizing "useful responses"
by Amplifying "useful responses"
by Recycling "useful responses"
by co-dependence of different immune responses

How to remember where you've been?

Figure 10. Points of comparison, analogy, and contrast between traditional immunity (T-cell/Antibody mediated responses) and proposed RNAi-based defense mechanisms.

as a trigger. Our cells don't normally use double stranded RNA to express our genes, they use single stranded RNA. Of course there may be cases where double stranded RNA is part of modulating gene expression, but for the most part, cells can avoid it if they need to. The interesting part of this avoidance is that it is evolutionary in nature. We presume that once the RNAi mechanism is in place, cells would evolve very diligently to avoid producing dsRNA in amounts that would shut off important endogenous genes. Any deviation from this could decrease the fitness of the organism, so over evolutionary time we expect a very effective avoidance of self-detrimental RNAi. This long-term mechanism differs from classic immunity in that the classic immune response avoids self-inflicted damage by a surveillance mechanism that (when everything is working properly) removes self-directed recognition elements continuously during the life of an organism. The consequence of this difference is that for RNA-based immunity it may be easier in real time to "trick" the system into targeting an endogenous component, something that could be an encouragement to the development of therapeutic strategies involving RNAi.

Breaking of the initial dsRNA trigger into small fragments reveals a third immune-related logic to the process. Certainly the dicing of the trigger serves to increase the number of independent molecules (and specificities) in the response, potentially providing a more effective trigger::target ratio for surveillance. In addition to this, the focus on short segments allows the system to respond to viruses that have mutated elsewhere in the genome but kept one or more essential sequences of greater than 20 bases. Finally, there is a benefit to breaking the infectivity of the effector molecules before disseminating them around the organism. I usually describe this by analogy to antiviral software: If you are worried about viruses infecting your computer, you will buy an antiviral software package that carries (i) a database of information about viruses (computer viruses in that case), (ii) a series of routines to establish which files are infected, and (iii) a series of remedies which either correct or delete the infected filed. The virus database that is part of this package doesn't need to have complete sequences for each virus, and indeed it would be a mistake for the antiviral software company to distribute such a database, as some of the components from the database might end up initiating infections. By taking from each virus only a set of relatively short signature sequences, it becomes possible to distribute identifying information without distributing the potential for infectivity. Breaking the double stranded RNA into 21–25 nt segments may serve the same role in cellular responses to unwanted RNA.

Dissemination of immune effector information is another feature of both classical and RNA-based immune mechanisms. For classical immunity this involves hitch-hiking with the blood circulation that permeates the body, as well as some very highly choreographed lymphocyte migration processes. For RNA-based immunity, the mechanisms of information dissemination are still being unveiled. Results demonstrating a concerted protein-based machinery that mediates dissemination of the RNAi response in *C. elegans* [e.g., 107] are

certainly exciting; understanding this machinery will be of great interest in
designing and planning applications of RNAi.

RNAi, like any cellular mechanism, requires use of energy and metabolic
resources. Balancing those resources with the current needs of the organism,
and focusing the resources available for this purpose on the most pressing
dangers, are essential for the system to fulfill its worth. For classic immunity,
there are mechanisms that manage the population of effector molecules
involved in surveillance (T and B cell repertoire), both by subtracting out
specificities that are not engaging targets and by amplifying specificities that
engage their targets. One expects, perhaps, to find similar overall management of specificities that guide the RNAi machinery. An enticing example of
such management comes from the involvement of RNA-directed RNA polymerases in the silencing process for plants, worms, and some single-celled
organisms (See Figure 11). First characterized in plant systems in the 1970s,
cellular enzymes that can copy RNA to RNA [e.g., 108] had little place in
the central dogma of molecular biology (DNA makes RNA makes Protein).
Considerable doubt regarding the source of such enzymes inhibited research
until they were purified and shown to be encoded by the cellular genome
[109–110] and subsequently shown to play key roles in RNAi in *Neurospora*,
worms, and plants [111–115]. One of the striking aspects of RdRP-based trigger amplification that has been described is that amplification only occurs
when a target has been engaged. The consequences of this guidance mechanism [116–122] are (i) that amplification of the effector signal is limited to
cases in which there is a real target, and (ii) that the spectrum of RNA silencing triggers can spread outside of the original area to encompass a broader
segment of a target that has been recognized as foreign/unwanted. The
RdRP-based amplification mechanism thus provides an example of honing
the immune activity of the RNAi system to "clear and present" dangers.

The immune system analogy to RNA-based surveillance brings up a final
question of how the system can remember prior challenges to provide optimal immunity. For the majority of RNA interference experiments done in *C.
elegans*, the visible effect disappears after a generation or so [e.g., 40]. This is
not always the case however, and there are instances in which gene-specific
effects of RNAi can last for numerous generations [e.g., 123–124]. Similar
long term effects have been studied in plant systems [e.g., 125]. Such effects
would not be expected from the simple model in Figure 9. Instead, a current
model (see Figure 12) is that the initial interaction between effector and target sequences might have a combination of short term consequences (e.g. inhibition of translation and degradation of the target mRNA), medium term
consequences (such as production of additional small RNA effectors complementary to the target) and long term consequences (including changes in
the physical conformation [chromatin context; 46,47] of the cellular DNA
that encodes the target transcript). This variety of responses to a similar
initial interaction event is in many ways analogous to the classic immune
system, where an initial target recognition interaction can lead to a plethora
of downstream consequences. In each case, the initial interaction complexes

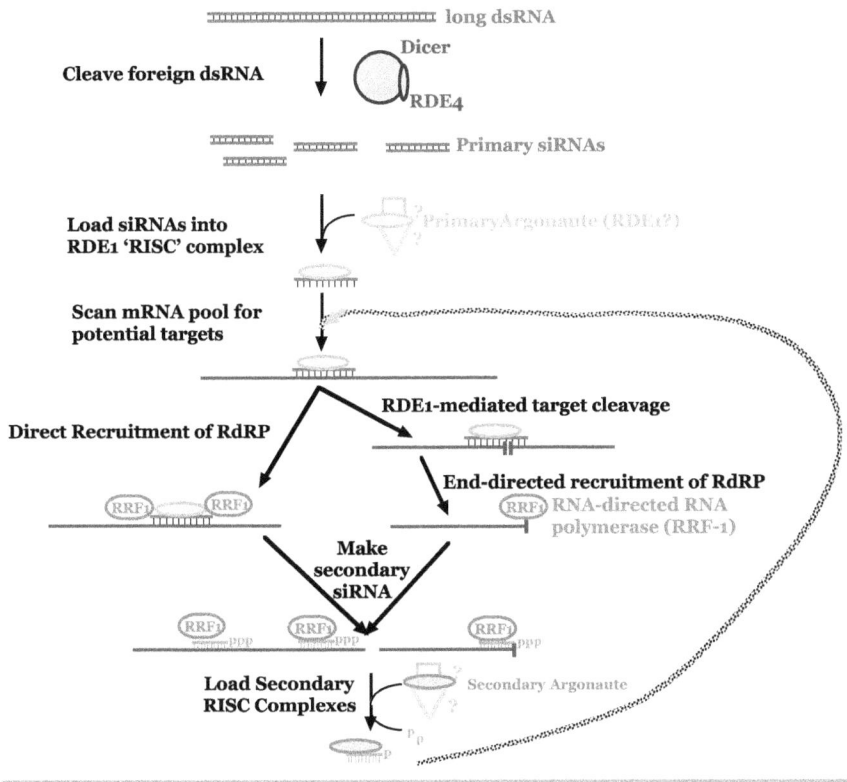

Figure 11. A model for amplified RNA interference in *C. elegans* somatic tissue. Based on discussion and references in the text, long dsRNA introduced into cells is initially attacked by a complex of a nuclease (Dicer) and a recognition component (RDE4) that "dice" the long dsRNA into short fragments. Loading of these fragments into a second protein complex results in a silencing complex that can scan the message population of the cell for matching sequences. These are then subject to two different consequences: cleavage (which should inactivate the message) and/or synthesis of short complementary RNAs [116–122]. The short complementary RNAs can join their own effector complexes (possibly including a different Argonaute family member, see [143]), resulting in a target-dependent amplification of the foreign dsRNA response. See text and references for further discussion of this proposed mechanism.

(RISC-mediated nucleic acid hybridization in the case or RNAi, antibody: antigen or T-Cell-Receptor::antigen in the case of the classic immune system) appear capable of recruiting a diversity of suppressive mechanisms based on the circumstances, with the duration of any given response (and subsequent memory) depending on a balance between longer and shorter term consequences.

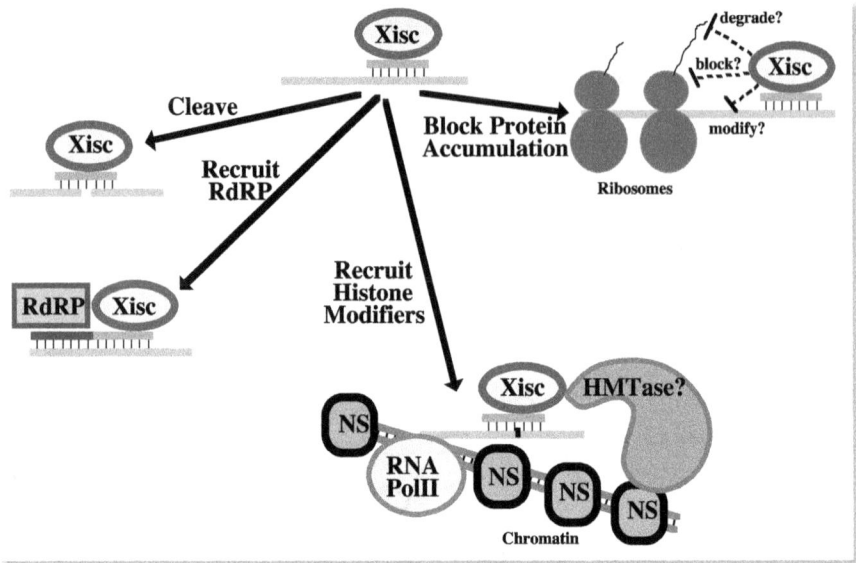

Figure 12. A model for multi-modal gene silencing as a result of siRNA effector recognition of RNA transcripts. A generic Argonaute:siRNA:target ternary complex is shown at the top, giving rise in principal to several different complexes in which silencing factors have been recruited. Left, top, cleavage of target transcript by Argonaute-like component or a recruited ally. Left, middle, recruitment of an RNA-directed RNA polymerase that might synthesize complementary RNA either primed by the initial siRNA or (as appears to be the case in *C. elegans*) with de-novo initiation. Bottom, a rough schematic diagram of tethered chromatin modification components acting on nearby nucleosomes and/or other DNA-associated factors (drawn here as a silencing HMT-ase = "Histone methyltransferase" [see 145] although numerous other epigenetic modifying activities could function equivalently). Note that this process would likely occur on a nascent RNA transcript still associated physically with the DNA template [144]. Right, recruitment of factor(s) that might block translation of the message [e.g., 48].

GOING FORWARD: PUZZLES AND CHALLENGES

RNAi is an extremely active field of current investigation and will certainly remain so for some time. Many of the central questions relate to basic mechanisms; many others relate to potential applications. From the perspective of understanding RNAi as a potential immune-type surveillance mechanism, several questions currently occupy the forefront (I have cobbled together a list in Figure 13). One question concerns possible roles for RNAi as an antiviral response outside of the plant kingdom. Several recent studies in invertebrate animals (worms and flies) rather clearly show the capability of RNAi to function in surveillance against viruses (and other selfish information such as transposons) in simple animals [e.g., 85,126–128]. That the issue has not yet been resolved for higher animals (mammals) could conceivably reflect the complexity of teasing apart specific and non-specific responses of mammals

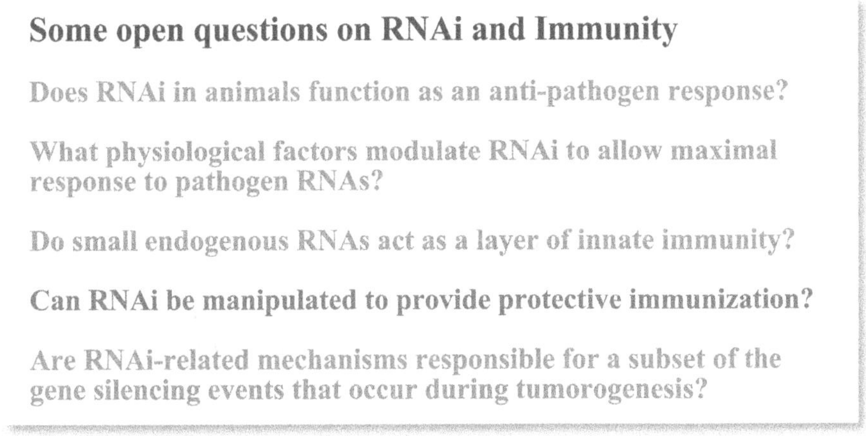

Some open questions on RNAi and Immunity

Does RNAi in animals function as an anti-pathogen response?

What physiological factors modulate RNAi to allow maximal response to pathogen RNAs?

Do small endogenous RNAs act as a layer of innate immunity?

Can RNAi be manipulated to provide protective immunization?

Are RNAi-related mechanisms responsible for a subset of the gene silencing events that occur during tumorogenesis?

Figure 13. Some open questions on RNAi and immunity.

to dsRNA. Alternatively, it is certainly conceivable at this point that the virus-protective role of RNAi has been lost in mammals.

One exciting development over the last several years has been the appearance in the literature of detailed structures of components parts of the RNA interference machinery [e.g., 129,130]. These structures have, both individually and in aggregate, led to an understanding of aspects of the mechanism would have only been dreams about during the early phases of the analysis of the system. With the emerging structural wisdom come a large number of thermodynamic and kinetic questions. For the less technically inclined reader, these challenge us to understand the contributions of energy and equilibrium to the natural system and to add the dimension of time to the static pictures such as those in Figures 9 and 11. Already it is clear that kinetic competition between different potential effectors at each stage of the RNAi mechanism is a key determinant of how the RNA-based surveillance system is used [e.g., 131,132]. Likewise, kinetic competition between the RNAi machinery and other protein: RNA interactions (RNA synthesis and processing machinery, RNA storage and turnover machinery, and the translation machinery) will undoubtedly determine the spectrum of RNAi events that can actually occur during the life of a cell [e.g., 133].

At the same time as detailed biochemical and structural studies are likely to illuminate the forefront of RNAi, there is still much to be learned from genetic analysis. The original screens of Tabara *et al.* [85] found just two *C. elegans* genes with the idealized property that they eliminated almost all RNA interference with little or no effect on the organism. Similarly, limited sets of comparable genes (although different individual components) were identified in the early genetic screens of plant and fungal systems [e.g., 93,94]. Vertebrate cells that lack the major Argonaute component involved in dsRNA-based surveillance are intriguingly alive (and capable of growth in a Petri dish) but incapable of forming a viable organism [134].

Correspondingly, some mutants in other systems that may have superficially appeared specific to the dsRNA response also exhibit intriguing variations in growth and/or physiology even in the absence of known pathogenic challenges [e.g., 135,136]. In addition to these observations, several biological forces which were limiting the original genetic screens are now clear. In some cases, the failure to recover mutants affecting a given stage in the process reflected a degree of genomic redundancy, with several different gene products each sufficient (at least partially) to execute a single reaction step [e.g., 85]. Conversely, some RNAi components were not identified in the early screens due to their shared involvement in RNAi-related (but distinct) processes which use similar molecular machineries and which are essential for organismal viability. In addition to the well characterized micro RNA regulatory system [128,137], the portfolio of RNAi-related processes will almost certainly include surveillance and regulatory roles within cells which we have yet to understand [e.g., 128]. As the expanding toolkit for analyzing essential and redundant genes in genetic model systems is applied, we should be able to open more than a few doors toward illumination of both the natural roles of RNAi and of numerous yet-to-be-elucidated cellular regulatory and surveillance functions.

RNA INTERFERENCE AS A TOOL IN MEDICINE?

A question that has generated considerable excitement beyond the research lab is whether effector dsRNAs might be used as a direct intervention to treat human disease. Indirect applications of RNAi in medicine have certainly jumped forward: RNAi takes its place among many different tools to understand gene regulation, assign functions to individual genes, and facilitate the discovery of potential therapeutic targets in disease systems.

Will direct administration of interfering RNA be a useful clinical tool? If a person has a virus infection, why not use double stranded RNA corresponding to that viral sequence as a drug to treat the person? If a person has a tumor, why not take a gene that's essential for that tumor and administer double stranded RNA corresponding to that gene to shut down growth of that tumor? If a person has a disease caused by an altered or out-of-control gene, why not try double stranded RNA corresponding to that gene as a potential therapeutic? There are many challenges and many conceivable benefits to this approach. There are scores of potential applications, all of which will require negotiating the thicket of delivery, safety, and efficacy in the complex circumstance of a genetically diverse target population and with the need to understand and anticipate host (and in some cases pathogen) responses to the specific dsRNA. Maybe the time frame in testing these approaches will be years, maybe tens of years, and maybe more. With all of the trepidation and caution that goes into such an enterprise, I still look forward to seeing research in this area progress as a future endeavor in both the public and the private sector.

I expect that there will be additional areas (beyond the gene discovery and therapeutic RNAi applications discussed above) in which understanding of RNA-triggered gene silencing will provide therapeutic opportunities and augment to our capacity to mitigate disease. Any potent and specific biological process (even if it is generally beneficial to the organism) comes with consequences to the organism if abnormalities in specificity or regulation occur. Aberrations in genetic silencing (both positive and negative) are certainly a major component of many human diseases, including most prominently cancer. Intensive investigations of dysregulation in cancer and other disease have turned up cases of defects in virtually every known cellular regulatory pathway. Regulation by small RNAs has rapidly joined this group [e.g., 139,140], with currently available data likely accounting for only a small fraction of such effects. As the potential contributions of RNA-triggered genetic silencing processes to both disease and the human response to disease continue to be characterized, it is conceivable that there will be clear cases in which manipulation of the RNAi machinery itself, either in a global manner or in a small subset of cells or effector functions, will become an attractive therapeutic strategy. As such situations arise, the availability of therapeutic interventions to manipulate aspects of the RNAi machinery such as small molecule drugs [e.g., 141] and biologically-based modulatory strategies [e.g., using viral anti-silencing components] will certainly provide worthy leads for potential treatment.

SCIENCE DOESN'T GROW ON TREES, EVEN IN SANTA CLARA COUNTY...

I want to finish with a few thanks. I have been fortunate to be associated with a family, a group of friends, a set of co-workers, and a number of institutions for which scientific inquiry and humanity have been equally highly valued. This has made it a joy to do science.

Since this article focuses most directly on experiments from the 1990s on the structural trigger for RNAi, I want to first specifically acknowledge the members of my lab and some of our collaborators that were directly involved in this work. The crew in my lab that were involved most directly in this particular effort were SiQun Xu, Mary Montgomery, Steve Kostas, Lisa Timmons, Susan White-Harrison, Jamie Fleenor, and Susan Parrish. Collaborations with Craig Mello and his group, particularly Sam Driver and Hiroaki Tabara likewise drove of the effort in wonderful ways, as did collaborations with Natasha Caplen and Rick Morgan at NIH, Farhad Imani at Johns Hopkins, and Titia Sijen, Femke Simmer, Karen Thijsen, and Ronald Plasterk at Utrecht. I hope you realize that even this rather substantial scientific consortium was just a piece of a very large puzzle that involves also many other scientists and groups around the world.

Two institutions deserve particular attribution for any contributions from myself and my lab. The Carnegie Institution of Washington's Department of Embryology is where we did our work on the structural characterization of the trigger, and the National Institutes of Health (in addition to supporting all of our colleagues and collaborators) funded all of the work that has gone on in my lab on this question.

For anybody to make their way in the world, there need to be inputs and contributions... and a lot of influences. When I sat down to put a few names down of people that at one point or another have had a positive influence, Figure 14 emerged. I apologize in advance for any inadvertent omissions (and numerous spelling errors)... you know who you are.

In randomized order: Lilly Lerner, Maria Esquela, Lynne Corboy, Yixian Zheng, Jenny Pang, Jim Manley, Robert Weinberg, Guy Rudin, Steven Siegel, Claire Craddock, John Hennessey, Andrew Godbey, Josh Glassman, Kevin O'Connell, Mark Lorell, Jim Kiessling, Benjamin Glass-Siegel, Ziva Reuveny, Gesine Dingkuhn, Vivian Hou, MarketBiology Students, Joe Robertson, Patrick Masson, Massachusetts Institute of Technology, Gabriel Chaen, Harold Smith, Caroline Mararah, Dina Goren, Sharon Long, Grace Pagalde, Rose Sherak, Mike Leong, Arend Sidow, Joan Miller, Metav Arusha, Peter Okkema, Elliot Meyrowitz, Aviva Richman, Robert Schleif, David Postman, Ursula Vogel, Ann Thompson, Barry Levine, Nathan Krantz-Fire, Michael Jantsch, David Remondini, Ed Hedgecock, Fred Tan, Mehrangiz Kamyab, Shira Lander, Sondra Lazarowitz, Gilbert Chu, John Gage, Karen Rosenfeld, Allie Liu, Min Kim, Ann Crowden, Richard Meserve, Mike Cleary, Sonya Palmer, Art Barnes, Mike Krause, Ashley Chi, Ann Corsi, Nipam Patel, Parmjit Jat, Mark Eaton, Michael Shen, Ben Hwang, Lucy Sherill, Linda Breeden, David Finkel, Gregory Fisher, Irv Weissman, Judith Greenberg, Kerstin Arusha, Lisa Steiner, Peter Sarnow, David Baillie, Lisa Weymouth, Toshi Oyama, Ann Halvorsen, Richard Durbin, Naomi Richman, Helen Sherak, Helen Reichel, Claudia Lipschultz, Terence Murphy, Jade Li, Won Kyu Pak, Ruth Rose, Poornima Parameswaran, Susan Michaelis, Chip Ferguson, Ed Ziff, Donald Moerman, Miriam Goodman, Bill Tiefenwerth, Kathy Meyrowitz, Chris Halvorsen, Mary Montgomery, Paul Chernoff, Josianne Eid, David Housman, Ben Hole, Cliff Tabin, Sarina Schwartz, The Fremont Chess Club, Betty Dyer, Garfield Moore, Julia Kay, Robin Kulakow, Kevin VanDoren, William Hurlbut, Charles Long, Daniel Blanchard, Sudha Mitra, Brittany Little, Peter Waterhouse, Casey Inman, Sara Eisenberg, Bill Fixen, Rachel Krantz, Mark Samuels, Maria Jasin, Geraldine Seydoux, Victor Ambros, Inder Verma, Chang Zhang Chen, Titia Sijen, Barbara Elspas, Jorge Mancillas, Elana Lubit, Steve Johnson, David Schwartz, Barbara Levine, Joyce Rosenfeld, Blake Hill, Hollenbeck Elementary School, Ruth Shinn, Phillip Sharp, Paul Englund, John Burke, Scott Oliver, Yvonne Pon, Peter Hahn, Amy Locks, Ben Stern, Marvin Sherak, Paul Kokulis, Clifford Locks, Robert Tjian, Siva Ophir, Beth Johnston, John Gearhart, Hee Young Kim, Denise Montell, Tory Prestera, Susan White-Harrison, Adil Ophir, Phil Beachy, Sam Halvorsen, Julia Pak, Alan Klotz, Hugh Rienhoff, Robert Waterston, Stuart Kim, Barbara Postman, Mary Dusso, Gina Fisher, Maurice Fox, Sarah Katz, Anita Finkel, Morgan Park, Maxine Singer, Gerry Crabtree, Mitch Kostich, Rick Myers, Carol Alberts, Peter Jackson, Ira Dyer, Levana Ruthschild, Daniel Nathans, Cynthia Kenyon, The National Institute of Child Health and Development, Joe Heilig, Sylvia O'Neill, The Balzano Family, Gary Otto, Ellen Zucker, The Medical Research Council, Yvette Goren, Simon Xu, Nancy Paiva, Andy Golden, Nancy Craig, Philip Anderson, Ellie Krantz, Mary Beth Shinn, Donna Rae Machado, Nina Federoff, Ann Sharp, Genevieve Pire-Halvorsen, Eleanor Maine, John Eichemendy, Richard Henderson, Gary Ruvkun, Ichi Maruyama, Kamiko Cangelosi, Shou Wei Ding, Mel Goudy, Hank Greely, Randall Kaufman, Geeta Narlikar, Haifan Lin, Victor Corces, Matt Kowitt, Alyssa Zucker, Alan Wolffe, Arielle Goren, Lynne Spencer, Joey Pinkel, Christine Norman, Joe Adler, Sam Fire, Nichol Thompson, Kathy Sherak-Chen, Kam Ophir, Richard Calendar, Judith Geller, Ken Kemphues, Mark Benvenuto, Judith Kimble, Nancy Hopkins, Susan Strome, Michael McCaffery, Kathy Berkner, Rich Breyer, Ann Brunet, Kirsten Crossgrove, Richard Jorgensen, Greg Wiederrecht, Ellen Cammon, Allan Shearn, 7_03 Students, Wendy Locks, Jim Darnell, Rebecca Raitzig, Kwok Han Lian, William Pavao, Baltimore-Washington Worm Club, Karla Kirkegaard, Ihor Lemischka, Miriam Fire, Ron Millar, Laura Loveland, Linda Henry, Lewis Chodosh, Mr. Steffen, Harold Weintraub, Path_218 Students, Joe Vokroy, Lois Edgar, Bill Reichel, Natasha Caplen, Ms. Escolar, the Fremont Math Club, Ms. Wilson, Nancy Blachman, Andy Hopkins, Jim McGhee, Hung Hsi Wu, Charles Yanofsky, Felix Khuner, De Anza College, Craig Mello, Steven Leong, Ken Lorell, Tim Schedl, Marcus Thompson, Robert Herman, Ilil Carmi, Heather McCullough, Gideon Eisenberg, Frederica Postman, Anne Villeneuve, Don Doering, Dan Donoghue, Margalit Krantz-Fire, Sarah Sheaffer, The Shapiro Family, Elias Spelioros, Peter Sklar, Ms. Beaufenkamp, Kim Kent, Nelson Blachman, Allan Spradling, Scott Hammond, Geil Shokat, David Baulcombe, Helen Hart, Jerry Fox, Menachem Ophir, Karen Lamarco, Michel Goedert, Ken Fisher, Zeke Gluzband, Molly Marcus, Don Riddle, Matt vanderRijn, The Helen Hay Whitney Foundation, Sim Esquela, The National Science Foundation, Diane Chaen, Tamir Ophir, Elizabeth Lincoln, Gavi Swerling, Jay Maniar, Lisa Sklar, Cori Bargmann, Sarah Barker, Mark Edgley, David Botstein, Jonathan Khuner, James Duc, Jim Lewis, Al Rosenfeld, Rachel Grossman, Doris Lorell, Richard Sutch, Jeff Shamma, Hannes Vogel, Bernie Elspas, Bruce Hoover, Jeff Yuan, Ky Sha, Suzy Halvorsen, Gloria Brienza, Chris Walsh, Forrest Foor, Sydney Brenner, Mike Pinney, Georgia Rosenblatt, Marvin Rosenfeld, Shrage Ophir, Micah Glass-Siegel, Rainer Sachs, Mark Kay, Greg Robinson, Connie Clay, Andrea Swerling, Greg Hannon, Tom Leong, Femke Simmer, Shin Lin, Dan Riordan, Tom Lee, Samantha Glassman, Mark Brencher, Doug Vollrath, Steve Galli, Louise Pape, Nadia Rosenfeld, Megan Jacoby, Florence Locks, Mary Lou Pardue, Jeff Levinsky, The National Institute of General Medical Sciences, Ms. Benevides, Dan Stinchcomb, Steve Carr, David Baltimore, Aaron Mitchell, Prank Solomon, Aurora Kerscher, Saul Roseman, Arnab Aryana, Timothy Bach, Roger Kornberg, Theresa Fritchle, Jamie Fleenor, Gladys Sherak, Sherrie Rakvin, Julie Baker, Joel Postman, Arlene Oyama, Fremont High School, Jim Priess, Tom McDonough, Beth Hare, Don Katz, Karen Beemon, Nelson Tandoc, Antoinette Glumac, Diane Leong, Nancy Maizels, Virginia Walbot, Phil Zamore, Mike Cherry, David Rosenfeld, Marc Shinn-Krantz, Hugh Tyson, Doug Fambrough, Michael Wassenegger, David Shore, Robert Horvitz, Matthew Grossman, Jack Lorell, Izzy Goren, Gary Struhl, Roy Halvorsen, Colin Hampton, David Lipman, Lester Dubins, Carl Baker, Jimo Borjigin, Erika Matunis, David Krantz, Bob Kingston, Ciel Berman, Tamara Doering, Martha Elspas, Teymour Boutros Ghalli, Jim Barsoum, Paul Miller, SuSun Slatky, Norma Fire, The University of California Santa Cruz, Shinting Chen, Gunther Stent, Steve Elloge, David Hirsch, Miriam Adler, Doug Koshland, Sherman Elspas, Ms. Pucile, Marcia Glass-Siegel, Alan Zahler, William Kelly, Connie Jewell, The University of California Berkeley, Jocelyn Shaw, Dennis Dixon, William Rossi, Peter Lawrence, Steve Kostas, Al Scott, The Rita Allen Foundation, Rick Irvin, Silas Xu, David Nelson, Kathy Wilson, Peter Kulakow, Tom Tullius, Jenny Hsieh, Uttam Rajbhandary, Susan Dymecki, Chelsea Glassman, Iva Greenwald, Judith Yanowitz, William Wood, Steve Wasserman, Ann Rose, Clarissa Cangelosi, Ann-Marie Macdonald, Weng Onn Lui, Tim Hunt, Michael Sklar, Genetics 235 Students, Mark Palm, Spice Kleinman, Mia Horowitz, Ellen Henderson, Michael Shokat, Erich Jarvis, Ann Marie Murphy, Jeff Axelrod, John Pringle, Alejandro Sanchez, Ulla Hansen, Deborah Robbins, Lisa Timmons, William Sherak, Lincoln Stein, Casonya Johnson, Nancy Locks, Bill Courchesne, Greg Barsh, Ronald Plasterk, David Meyrowitz, Orit Ophir, Trelligans Inc, Chris Phillips, Becky Hohman, Stan Finkel, Steven Tschantz, Scott Rosenfeld, Robert Brown, Phil Cormier, Hilla Keren, Rachel Meyrowitz, Laurie Kovens, Nina Sherak, Brian Horblit, Howard Sklar, Ben Sher, Susan Parrish, Peter Sorger, Philip Fire, Miles Greiner, Boris Magasanik, Frank Laski, John White, Jake Raitzig, Chen Ming Fan, Carrie Athans, Paul Shimmel, Daniel Shinn-Krantz, Tom Changnan, Sam Dedras, Susan Kern, Kyung Lee, Kiyoshi Nagai, Verena Jantsch, Jonathan Adler, Nigel Crawford, Sam Ward, HsuTze Lee, Bruce Vogel, Robert Ely, David Miller, Joe Gall, Jonathan Gent, Jonathan Hodgkin, Susan Blachman, Miriam Fox, Margarita Siafaca, Michael Senturia, Steve L'Hernault, Marian Barber, Michael Adler, Diane Shakes, Naudia Lauder, Saul Glass-Siegel, Donald Brown, Stanford University, Marnie Halpern, Pat Englar, Judith Swan, Malcolm Gefter, Ari Krantz, Maitreya Levanchild, Barbara Messenger, The National Institutes of Health, Michael Brent, Libsia Chen, Elizabeth Glassman, John Sulston, Paul Sternberg, Susanna Lewis, Bingwei Lu, Man Wah Tan, Gary Tanegawa, Ann Blachman, Hiroaki Kagawa, Tina Trapane, Cecilia Mello, Vaun Loveland, Jan Riordan, Eric Fyrberg, Tom Loveland, Dan Blachman, Barbara Sollner-Webb, Dennis Balinger, David Friedheim, Bill Kupiec, Zhou Wang, Earl Potts, Narry Kim, Mike Fuller, Stella Kerscher, Carol Berktower, Carolyn Norris, Mary Waye, Russell Rarity, Paina Ophir, Guy Benian, Lucy Sherak, Chalermporn Ongvarrasopone, Janet Fire, Rich Mulligan, Robert Rosenfeld, Michael Klass, Richard Morgan, Iris Martinez, The Carnegie Institution of Washington, Rachel Hertzman, Witold Filipowicz, Andy Hoyt, Janet Williams, Doug Melton, Kaja Arusha, Joel Schildbach, Johns Hopkins University, Calvin Moore, Mortimer Locks, Yossi Gruenbaum, Connie Cepko, 020_731 Students, Irving Zucker, RuChih Huang, Sam Driver, Ian Purse, JooHong Ahnn, Maurice Bessman, Gene Brown, Lenny Brand, The Medical Research Council Laboratory of Molecular Biology, Karen Perry, Oliver Kerscher, Brian Harfe, Glenese Johnson, Rose Stern, Rose Kass, Sam Katz, Chaya Krishna, SiQun Xu, Ruth Starczyk, Elsabetta Ullu, Donna Albertson, Math_51C Students, Harvey Lodish, Stan Balazar, Tom Fulton, Phil Pizzo, 020_348 Students, Rosa Alcazar, Sarah Hammonstree, Karen Chapman, Adam Rosenblatt, Louisa Ho, Richard Pagano, Marianne Bienz, Nicholas Hammonstree, The American Cancer Society, Tom Blumenthal, Anna Esquela, Constance Sherak, Hugh Pelham, Monty Lerner, Karin Kopciak, Corwin Shokat, Paula Grabowski, Michael Wilcox, Mohammed Islam, Allen Strause, Nathan Sato, Rebecca Krantz, Magda Konarska, Sally Robinson-Seaver, Marry Chalfie, Bino Palmer, David Landsman, Steve Hardy, Michael Raitzig, Seth Alberts, Ilana Sherak, Jonathan Zucker, Mary Esteve, Richard Craddock, Eric Haag, Lori Steffy, Mango Junior High School, Crystal Myers, Donald Sherak, Mei Hsu, Elizabeth Lee, Tabitha Doniach, William Whitson, Steven Locks, Micro 233 Students, Stan Cohen, Richard Padgett, Larry Lasky, Anthony Hyman, Robert Sherak, Tom Tuschl, Charles Vinson, Bruce Reynolds, Doug Harrison, Lia Gracey, Patty Winningham, Mike Sepanski, Pat Cammon, Mike Cangelosi, Richard Swerling, Marina Ratner, Kelly Liu, Amy Alper, David Eisenmann, Joe Lipsick, Monroe Postman, Phil Newmark, Steve McKnight, Richard Field, Ms. Mayfield, Allan Kensky, Bernard Sherak, William Matlack, Trina Shroer, Kyle Cunningham, Steve Winans, Eric Barklis, more...

Figure 14. A few of the people and groups that the author would like to acknowledge for their help, support, encouragement. The list is in computationally randomized order with a few omissions (apologies) and misspellings (apologies).

REFERENCES

1. Watson, J. and Crick, F. (1953) Molecular structure of nucleic acids; a structure for deoxyribose nucleic acid. Nature 171: 737-738.
2. Rich, A., and Davies, D. (1956) A new two-stranded helical structure: polyadenylic acid and polyuridylic acid. J. Am. Chem. Soc. 78, 3548.
3. Volkin, E. and Astrachan, L. (1956) Phosphorus incorporation in Escherichia coli ribo-nucleic acid after infection with bacteriophage T2. Virology 2:149-161.
4. Hall, B. and Spiegelman, S. (1961) Sequence complementarity of T2-DNA and T2-specific RNA. PNAS 47:137-163.
5. Spencer, M., Fuller, W., Wilkins, M., and Brown, G. (1962) Determination of the helical configuration of ribonucleic acid molecules by X-ray diffraction study of crystalline amino-acid-transfer ribonucleic acid. Nature 194:1014-1020.
6. Rosenberg, J., Seeman, N., Kim, J., Suddath, F., Nicholas, H., and Rich, A. (1973) Double helix at atomic resolution. Nature 243:150-154.
7. Montagnier, L. and Sanders, F. (1963) Replicative form of encephalomyocarditis virus Ribonucleic Acid. Nature 199:664-667.
8. Weissmann, C., Borst, P., Burdon, R., Billeter, M., and Ochoa, S. (1964) Replication of viral RNA, III. Double-stranded replicative form of MSW phage RNA. PNAS 51:682-690.
9. Baltimore, D., Becker, Y., and Darnell, J. (1964) Virus-specific double-stranded RNA in poliovirus-infected cells. Science 143:1034-1036.
10. Hoskins, M. (1935) A protective action of neurotropic against viscerotropic yellow fever virus in Macacus rhesus. American Journal of Tropical Medicine, 15, 675-680.
11. Findlay, G. and MacCallum, F. (1937) An interference phenomenon in relation to yellow fever and other viruses. J. Path. Bact. 44, 405-424.
12. Isaacs, F. and Lindenmann, J. (1957) Virus Interference. I. The Interferon. Proc. Royal Soc. B 147, 268-273.
13. Shope, R. (1953) An antiviral substance from Penicillium funiculosum. J. Exp. Med. 97:601-650.
14. Lampson, G., Tytell, A., Field, A., Nemes, M., and Hilleman, M. (1967) Inducers of interferon and host resistance. I. Double-stranded RNA from extracts of Penicillium funiculosum. PNAS 58:782-789.
15. Field, A., Tytell, A., Lampson, G., and Hilleman, M. (1967) Inducers of interferon and host resistance. II. Multistranded synthetic polynucleotide complexes. PNAS 58:1004-1010.
16. Tytell, A., Lampson, G., Field, A., and Hilleman, M. (1967) Inducers of interferon and host resistance. 3. Double-stranded RNA from reovirus type 3 virions [reo 3-RNA]. PNAS 58:1719-1722.
17. Field, A., Lampson, G., Tytell, A., Nemes, M., and Hilleman, M. (1967) Inducers of interferon and host resistance, IV. Double-stranded replicative form RNA [MS2-Ff-RNA] from E. coli infected with MS2 coliphage. PNAS 58:2102-2108.
18. Wingard, S. (1928) Hosts and symptoms of ringspot, a virus disease of plants. J. agric. Res. 37 127-153.
19. Price, W. (1936) Virus concentration in relation to acquired immunity from tobacco ringspot. Phytopathology 26: 503-529.
20. Brenner, S. (2003) Nature's gift to science [Nobel lecture]. Chembiochem 4:683-687.
21. Kimble, J., Hodgkin, J., Smith, T., and Smith, J. (1982) Suppression of an amber mutation by microinjection of suppressor tRNA in C. elegans. Nature 299:456-458.
22. Stinchcomb, D., Shaw, J., Carr, S., and Hirsh, D. (1985) Extrachromosomal DNA transformation of Caenorhabditis elegans. Mol. Cell. Biol. 5:3484-3496.
23. Fire, A. (1986) Integrative transformation of Caenorhabditis elegans. EMBO J. 5:2673-2680.

24.   Mello, C., Kramer, J., Stinchcomb, D., and Ambros, V. (1991) Efficient gene trans-
       fer in *C.elegans:* extrachromosomal maintenance and integration of transforming
       sequences. EMBO J 10:3959-3970.
25.   Mello, C. and Fire, A. (1995) DNA transformation. Methods Cell Biol 48:451-482.
26.   Brenner, S. (1974) The genetics of *Caenorhabditis elegans.* Genetics 77:71-94.
27.   Moerman, D. and Baillie, D. (1979) Genetic Organization in *Caenorhabditis elegans:*
       Fine-Structure Analysis of the *unc-22* Gene. Genetics 91:95-103.
28.   Moerman, D., Benian, G., and Waterston, R. (1986) Molecular cloning of the mus-
       cle gene *unc-22* in *Caenorhabditis elegans* by Tc1 transposon tagging. PNAS 83:2579-
       2583.
29.   Fire, A. and Moerman, D. (1986) Transgenic Twitchers. Worm Breeder's Gazette
       9[2]:13-15.
30.   Stephenson, M. and Zamecnik, P. (1978) Inhibition of Rous sarcoma viral RNA
       translation by a specific oligodeoxyribonucleotide. PNAS 75:285-288.
31.   Izant, J. and Weintraub, H. (1984) Inhibition of thymidine kinase gene expression
       by anti-sense RNA: a molecular approach to genetic analysis. Cell 36:1007-1015.
32.   Fire, A. and Harrison, S. (1988) Sense and Antisense Disruption of Muscle Genes.
       Worm Breeder's Gazette 10[2]: 89-90.
33.   Fire, A., Albertson, D., Harrison, S., and Moerman, D. (1991) Production of anti-
       sense RNA leads to effective and specific inhibition of gene expression in *C. elegans*
       muscle. Development 113:503-514.
34.   Guo, S. and Kemphues, K. (1995) par-1, a gene required for establishing polarity in
       C. elegans embryos, encodes a putative Ser/Thr kinase that is asymmetrically distrib-
       uted. Cell 81:611-620.
35.   Rocheleau, C., Downs, W., Lin, R., Wittmann, C., Bei, Y., Cha, Y., Ali, M., Priess, J.,
       and Mello, C. (1997) Wnt signaling and an APC-related gene specify endoderm in
       early *C. elegans* embryos. Cell 90:707-716.
36.   Kuwabara, P. (1996) Interspecies comparison reveals evolution of control regions in
       the nematode sex-determining gene tra-2. Genetics 144:597-607.
37.   Seydoux, G. and Fire, A. (1994) Soma-germline asymmetry in the distributions of
       embryonic RNAs in *Caenorhabditis elegans.* Development 120:2823-2834.
38.   Soukup, G. and Breaker, R. (1999) Relationship between internucleotide linkage
       geometry and the stability of RNA. RNA 5:1308-1325.
39.   Mellits, K., Pe'ery, T., Manche, L., Robertson, H., and Mathews, M. (1990) Removal
       of double-stranded contaminants from RNA transcripts: synthesis of adenovirus VA
       RNAi from a T7 vector. Nucleic Acids Res 18:5401-5406.
40.   Fire, A., Xu, S., Montgomery, M., Kostas, S., Driver, S., and Mello, C. (1998) Potent
       and specific genetic interference by double-stranded RNA in *Caenorhabditis elegans.*
       Nature 391:806-811.
41.   Montgomery, M., Xu, S., and Fire, A. (1998) RNA as a target of double-stranded
       RNA-mediated genetic interference in *Caenorhabditis elegans.* PNAS 95:15502-
       15507.
42.   Tomizawa, J., Itoh, T., Selzer, G., and Som, T. (1981) Inhibition of ColE1 RNA
       primer formation by a plasmid-specified small RNA. PNAS 78:1421-1425.
43.   Chalker, D., and Yao, M. (2001) Nongenic, bidirectional transcription precedes and
       may promote developmental DNA deletion in Tetrahymena thermophila. Genes
       Dev 15:1287-1298.
44.   Mochizuki, K., Fine, N., Fujisawa, T., and Gorovsky, M. (2002) Analysis of a piwi-re-
       lated gene implicates small RNAs in genome rearrangement in tetrahymena. Cell
       110:689-699.
45.   Yang, V., Lerner, M., Steitz, J., and Flint, S. (1981) A small nuclear ribonucleopro-
       tein is required for splicing of adenoviral early RNA sequences. PNAS 78:1371-
       1375.
46.   Wassenegger, M., Heimes, S., Riedel, L., and Sanger, H. (1994) RNA-directed de
       novo methylation of genomic sequences in plants. Cell 76:567-576.

47. Ingelbrecht, I., Van Houdt, H., Van Montagu, M., and Depicker, A. (1994) Posttranscriptional silencing of reporter transgenes in tobacco correlates with DNA methylation. PNAS 91:10502-10506.
48. Olsen, P. and Ambros, V. (1999) The lin-4 regulatory RNA controls developmental timing in *Caenorhabditis elegans* by blocking LIN-14 protein synthesis after the initiation of translation. Dev. Biol. 216:671-680.
49. Gall, J., and Pardue, M. (1969) Formation and detection of RNA-DNA hybrid molecules in cytological preparations. PNAS 63:378-383.
50. Chalfie, M., Tu, Y., Euskirchen, G., Ward, W., and Prasher, D. (1994) Green fluorescent protein as a marker for gene expression. Science 263:802-805.
51. Heim, R., Cubitt, A., and Tsien, R. (1995) Improved green fluorescence. Nature 373:663-664.
52. Kelly, W., Xu, S., Montgomery, M., and Fire, A. (1997) Distinct requirements for somatic and germline expression of a generally expressed *Caenorhabditis elegans* gene. Genetics 146:227-238.
53. Timmons, L. and Fire, A. (1998) Specific interference by ingested dsRNA. Nature 395:854.
54. Tabara, H., Grishok, A., and Mello, C. (1998) RNAi in *C. elegans*: soaking in the genome sequence. Science 282:430-431.
55. Kennerdell, J. and Carthew, R. (1998) Use of dsRNA-mediated genetic interference to demonstrate that frizzled and frizzled 2 act in the wingless pathway. Cell 95:1017-1026.
56. Misquitta, L. and Paterson, B. (1999) Targeted disruption of gene function in *Drosophila* by RNA interference [RNA-i]: a role for nautilus in embryonic somatic muscle formation. PNAS 96:1451-1456.
57. Ngo, H., Tschudi, C., Gull, K., and Ullu, E. (1998) Double-stranded RNA induces mRNA degradation in Trypanosoma brucei. PNAS 95:14687-14692.
58. Waterhouse, P., Graham, M., and Wang, M. (1998) Virus resistance and gene silencing in plants can be induced by simultaneous expression of sense and antisense RNA. PNAS 95:13959-13964.
59. Ui-Tei, K., Zenno, S., Miyata, Y., and Saigo, K. (2000) Sensitive assay of RNA interference in *Drosophila* and Chinese hamster cultured cells using firefly luciferase gene as target. FEBS Lett. 479:79-82.
60. Wianny, F. and Zernicka-Goetz, M. (2000) Specific interference with gene function by double-stranded RNA in early mouse development. Nat. Cell. Biol. 2:70-75.
61. Svoboda, P., Stein, P., Hayashi, H., and Schultz, R. (2000) Selective reduction of dormant maternal mRNAs in mouse oocytes by RNA interference. Development 127:4147-4156.
62. Caplen, N., Fleenor, J., Fire, A., and Morgan, R. (2000) dsRNA-mediated gene silencing in cultured *Drosophila* cells: a tissue culture model for the analysis of RNA interference. Gene 252:95-105.
63. Selker, E., Cambareri, E., Jensen, B., and Haack, K. (1987) Rearrangement of duplicated DNA in specialized cells of *Neurospora*. Cell 51:741-752.
64. Goyon, C. and Faugeron, G. (1989) Targeted transformation of *Ascobolus* immersus and de novo methylation of the resulting duplicated DNA sequences. Mol. Cell. Biol. 9:2818-2827.
65. Napoli, C., Lemieux, C., and Jorgensen, R. (1990) Introduction of a Chimeric Chalcone Synthase Gene into Petunia Results in Reversible Co-Suppression of Homologous Genes in trans. Plant Cell 2:279-289.
66. van der Krol, A., Mur, L., Beld, M., Mol, J., and Stuitje, A. (1990) Flavonoid genes in petunia: addition of a limited number of gene copies may lead to a suppression of gene expression. Plant Cell 2:291-299.
67. Romano, N. and Macino, G. (1992) Quelling: transient inactivation of gene expression in *Neurospora* crassa by transformation with homologous sequences. Mol. Microbiol. 6:3343-3353.

68. Cameron, F. and Jennings, P. (1991) Inhibition of gene expression by a short sense fragment. Nucleic Acids Res. 19:469-475.

69. Cambareri, E., Jensen, B., Schabtach, E., and Selker, E. (1989) Repeat-induced G-C to A-T mutations in *Neurospora*. Science 244:1571-1575.

70. deCarvalho, F., Gheysen, G., Kushir, S., Montagu, M., Inzé, D., and Castensana, C. (1992). Suppression of ß-1,3-glucanase transgene expression in homozygous plants. EMBO J. 11, 2595-2602.

71. Cogoni, C., Irelan, J., Schumacher, M., Schmidhauser, T., Selker, E., and Macino, G. (1996) Transgene silencing of the al-1 gene in vegetative cells of *Neurospora* is mediated by a cytoplasmic effector and does not depend on DNA-DNA interactions or DNA methylation. EMBO J. 15:3153-3163.

72. Palauqui, J., Elmayan, T., Pollien, J., and Vaucheret, H. (1997) Systemic acquired silencing: transgene-specific post-transcriptional silencing is transmitted by grafting from silenced stocks to non-silenced scions. EMBO J. 16:4738-4745.

73. Voinnet, O. and Baulcombe, D. (1997) Systemic signalling in gene silencing. Nature 389:553.

74. Abel, P., Nelson, R., De, B., Hoffmann, N., Rogers, S., Fraley, R., and Beachy, R. (1986) Delay of disease development in transgenic plants that express the tobacco mosaic virus coat protein gene. Science 232:738-743.

75. Lindbo, J., Silva-Rosales, L., Proebsting, W., and Dougherty, W. (1993) Induction of a Highly Specific Antiviral State in Transgenic Plants: Implications for Regulation of Gene Expression and Virus Resistance. Plant Cell 5:1749-1759.

76. Kumagai, M., Donson, J., della-Cioppa, G., Harvey, D., Hanley, K., and Grill, L. (1995) Cytoplasmic inhibition of carotenoid biosynthesis with virus-derived RNA. PNAS 92:1679-1683.

77. Anandalakshmi, R., Pruss, G., Ge, X., Marathe, R., Mallory, A., Smith, T., and Vance, V. (1998) A viral suppressor of gene silencing in plants. PNAS 95:13079-13084.

78. Kasschau, K. and Carrington, J. (1998) A counterdefensive strategy of plant viruses: suppression of posttranscriptional gene silencing. Cell 95:461-470.

79. Beclin, C., Berthome, R., Palauqui, J., Tepfer, M., and Vaucheret, H. (1998) Infection of tobacco or *Arabidopsis* plants by CMV counteracts systemic post-transcriptional silencing of nonviral [trans]genes. Virology 252:313-317.

80. Brigneti, G., Voinnet, O., Li, W., Ji, L., Ding, S., and Baulcombe, D. (1998) Viral pathogenicity determinants are suppressors of transgene silencing in Nicotiana benthamiana. EMBO J. 17:6739-6746.

81. Ratcliff, F., Harrison, B., and Baulcombe, D. (1997) A similarity between viral defense and gene silencing in plants. Science 276:1558-1560.

82. Metzlaff, M., O'Dell, M., Cluster, P., and Flavell, R. (1997) RNA-mediated RNA degradation and chalcone synthase A silencing in petunia. Cell 88:845-854.

83. Parrish, S., Fleenor, J., Xu, S., Mello, C., and Fire, A. (2000) Functional anatomy of a dsRNA trigger: differential requirement for the two trigger strands in RNA interference. Mol. Cell. 6:1077-1087.

84. Yang, D., Lu, H., and Erickson, J. (2000) Evidence that processed small dsRNAs may mediate sequence-specific mRNA degradation during RNAi in *Drosophila* embryos. Curr. Biol. 10:1191-1200.

85. Tabara, H., Sarkissian, M., Kelly, W., Fleenor, J., Grishok, A., Timmons, L., Fire, A., and Mello, C. (1999) The rde-1 gene, RNA interference, and transposon silencing in *C. elegans*. Cell 99:123-132.

86. Tabara, H., Hill, R., Mello, C., Priess, J., and Kohara, Y. (1999) pos-1 encodes a cytoplasmic zinc-finger protein essential for germline specification in *C. elegans*. Development 126:1-11.

87. Tabara, H., Yigit, E., Siomi, H., and Mello, C. (2002) The dsRNA binding protein RDE-4 interacts with RDE-1, DCR-1, and a DExH-box helicase to direct RNAi in *C. elegans*. Cell 109:861-871.

88. Green, S. and Mathews, M. (1992) Two RNA-binding motifs in the double-stranded RNA-activated protein kinase, DAI. Genes Dev. 6:2478-2490.

89. Cox, D., Chao, A., Baker, J., Chang, L., Qiao, D., and Lin, H. (1998) A novel class of evolutionarily conserved genes defined by piwi are essential for stem cell self-renewal. Genes Dev. 12:3715-3727.

90. Bohmert, K., Camus, I., Bellini, C., Bouchez, D., Caboche, M., and Benning, C. (1998) AGO1 defines a novel locus of *Arabidopsis* controlling leaf development. EMBO J. 17:170-180.

91. Lynn, K., Fernandez, A., Aida, M., Sedbrook, J., Tasaka, M., Masson, P., and Barton, M. (1999) The Pinhead/Zwille gene acts pleiotropically in *Arabidopsis* development and has overlapping functions with the Argonaute1 gene. Development 126:469-481.

92. Zou, C., Zhang, Z., Wu, S., and Osterman, J. (1998) Molecular cloning and characterization of a rabbit eIF2C protein. Gene 211:187-194.

93. Elmayan, T., Balzergue, S., Beon, F., Bourdon, V., Daubremet, J., Guenet, Y., Mourrain, P., Palauqui, J., Vernhettes, S., Vialle, T., Wostrikoff, K., and Vaucheret, H. (1998) *Arabidopsis* mutants impaired in cosuppression. Plant Cell 10:1747-1758.

94. Cogoni, C. and Macino, G. (1997) Isolation of quelling-defective [qde] mutants impaired in posttranscriptional transgene-induced gene silencing in *Neurospora* crassa. PNAS 94:10233-10238.

95. Liu, Q., Rand, T., Kalidas, S., Du, F., Kim, H., Smith, D., and Wang, X. (2003) R2D2, a bridge between the initiation and effector steps of the *Drosophila* RNAi pathway. Science 301:1921-1925.

96. Catalanotto, C., Azzalin, G., Macino, G., and Cogoni, C. (2002) Involvement of small RNAs and role of the qde genes in the gene silencing pathway in *Neurospora*. Genes Dev. 16:790-795.

97. Okamura, K., Ishizuka, A., Siomi, H., and Siomi, M. (2004) Distinct roles for Argonaute proteins in small RNA-directed RNA cleavage pathways. Genes Dev. 18:1655-1666.

98. Hamilton, A. and Baulcombe, D. (1999) A species of small antisense RNA in post-transcriptional gene silencing in plants. Science 286:950-952.

99. Tuschl, T., Zamore, P., Lehmann, R., Bartel, D., and Sharp, P. (1999) Targeted mRNA degradation by double-stranded RNA in vitro. Genes Dev. 13:3191-3197.

100. Hammond, S., Bernstein, E., Beach, D., and Hannon, G. (2000) An RNA-directed nuclease mediates post-transcriptional gene silencing in *Drosophila* cells. Nature 404:293-296.

101. Zamore, P., Tuschl, T., Sharp, P., and Bartel, D. (2000) RNAi: double-stranded RNA directs the ATP-dependent cleavage of mRNA at 21 to 23 nucleotide intervals. Cell 101:25-33.

102. Bernstein, E., Caudy, A., Hammond, S., and Hannon, G. (2001) Role for a bidentate ribonuclease in the initiation step of RNA interference. Nature 409:363-366.

103. Elbashir, S., Lendeckel, W., and Tuschl, T. (2001) RNA interference is mediated by 21- and 22-nucleotide RNAs. Genes Dev 15:188-200.

104. Manche, L., Green, S., Schmedt, C., and Mathews, M. (1992) Interactions between double-stranded RNA regulators and the protein kinase DAI. Mol. Cell. Biol. 12:5238-5248.

105. Elbashir, S., Harborth, J., Lendeckel, W., Yalcin, A., Weber, K., and Tuschl, T. (2001) Duplexes of 21-nucleotide RNAs mediate RNA interference in cultured mammalian cells. Nature 411:494-498.

106. Caplen, N., Parrish, S., Imani, F., Fire, A., and Morgan, R. (2001) Specific inhibition of gene expression by small double-stranded RNAs in invertebrate and vertebrate systems. PNAS 98:9742-9747.

107. Winston, W., Molodowitch, C., and Hunter, C. (2002) Systemic RNAi in *C. elegans* requires the putative transmembrane protein SID-1. Science 295:2456-2459.

108. Astier-Manifacier, S. and Cornuet, P. (1971) RNA-dependent RNA polymerase in Chinese cabbage. Biochim. Biophys. Acta 232:484-493.

109. Schiebel, W., Haas, B., Marinkovic, S., Klanner, A., and Sanger, H. (1993) RNA-directed RNA polymerase from tomato leaves. I. Purification and physical properties; II. Catalytic in vitro properties. J. Biol. Chem. 268:11851-11867.

110. Schiebel, W., Pelissier, T., Riedel, L., Thalmeir, S., Schiebel, R., Kempe, D., Lottspeich, F., Sanger, H., and Wassenegger, M. (1998) Isolation of an RNA-directed RNA polymerase-specific cDNA clone from tomato. Plant Cell 10:2087-2101.

111. Cogoni, C. and Macino, G. (1999) Gene silencing in *Neurospora* crassa requires a protein homologous to RNA-dependent RNA polymerase. Nature 399:166-169.

112. Smardon, A., Spoerke, J., Stacey, S., Klein, M., Mackin, N., and Maine, E. (2000) EGO-1 is related to RNA-directed RNA polymerase and functions in germ-line development and RNA interference in *C. elegans*. Curr. Biol. 10:169-178.

113. Mourrain, P., Beclin, C., Elmayan, T., Feuerbach, F., Godon, C., Morel, J., Jouette, D., Lacombe, A., Nikic, S., Picault, N., Remoue, K., Sanial, M., Vo, T., and Vaucheret, H. (2000) *Arabidopsis* SGS2 and SGS3 genes are required for posttranscriptional gene silencing and natural virus resistance. Cell 101:533-542.

114. Dalmay, T., Hamilton, A., Rudd, S., Angell, S., and Baulcombe, D. (2000) An RNA-dependent RNA polymerase gene in *Arabidopsis* is required for posttranscriptional gene silencing mediated by a transgene but not by a virus. Cell 101:543-553.

115. Sijen, T., Fleenor, J., Simmer, F., Thijssen, K., Parrish, S., Timmons, L., Plasterk, R., and Fire, A. (2001) On the role of RNA amplification in dsRNA-triggered gene silencing. Cell 107:465-476.

116. Vaistij, F., Jones, L., and Baulcombe, D. (2002) Spreading of RNA targeting and DNA methylation in RNA silencing requires transcription of the target gene and a putative RNA-dependent RNA polymerase. Plant Cell 14:857-867.

117. Makeyev, E. and Bamford, D. (2002) Cellular RNA-dependent RNA polymerase involved in posttranscriptional gene silencing has two distinct activity modes. Mol. Cell. 10:1417-1427.

118. Alder, M., Dames, S., Gaudet, J., and Mango, S. (2003) Gene silencing in *Caenorhabditis elegans* by transitive RNA interference. RNA 9:25-32.

119. Nicolas, F., Torres-Martinez, S., and Ruiz-Vazquez, R. (2003) Two classes of small antisense RNAs in fungal RNA silencing triggered by non-integrative transgenes. EMBO J. 22:3983-3991.

120. Pak, J. and Fire, A. (2007) Distinct populations of primary and secondary effectors during RNAi in *C. elegans*. Science 315:241-244.

121. Sijen, T., Steiner, F., Thijssen, K., and Plasterk, R. (2007) Secondary siRNAs result from unprimed RNA synthesis and form a distinct class. Science 315:244-247.

122. Dougherty, W. and Parks, T. (1995) Transgenes and gene suppression: telling us something new? Curr. Opin. Cell Biol. 7:399-405.

123. Grishok, A., Tabara, H., and Mello, C. (2000) Genetic requirements for inheritance of RNAi in *C. elegans*. Science 287:2494-2497.

124. Vastenhouw, N., Brunschwig, K., Okihara, K., Muller, F., Tijsterman, M., and Plasterk, R. (2006) Gene expression: long-term gene silencing by RNAi. Nature 442:882.

125. Park, Y., Papp, I., Moscone, E., Iglesias, V., Vaucheret, H., Matzke, A., and Matzke, M. (1996) Gene silencing mediated by promoter homology occurs at the level of transcription and results in meiotically heritable alterations in methylation and gene activity. Plant J. 9:183-194.

126. Lu, R., Maduro, M., Li, F., Li, H., Broitman-Maduro, G., Li, W., and Ding, S. (2005) Animal virus replication and RNAi-mediated antiviral silencing in *Caenorhabditis elegans*. Nature 436:1040-1043.

127. Wilkins, C., Dishongh, R., Moore, S., Whitt, M., Chow, M., and Machaca, K. (2005) RNA interference is an antiviral defence mechanism in *Caenorhabditis elegans*. Nature 436:1044-1047.

128. Ketting, R., Haverkamp, T., van Luenen, H., and Plasterk, R. (1999) Mut-7 of *C. elegans*, required for transposon silencing and RNA interference, is a homolog of Werner syndrome helicase and RNaseD. Cell 99:133-141.

129. Song, J., Smith, S., Hannon, G., and Joshua-Tor, L. (2004) Crystal structure of Argonaute and its implications for RISC slicer activity. Science 305:1434-1437.

130. Macrae, I., Zhou, K., Li, F., Repic, A., Brooks, A., Cande, W., Adams, P., and Doudna, J. (2006) Structural basis for double-stranded RNA processing by Dicer. Science 311:195-198.

131. Schwarz, D., Hutvagner, G., Du, T., Xu, Z., Aronin, N., and Zamore, P. (2003) Asymmetry in the assembly of the RNAi enzyme complex. Cell 115:199-208.

132. Khvorova, A., Reynolds, A., and Jayasena, S. (2003) Functional siRNAs and miRNAs exhibit strand bias. Cell 115:209-216.

133. Tonkin, L. and Bass, B. (2003) Mutations in RNAi rescue aberrant chemotaxis of ADAR mutants. Science 302:1725.

134. Liu, J., Carmell, M., Rivas, F., Marsden, C., Thomson, J., Song, J., Hammond, S., Joshua-Tor, L., and Hannon, G. (2004) Argonaute2 is the catalytic engine of mammalian RNAi. Science 305:1437-1441.

135. Deshpande, G., Calhoun, G., and Schedl, P. (2005) *Drosophila* argonaute-2 is required early in embryogenesis for the assembly of centric/centromeric heterochromatin, nuclear division, nuclear migration, and germ-cell formation. Genes Dev. 19:1680-1685.

136. Kalmykova, A., Klenov, M., and Gvozdev, V. (2005) Argonaute protein PIWI controls mobilization of retrotransposons in the *Drosophila* male germline. Nucleic Acids Res. 33:2052-2059.

137. Lee, R., Feinbaum, R., and Ambros, V. (1993) The *C. elegans* heterochronic gene lin-4 encodes small RNAs with antisense complementarity to lin-14. Cell 75:843-854.

138. Grishok, A., Pasquinelli, A., Conte, D., Li, N., Parrish, S., Ha, I., Baillie, D., Fire, A., Ruvkun, G., and Mello, C. (2001) Genes and mechanisms related to RNA interference regulate expression of the small temporal RNAs that control *C. elegans* developmental timing. Cell 106:23-34.

139. He, L., Thomson, J., Hemann, M., Hernando-Monge, E., Mu, D., Goodson, S., Powers, S., Cordon-Cardo, C., Lowe, S., Hannon, G., and Hammond, S. (2005) A microRNA polycistron as a potential human oncogene. Nature 435:828-833.

140. Lu, J., Getz, G., Miska, E., Alvarez-Saavedra, E., Lamb, J., Peck, D., Sweet-Cordero, A., Ebert, B., Mak, R., Ferrando, A., Downing, J., Jacks, T., Horvitz, H., and Golub, T. (2005) MicroRNA expression profiles classify human cancers. Nature 435:834-838.

141. Chiu, Y., Dinesh, C., Chu, C., Ali, A., Brown, K., Cao, H., and Rana, T. (2005) Dissecting RNA-interference pathway with small molecules. Chem. Biol. 12:643-648.

142. Draper, B., Mello, C., Bowerman, B., Hardin, J., and Priess, J. (1996) MEX-3 is a KH domain protein that regulates blastomere identity in early *C. elegans* embryos. Cell 87:205-216.

143. Yigit, E., Batista, P., Bei, Y., Pang, K., Chen, C., Tolia, N., Joshua-Tor, L., Mitani, S., Simard, M., and Mello, C. (2006) Analysis of the *C. elegans* Argonaute family reveals that distinct Argonautes act sequentially during RNAi. Cell 127:747-757.

144. Bao, N., Lye, K., and Barton, M. (2004) MicroRNA binding sites in *Arabidopsis* class III HD-ZIP mRNAs are required for methylation of the template chromosome. Dev. Cell 7:653-662.

145. Volpe, T., Kidner, C., Hall, I., Teng, G., Grewal, S., and Martienssen, R. (2002) Regulation of heterochromatic silencing and histone H3 lysine-9 methylation by RNAi. Science 297:1833-1837.

Portrait photo of Dr. Fire by photographer Linda Cicero.

CRAIG C. MELLO

I recall a sunny September morning in Virginia. I remember the sound of the school bus taking away the older kids, including my two siblings Jean and Frank. My mother, no doubt, was busy with my baby brother Roger. I was playing in the creek as I often did, turning over stones, looking for small animals. I remember a mourning dove cooing on the telephone wire, and the way the sunlight felt on my red sweatshirt and my rolled up hand-me-down blue jeans. I remember a sense of contentment with being alive, a feeling that infuses many of my early memories in a general, fuzzy, unfocused way. However, this memory is different. It was etched with stunning clarity in my mind by adrenalin and other sharper emotions. That morning, a box turtle decided to choose this peaceful moment to make its way across the street adjacent to the field where I was playing. My attention was drawn to the road by the sound of an oncoming car, and I remember my excitement with seeing the turtle changing to shock as I watched the car swerve with clear intention toward the turtle. I remember a smirking teenage boy driving off, leaving the turtle, his shell broken, still struggling to move to the edge of the road. The turtle died before my eyes, etching this scene deeply into my mind. Even though this might seem a sad memory, the fact is that I'm grateful in a sense. That morning of my youth seems timeless now. I can see in my heart that the child playing in the creek is me, and that I haven't changed much really in the intervening years. I'm still turning over stones, hoping to find something new. I'm still struggling to understand what drives us humans to cruelty and hoping that knowledge of our place in the world can help us to achieve a higher purpose.

I was born in New Haven, Connecticut on October 18th 1960, the third child of a paleontologist father and artist mother (James and Sally Mello). In 1962 my father completed his doctorate in paleontology at Yale University, and my family moved to Falls Church in northern Virginia so that he could take a position with the US Geological Survey (USGS) in Washington, DC. My parents met while attending Brown University and were the first children in their respective families to attend college. My grandparents on both sides withdrew from school as teenagers to work for their families. My paternal grandfather, Frank Mello, was of Azorean descent although he was born in Warren, Rhode Island. He was an outstanding athlete nicknamed "Bullet" Mello for his speed. He played semi-pro baseball and football. He worked a variety of jobs including delivering grain for many years and operating trucks for the town. My grandmother, Elena (Primiano) Mello, was of Italian descent, but was also born in Warren, Rhode Island, she worked in local textile factories. They both worked for their families for close to ten years before they were able to marry and start their own household at the age of 24. On my mother's side I have English and Scottish roots dating to colonial times and including a distant link to Lyman Hall who signed the declaration of independence. My maternal grandfather, William Cameron, ran a very successful plumbing business in Middletown, CT. My grandmother, Ida (Hall) Cameron, was a home- maker. I'm proud of my melting pot origins, and of the accomplishments of my grandparents. They worked hard, and sacrificed so that their children could go to college. They were wonderful, creative, thoughtful and extremely loving people who gave me a refreshing perspective on what's important in life.

After a brief stay in Falls Church, we moved to Fairfax, VA, when my father switched from the USGS to a position as assistant director at the Smithsonian Museum of Natural History. Among my fondest early memories are field trips with my father and the whole family to Colorado and Wyoming and more frequent trips to the Blue Ridge mountains in Virginia. I remember searching for fossils, hiking, exploring, and wonderful family discussions around the campfire.

My family had a very strong tradition of discussions around the dinner table. This experience was extremely important to me. I learned to argue, to listen, and to admit it (sometimes grudgingly) when I was wrong about something. These were often lively discussions, and my parents did a great job of allowing each of us to be heard. At a time when I was not performing so well in school, these daily discussions helped to build my confidence and self esteem. I struggled during the first few years of grade school. I started first grade at the age of 5 in a local private school because I was too young to enter first grade in the public system. I don't know if I was a slow learner, or just not interested, but I did not do well in school until the 7th grade. In second grade, I remember faking that I could read and the embarrassment of being called on in class. I much preferred playing outdoors, in the woods and creeks, to time spent in the classroom. Meanwhile, my older siblings were model students, raising the teacher's expectations for me. If not for the family discussions, where I was respected and could hold my own in arguments, I might have been discouraged with my academic prospects.

During these early years, I remember having no doubt that I would be a scientist when I grew up. I was amazed that so few adults (including my teachers) understood basic concepts such as deep (geologic) time, the vastness of the universe, and the common evolutionary origins of life. In first grade, the private school I attended had a Bible session each day, and I remember being shocked that the teacher presented the story of Noah and his ark as fact. Similarly, I was exposed to religious instruction in Sunday school. My father had agreed to raise us as Catholics. My mother was a Methodist by birth but did not practice her religion and did not attend Catholic services with the family. I remember learning the argument of intelligent design in Sunday school, as a counterargument to evolution. Given my own exposure to my dad's museum, and our family discussions about evolution and the history of the earth, these exposures to religious dogma actually had the effect of intensifying my interest in science as a way of knowing about the world.

By the time I was in middle school, I had decided to reject religious dogma altogether. The 'absolute knowledge' offered, was in my view, inadequate to explain the world around me. Furthermore, it seemed wrong to claim knowledge based on ones culture or upbringing. I saw the leap of faith involved in religion as smothering dialogue, closing the door on non-believers and walling them out of one's society. In contrast, the scientific method with its focus on asking questions and admitting no absolutes, was and continues to be refreshing to me. Science is grounded on, and values, dialogue. It is a human enterprise that breaks down walls and challenges its practitioners to admit ignorance and to question all ideas. However, we must all arrive at and defend our moral choices of right and wrong. Science can't touch these issues and shouldn't try. I believe that there is no more spiritual and worthwhile undertaking than that of trying to understand the world around us, and our place in it. The world is a far more remarkable place than we can imagine. Its mysteries define the human condition; to exist without knowing why.

My first exposure to academic science came in 7th grade, and during that year I can remember for the first time applying myself to my studies. I became an avid reader of science fiction, an amateur astronomer, and a serious student. I remember organizing my desk at home and doing homework, with music blasting, for at least a couple hours every night. I attended Fairfax High School, where I took all of the science courses offered except advanced physics. My earth science, chemistry and biology teachers were excellent. My biology teacher, Randy Scott, was also my wrestling, football and track coach. He was a wonderful man, who had a large role in fostering my interest in biology. I reconnected with him and was able to thank him after the news of October, but tragically, he has since lost his battle with cancer.

In 1978, I learned about molecular biology from a newspaper article in the Washington Post. The article described the cloning of the human insulin gene in bacteria, and described how the bacterial cells were able to read the human genetic code and produce functional human insulin. I found this concept incredible and extremely exciting. Incredible because the bacterial cells were able to speak the same language as the human cells, reading out the genetic code to make functional, life-giving, human protein for diabetic patients. Prior to that time, diabetics used animal insulin. I found this extremely exciting because I could see the potential for understanding disease at the genetic level and for treating it with molecular medicines, like insulin, and with gene therapy.

At Brown University, I pursued biochemistry and molecular biology as my major and had inspiring teachers, including Frank Rothman, Ken Miller, Susan Gerbi and Nelson Fausto. Brown provided a wonderful environment for learning, and had the added benefit of being close to my grandparents' home in Warren, RI, and to my small sailboat on the Warren River, a tributary to the upper Narragansett Bay. Sailing continues to be an important part of my life. It gives me a sense of place that settles and refreshes my mind. I sail in a wide range of conditions, for hours and hours (if possible). I don't prefer to race, but rather to explore. I gradually trained my parents and grandparents to accept the fact that if the wind died, then I might not get back until long after dark. Eventually, they even agreed to my taking camping gear and doing overnight trips ranging along the coast from Martha's Vineyard to half way up Long Island Sound.

After Brown, I went to Colorado for graduate school, where I enjoyed the mountains again and a really fantastic and inspiring course in molecular, cellular and developmental biology. The course consisted of a small group of 15 or so students with outstanding instructors, including Drs. Dick McIntosh, Mike Yarus, Larry Gold, Bill Wood and others. At Boulder, I was introduced to *C. elegans* in the laboratory of Dr. David Hirsh. David's lab was fantastic – filled with people who would prove to be really important in my future training. These included Dan Stinchcomb, who introduced me to the practice of molecular biology; Mike Krause, Jim Kramer, and Ken Kemphues, with whom I collaborated; and Jim Priess with whom I did my postdoctoral work.

When I joined David's lab in 1982, no one had succeeded in introducing DNA back into *C. elegans* (a method referred to as "DNA transformation"). Work in yeast had identified functional DNA elements that direct the replication and partitioning of chromosomes (replication origins and centromeres, respectively). Working with Dan Stinchcomb, my project was to identify such elements from the worm, with the goals of 1) understanding these essential functional chromosomal elements, and 2) of using them to produce stable artificial chromosomes for worm molecular genetics. During my first year in Boulder, David Hirsh decided to take a position in industry, and so I chose to move to Harvard University where I could continue my research with Dan Stinchcomb, who was starting up an independent lab there.

I thoroughly enjoyed Harvard! Dan set up his lab at the Biolabs in Cambridge next to Victor Ambros, another brand new, junior faculty member at Harward working on *C. elegans*. Dan and Victor integrated their labs to make a single "wormlab", and both served as advisors to me during my studies. I loved my project and worked long hours in the lab, never going home until I had a gel running or something incubating, so as to use the overnight hours. I took advantage of opportunities to attend lectures on a wide range of subjects. I obtained permission to use the large refracting telescope located atop the Science center, which was, surprisingly, available for individual use. I got to meet and teach with Stephen J. Gould, who's essays on natural history and the philosophy of science had inspired me over the years. Gould's, *"The Freezing of Noah,"* is one of my favorites as it captures the essence of good science; admitting when your theory is wrong and developing a new theory.

I learned an important lesson in graduate school; that it's not enough to be persistent and to work hard, it's also important to attack the question you wish to address from every conceivable angle. By focusing on identifying worm centromere activities using yeast as a model system, I ended up learning about the yeast centromere, not the worm centromere. While this project was fulfilling and interesting to me, it was flawed. To study the yeast centromere, I should have been working with the yeast sequences directly. To study the worm centromere, I should have been injecting DNA into the worm. Only after I began to experiment directly with the worm did my project really take off.

Technology is what drives science, and yet, developing new technology is often a thankless task. Getting something to work that has never been done before can be exceedingly frustrating because you may never know how close you were to success, and failures quite often teach you nothing. Partly because of this, those working on technology development often tend to band together and share ideas more than would otherwise be common among scientists. This was certainly the case for Andrew Fire and me. We were both working on developing techniques for DNA transformation in worms. Andy had some early success and developed a number of clever methods. I followed up with some improvements. And together we made DNA transformation a routine procedure for the worm. In the course of these studies, we became frequent correspondents, spending hours on the phone (before email was invented). We developed the mutual trust and respect that ultimately led to our collaboration on RNAi.

After graduating from Harvard, I joined the lab of Jim Priess at the Fred Hutchinson Cancer Research Center in Seattle Washington. Jim is one of those rare scientists who has "a feeling for the organism" as E.F. Keller put it when describing Barbara McClintock. Jim put me in touch with my own feelings for the worm. Through Jim, I was able to learn genetics, without which our later work on RNAi would have remained entirely descriptive. In Jim's lab, we identified genes that act as regulators of early development in *C. elegans*. It turns out that some of these genes are connected to RNAi-related mechanisms in ways that we are still trying to understand.

In Seattle, my daughter Melissa was born in 1992. I wish she could remember those first two years of her life in Seattle. We hiked and biked together regularly and had a wonderful time. However, her mother and I struggled to find enough time together as a family. Melissa's mother, Margaret Hunter, worked mornings and weekends as a chef at a Café in Seattle. Because my schedule often demanded late nights and weekends, we handed Melissa off from one to the other and rarely had enough time to be together as a family. Shortly after we moved to Massachusetts in 1994, we separated and divorced. Fortunately, we remain respectful and friendly to this day. I focused on my work and continued to have Melissa with me half of each week.

Shortly before I started my lab in 1994, I learned from Ken Kemphues and his student Sue Guo about an "antisense" RNAinjection technique that surprisingly well in *C. elegans* silenced target genes. I began using this method to study the genes we had identified during my genetic studies with Jim Priess. The genome-sequencing project for *C. elegans* had begun in earnest and had revealed dozens of genes in the sequence data base that were similar in DNA sequence to those that I had discovered in Jim's lab. These related genes (or homologs, as we call them) could have important developmental functions, and so I began using the RNA injection method described by Guo and Kemphues to silence them in order to identify those functions.

At that time, RNA injection was performed according to the same procedure that Andy and I had developed for DNA injection. A fine, sharp, glass needle was inserted with care through the cuticle of the worm and positioned inside the large shared cytoplasm of a gonad that contains hundreds of germline nuclei. After positioning the needle and injecting, the procedure was then carried out a second time on the other gonad arm, two injections per worm. The power of this gene-silencing approach accelerated our studies and we began to make rapid progress in understanding the developmental mechanisms that specify cell fate in the early embryo. However, we also became interested in the silencing phenomenon itself. The first observation that truly galvanized my interest occurred when, having injected RNA targeting *apx-1,* a gene essential for embryogenesis, I observed by chance that some embryos hatched and matured to adulthood only to produce 100% *apx-1* dead embryos. The silencing phenomenon had skipped a generation and had been passed on *via* the germline to the next generation! This was truly amazing and prompted further studies that demonstrated the transmission of silencing for multiple generations *via* both the sperm and the egg.

The first graduate student to work on RNAi in my lab, Sam Driver, discovered, in part by accident while learning to inject, that the RNA need not be delivered directly to the germline. Injection anywhere in the body was sufficient to induce interference that spread into the germline and was transmitted to progeny. These findings, along with the inheritance properties, and the lack of strand specificity (first noted by Guo and Kemphues), prompted us to recognize the silencing phenomenon as an active response in the organism to the RNA. To distinguish this mechanism from the earlier "antisense" methodology, we decided to give it the simple name RNAi (for RNA interference). We envisioned a mechanism where either strand could template the production of the other strand and could somehow build up silencing RNA levels. The specificity of the silencing indicated that ultimately, after amplification, the antisense strand must unwind from its complement to find its target RNA and induce silencing.

Throughout this period, Andy and I continued to correspond and collaborate. It was Andy's suggestion that dsRNA contaminating our preparations could be the actual trigger molecule underlying RNAi. At the time, I was still thinking of dsRNA as an amplification intermediate, rather than trigger. It was not until after Andy sent me purified double stranded RNA to test in my own hands that I became a believer in this molecule as a potent trigger for gene silencing. We now know that dsRNA is both a trigger and intermediate in RNAi. The concept of dsRNA as a trigger for sequence-specific gene silencing only makes sense if one recognizes that the organism is actively responding by unwinding the RNA strands both for amplification and to generate single strands capable of base pairing with targets. This concept of an active response in the animal prompted Hiroaki Tabara in my lab to undertake his exciting genetic studies that identified cellular gene products that mediate silencing. As discussed in my lecture, dsRNA is not the only trigger for this silencing mechanism. However, importantly, dsRNA turned out to be a highly conserved trigger that rapidly led to the application of RNAi in diverse species including humans.

1998 was a truly outstanding year. In January of that year, Andy and I published our paper on RNAi. In August, I married Edit Kiss and became the stepfather of two wonderful kids, David and Sarah Apotheker. In the year 2000 our daughter Victoria was born. In an unfortunate twist of fate, Victoria developed type-one diabetes in the fall of 2001. Suddenly, I had to learn how to inject into a human, my own daughter, for the first time. Ironically, human insulin, the same bacterially synthesized molecule that inspired me to pursue molecular biology, is now giving Victoria her very life. This experience has given me a new perspective on the importance of medical research. Edit, who is a wonderful nurse, is now taking care of Victoria, and serving as a diabetes counselor for newly diagnosed families.

With RNAi and the completion of the genome sequences for humans and numerous other organisms, we now have unprecedented opportunities to develop new, life saving therapies and to advance the basic understanding of our biology. We humans have a potentially very bright future. The biological mechanisms at work inside our cells are truly ancient and remarkably stable, more stable even than the positions of continents and oceans on the face of the Earth. However, in my view, our thriving global economy has engendered serious problems. Climate change and other forces beyond our control could easily disrupt our economies causing widespread human suffering at unprecedented levels. We are fishing out oceans, depleting our topsoils, and exhausting our sources of fossil fuel and fresh water. Scientists and policy makers must begin to work together to foster the development of technologies that are sustainable and resilient. As humans, we must work with common purpose around the world to prepare for the challenges and opportunities ahead. I hope that I can further that cause.

RETURN TO THE RNAi WORLD: RETHINKING GENE EXPRESSION AND EVOLUTION

Nobel Lecture, December 8, 2006

by

CRAIG C. MELLO

Howard Hughes Medical Institute and Program in Molecular Medicine, University of Massachusetts Medical School, 373 Plantation Street, Worcester, MA 01605, USA.

It's wonderful to be here today, I would like to start with the most important part, by saying thank you. First of all, I want to thank Andy Fire for being such a tremendous colleague, friend and collaborator going back over the years. Without Andy I definitely wouldn't be here today. I need to thank the University of Massachusetts for providing for my laboratory, for believing in me and for giving me not only a place and money, but great colleagues with whom to pursue my research. Without UMass and the great environment provided for me there, I probably would not be here today. And, of course my family; I'm not going to spend time now thanking them individually, but they know how important they are.

I'm going to talk today about *C. elegans* and the role of RNAi in *C. elegans* development. This animal is aptly named for its elegant simplicity (Figure 1). Only one millimeter in length and yet capable of producing 300 progeny in three days by self fertilization. One of the most beautiful things about *C. elegans*, immediately apparent upon viewing it in the microscope, is its transparency. Sydney Brenner recognized the importance of this

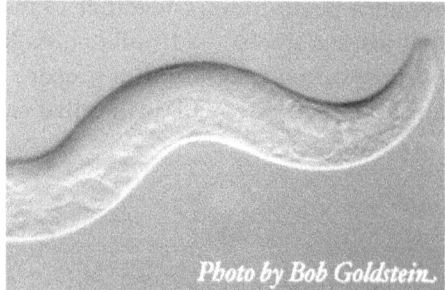

Photo by Bob Goldstein

Figure 1.

attribute when deciding what organism to work on. As animals go, *C. elegans* is relatively simple, having only about a thousand cells in the adult organism. Indeed, the origin and fate of every cell, both in the embryo and adult, has been determined – an amazing accomplishment. At any stage of development, you can look at a cell and know where that cell came from, tracing its origin back in time to the first division of the embryo.

It's a beautiful system. In fact, the researchers who work in *C. elegans* have their own lineage. Almost all of us can trace ourselves back to Sydney Brenner, who pioneered the modern genetic analysis of this organism. My particular ancestors, if you will, in the lineage of researchers are shown in Figure 2. I owe a tremendous amount of thanks to Dan Stinchcomb, for teaching me molecular biology and really being a fantastic mentor during my initial years in graduate school; Victor Ambros who, along with Dan, provided a wonderful joint laboratory at Harvard University where I did graduate work; and then Jim Priess, who taught me genetics, and was a tremendous mentor and a great friend out in Seattle where I conducted my postdoctoral research at the Fred Hutchinson Cancer Research Center. I owe a tremendous amount to these individuals. I'll show you more pictures of people I will need to thank as I go along.

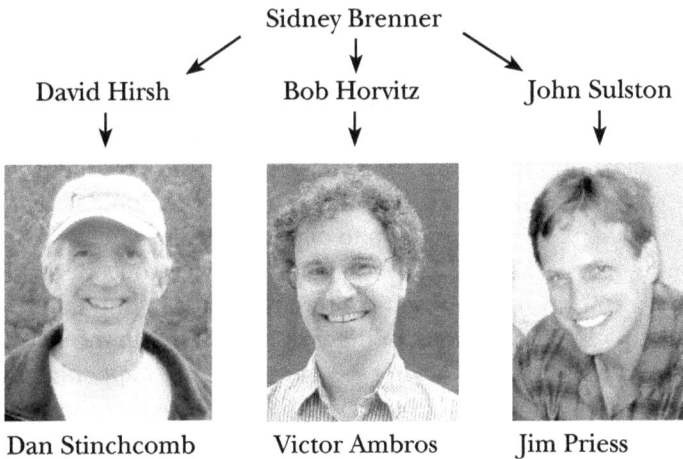

Figure 2.

I want to get down to the theme of my talk today, which really is, in part, about how we continually underestimate the complexity of life. It's the correction of these underestimations that is quite often what this prize is really recognizing. As science progresses, our knowledge expands, we think we understand, and too often we become overconfident. The fact is, I think we almost always underestimate the complexity of life and of nature. Today has been a true celebration of that beauty and complexity. I attended the Physics talks and the Chemistry talks and it was just spectacular to contemplate. An embryonic universe 13.7 billion years old, originally on the scale of inches across, expanding in seconds through a mysterious process of inflation to occupy the nearly infinite dimensions of space. And to explore the workings of a polymerase at the atomic level, whose origins derive from a common ancestor of all life on earth some 3.5 billion years ago.

These stories are so beautiful and stunning in their complexity. For every answer they provide they raise a thousand new questions. And so one thing I'd like to accomplish in my talk, is to raise questions that I can't answer, to talk about the unknown some more. Andy's done such a wonderful job of introducing the subject, giving me the luxury of spending some time talking about potential implications and some of the things we don't know but would love to understand in the future. It is the unknown that inspires us and sparks our curiosity, and so I'd like to try to focus more on telling you about what we don't know and on speculating on what is possible.

If one looks carefully, the complexity of living things becomes strikingly clear. Consider for example the natural environment of *C. elegans*. Figure 3 is an electron micrograph taken by George Barron, who works on nematophagous fungi. The unfortunate worm shown here has become ensnared in a trap set by a fungus that preys on nematodes. It really is a jungle out there for these poor little animals; they struggle to survive, just like the rest of us. The soil is filled with hundreds of different species of these fungi that prey on worms as they're swimming around in the soil. These fungi can sense the motion or contact of a worm and, after the worm has entered its lariat, the fungus inflates it to constrict the snare around the animal, trapping it. The fungus can then send hyphae into the worm to digest it. So, imagine that these poor elegant little animals are actually struggling to survive out there. Nature is filled with complexity that we don't appreciate. This is so tiny, that you would walk over millions, if not billions, of these little creatures in the soil every day on your way to work, never realizing the things that are happening there.

One of the great triumphs of biology was the discovery of the structure of DNA. The structure of DNA was first determined by Watson and Crick, who showed how two strands composed of four basic building blocks form polymers that intertwine in a beautiful helical staircase structure. This structure explains so much, really, about the basic biology of living things. It explains the segregation patterns, first described by Gregor Mendel, for certain genetic traits of pea plants. The structure alone, as Watson and Crick noted, suggests how the genetic material can be replicated. They stated in their famously brief paper in Nature [1] that, "It has not escaped our notice that the specific pairing we have postulated immediately suggests a possible copying mechanism for the genetic material." The DNA strands are wrapped around each other and each can template the production of a perfect copy of the strand to which it's bound simply by unwinding and allowing the polymerase to copy it. In Roger Kornberg's talk, we heard about an RNA polymerase that can transcribe the DNA to produce RNA copies of the genetic information. These copies provide templates for the polymerization of the proteins through another elaborate and really beautiful process, called translation, that I certainly don't have time to describe today. I would hope that, if you're interested in these basic workings of the cell, which certainly I think everyone should be interested in, you should look at the literature and do some searching on the Internet to learn more about this process – it's truly amazing.

But, one of the problems with a discovery like this one, of DNA, is that we tend to become overconfident in the explanatory power of the discovery.

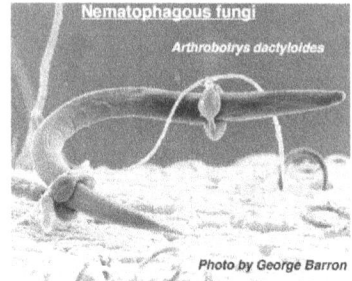

Does the DNA sequence information control all of the events in the cell? Cells are constantly responding to their environment and to surrounding cells, and these external influences can alter the cell in heritable ways that do not require changes in the primary sequence information in the DNA. Consider the early *C. elegans* embryo. During these early divisions, maternal mRNA and protein products that are stored in the egg direct numerous cell-cell signaling and differentiation events that give rise to the multicellular organism. These are exemplified by the distribution of the PIE-1 protein (Figure 4). PIE-1 tracks with, and is essential for, germline specification. As shown in this image from a movie, PIE-1 – in this case tagged with a glowing jellyfish protein – becomes localized after each division to the germ-line cell. In this two-cell embryo PIE-1 protein is localized exclusively to the posterior cell where it is concentrated in the nucleus. This occurs through a fundamental developmental process called asymmetric (unequal) cell division. As a result of this process, the two daughter cells differ with respect to their content of maternally provided products, like PIE-1. These products, in turn, can direct the subsequent development of these cells such that, once differentiated in this way, these cells remain committed to their specific tasks in the animal through numerous rounds of cell division. These remarkably stable differentiation events can be maintained for the entire life of an organism without any underlying changes in the DNA sequence. The germline cells, which in *C. elegans* inherit PIE-1 protein, are the only cells that retain the potential to launch the developmental program again in the next generation.

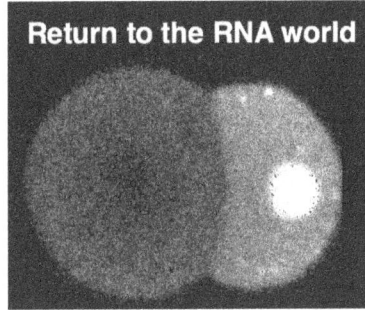

Figure 4.

How do developing cells, all with the same DNA content, lock in different programs of gene expression that are stable through so many rounds of cell division? One possibility, as I will discuss below, is that mechanisms related to those that mediate RNA interference have a role in this process. It has been suggested that the origin of life on Earth may have begun with self-replicating nucleic acid polymers that were more similar chemically to RNA than to DNA, a classic hypothesis referred to as the "RNA World" hypothesis. Hence the provocative title in Figure 4, "Return to the RNA World," a world in which RNA molecules may have carried, and may still carry, genetic information. The direct ancestors through cell division of the *C. elegans* germ cells were primordial germ cells in the common metazoan (probably worm-like) ancestor of worms and humans, and going even farther back are direct descendants of the hypothetical self-replicating RNA molecules that gave rise to all life on Earth some 3.5 billion years ago. We heard earlier today, in the physics lectures, that the temperature of the cosmic-background radiation is consistent with an age for the universe of 13.7 billion years. Thus life on earth is about a quarter of the age of the universe. Living things and these mechanisms that we are talking about today are incredibly ancient. RNAi itself is at least one billion years old. Biological mechanisms are far more constant than the positions of continents on our planet. That fact and the implicit concept of deep time are among the most profound discoveries of science.

Considering the possible origins of life in a world where information was stored in RNA polymers, and considering the remarkable sophistication of living things and the constancy of the basic and fundamental underlying mechanisms of biology, and finally, considering what we now know about RNA and RNA interference, it is perhaps a good time to reconsider the idea that genetic information is stored primarily in the nucleotide sequence of our DNA. In thinking about this, it is interesting to consider what previous scientists thought about the mechanism of inheritance before DNA and RNA were discovered. For example, in the late 1800s August Weismann, a famous naturalist and early thinker on mechanisms of inheritance, coined the term "biophore" to describe the hereditary agent [2]. Ernst Mayr, in describing Weismann's work in his book *The Growth Of Biological Thought*, characterizes Weismann's ideas as flawed. Weismann said that "1) there is a special particle, the biophore, for each trait; 2) that these particles can grow and multiply independent of cell division; 3) that both the nucleus and cytoplasm consist of these biophores; 4) that a given biophore may be represented by many replicas in a single nucleus, including the germ cell; and 5) that during cell division the daughter cells may receive different kinds and numbers of biophores through unequal cell division" [3]. Mayr concludes that "As we now know" (thanks to Mendel), "postulates (2) and (5) are wrong and are responsible for the fact that Weismann was not able to arrive at a correct theory of inheritance."

Well, are they really wrong? If you try to apply Weismann's concepts to all genes or genetic traits they are clearly not adequate to explain inheritance. For example, Weismann's biophores could not explain the striking segregation patterns first observed by Mendel for the genetic traits of pea plants. Thus, yes, it would be wrong to apply Weismann's theory to define all genetic inheritance. But let's consider applying Weismann's theory to some traits, and then replace the term "biophores" with the term "siRNAs." Andy introduced siRNAs as these small interfering RNAs, as we call them; the little chunks of RNA that go on and silence genes. If we put "siRNAs" into each place where "biophores" appears in Weismann's theory, we then have a very different situation: 1) there is a particle, containing siRNAs, for some traits; 2) these siRNAs can grow and multiply independent of cell division; 3) both the nucleus and the cytoplasm can contain the siRNAs; 4) a given siRNA may be represented by many replicas; and 5) that during cell division the daughter cells may receive different kinds and numbers of siRNAs through unequal cell division. And with these changes, and in light of what we now know about RNAi, (as will be discussed more below), it becomes clear that these postulates are not necessarily wrong. Weismann had some very good ideas and we shouldn't discard them out of hand. RNA may play a role in inheritance and evolution. I'll talk about a mechanism for RNA-directed inheritance toward the end of my talk. Furthermore, I'll suggest how natural variation in silencing levels could underlie heritable phenotypic variation upon which evolution could act.

To help introduce RNAi, I'm going to describe some movies that try to capture the essence of the RNAi process. Andy and I take hours to explain RNAi, but that won't do for today's television audience. It's a major problem for television programmers, as you may know. People watch with the remote control handy at all times, so you have to get your point across very quickly, before your viewer loses interest and clicks to another channel. Consequently, in the television industry, they're very good at making models and graphics that can show complex mechanisms like RNAi in just a few seconds.

So, here's what CBS Evening News came up with to try to explain RNAi. For the average viewer at prime time the attention span is about fifteen seconds. So, here's what CBS Evening News came up with (Figure 5). In the movie, the double stranded RNA flies onto the scene then opens at one end and begins to open and close as though it's chewing. Defective genes, shaped like colored cheese puffs, then begin to fly into

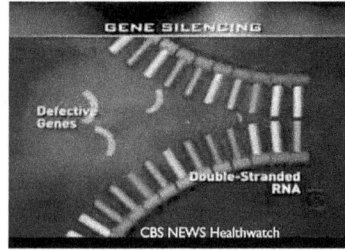

Figure 5. Image © 2005 CBS Evening News.

the mouth of the RNA from the left. The RNA is literally eating the DNA for lunch. Now, Andy and I knew that RNA interference was something incredible when we started working on it, but we really didn't have any idea that it was this incredible. Of course there are a few details that are glossed over in this explanation.

Public broadcasting has the luxury of an audience that tends to have a bit more patience, and they came up with a 15 minute segment and another strategy, "the cop", to explain RNAi (Figure 6). They describe a little policeman who looks out for viruses and other misbehavior in the cell. When he sees double-stranded RNA he realizes something is not right. He then goes on to use "enzymatic kung fu" to destroy not

Figure 6. Image © 2005 PBS NOVA ScienceNOW.

only the dsRNA with that sequence, but all RNAs with that sequence that he encounters in the cell.

I like both of these movies because they illustrate a really important concept; that is, that RNAi is an active process, that there is an organismal response to the dsRNA [4]. We realized this at an early stage, because, first of all, as Andy mentioned, the silencing was heritable. RNA injected into an animal resulted in silencing that was transmitted to progeny and even transmitted through crosses for multiple generations via the egg or the sperm. So, the interference mechanism can be initiated in one generation and then transmitted in the germline. And, interestingly, RNAi is also systemic; RNA injected anywhere in the body, or even delivered by ingestion, can get into all the tissues, including the germline. So RNAi involves a transport mechanism, meaning it can be transferred from cell to cell in the body.

The inheritance properties and systemic nature of RNAi, along with its remarkable potency in *C. elegans,* all pointed toward an active organismal response to the double-stranded RNA. What we wanted to do immediately, upon realizing that there was an active response in the organism, was to find the genes in the animal that encode that response. Therefore, we set out to use the powerful genetics of *C. elegans* to look for mutant strains defective in RNAi. We imagined that these mutants would define genes required for the recognition of the foreign double-stranded RNA, genes required for the transport of the silencing signal from cell to cell, genes required for the amplification of silencing, and genes required for the silencing apparatus itself. Hiroaki Tabara (Figure 7), was the first person doing RNAi genetics in the world. He was a courageous postdoc who came to my lab to study development, but was willing to tackle something as unusual and as odd as RNAi. The screen that he did was very simple. Basically, he mutagenized animals, let them grow for two generations until mutations that had been induced would become homozy-gous and then, using a trick first developed by Lisa Timmons in Andy's lab [5], he fed the worms *E. coli* expressing double-stranded RNA targeting an es-

Figure 7. Hiroaki Tabara

sential worm gene. According to this strategy, if the animals have an intact RNAi response, then RNAi would knock out the activity of the essential gene, causing lethality. Now, if by chance a mutant exists in the population that lacks an RNAi response, then RNAi would not occur, and the corresponding animal and its progeny would be viable. Hiroaki used this very powerful genetic selection to identify mutants defective in RNAi, and his screen worked very, very well.

Hiroaki was able to identify numerous mutants. Some of these lacked the RNAi response and had no other obvious phenotypes, like *rde-1* and *rde-4.* However, some of his mutants had additional defects, including a very striking phenotype in which the transposons, which are selfreplicating DNA elements present in the genomes of all organisms, became hyperactive, causing muta-tions by jumping from place to place in the genome. In addition, these same mutants had a reduced tendency to silence transgenes in the germline (trans-genes are genes that are experimentally introduced into the organism). In normal worms, transgenes have the vexing property of becoming silent after introduction into the animal experimentally. The same mutants with activated transposons also exhibited activation of transgenes in the germline.

These observations suggested that the normal physiological function of RNAi might be to defend cells against the potentially damaging effects of transposons and other foreign genetic elements (perhaps including viruses). However, there was a big problem with this relatively simple model. The *rde-
1* and *rde-4* mutants, as I indicated earlier, had no other phenotypes. They

were strongly deficient in RNAi in response to double-stranded RNA, but the transposon silencing and the transgene silencing mechanisms were still functioning in these mutant strains. These observations indicated to us, even at that very early stage of our analysis, the existence of some additional, very interesting complexity. The rest of the science that I will discuss below really relates to our further investigation of this complexity, and to how these investigations led to the realization that related silencing pathways with distinct triggering mechanisms are at work in *C. elegans*. Keeping in mind that the Nobel committee was careful to recognize, very specifically, the initiation of gene silencing by double-stranded RNA, I hope that we can look forward to future recognition of silencing that is triggered in other ways. For example, silencing driven from endogenous dsRNA-encoding genes, microRNAs, or silencing triggered by the introduction of transgenes.

Hiroaki cloned the *rde-1* gene and showed that it encodes a highly conserved protein that we now refer to as an Argonaute protein [6]. RDE-1 was an interesting protein for a couple of reasons. It had highly conserved domains found in related genes in organisms as diverse as plants and humans, and yet nothing was known about the enzymatic activities or the biological functions of these domains. This was a very exciting time in the laboratory. We at last had a gene that we knew was involved in the mechanism. Furthermore, previous work on one gene closely related to RDE-1 from *Drosophila* had linked this gene family to an epigenetic silencing pathway in the fruit fly [7, 8], and work in plants had linked a member of the family to the control of development [9]. Very shortly after our paper was published, Carlo Cogoni and Giuseppe Macino [10] published a very nice paper implicating an RDE-1 family member in silencing triggered by the introduction of a transgene in the fungus *Neurospora*. So from these findings in other organisms, and from Hiroaki's genetics, we hypothesized that there may be other types of triggers that initiate related silencing pathways either through natural developmental mechanisms or in response to transposons and transgenes.

A very exciting possibility occurred to us after we cloned *rde-1*. To explain this possibility, I first have to describe some fundamental facts about genes and how the amazingly successful genome-sequencing projects around the world have impacted biological research. Genes are composed of long sequences of nucleotides that specify the protein-coding potential and/or other functions of their gene products. The relationships between genes can be inferred by looking at the nucleotide sequence of the gene. For example, by using the nucleotide sequence to infer the protein-coding potential of all the known genes related to *rde-1*, it was possible to build what is called a phylogenetic tree (Figure 8), in which the most similar members of the gene family (often referred to as homologs) are closest to each other on the tree. Interestingly, it turns out that *rde-1* is a member of a large gene family, with 26 related genes in *C. elegans*. Similarly, there are multiple Argonaute genes in almost every organism. The organisms from which each gene in the tree is derived are indicated by a prefix as follows: *C. elegans* (Ce), humans (Hs), the plant *A. thaliana* (At), fruit fly (Dm), and fission yeast (Sp).

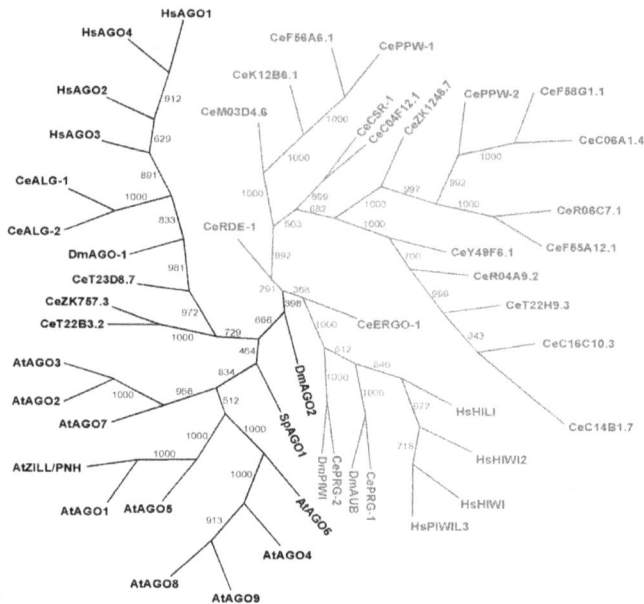

Figure 8.

The genes named in black represent the most highly conserved branch of the family, with members in plants, animals and fungi. The green branch, often referred to as the Piwi family after its founding member, has family members in all animals (but not in plants or fungi). Finally, the red branch of the tree represents a *C. elegans*-specific subfamily of genes that are equally divergent from both the black and green families. This remarkable diversity of Argonaute genes raised the exciting possibility that different members of the family have become specialized in each organism to perform distinct functions. For example, RDE-1 according to our genetic studies is required for gene silencing in response to foreign dsRNA. Perhaps other members of the *C. elegans* Argonaute gene family mediate transposon and transgene silencing. Still others may function in developmental pathways related to RNAi. An outstanding graduate student, Alla Grishok (Figure 9), took on the task of trying to test these ideas.

To do this, Alla set out to inactivate members of the *C. elegans* Argonaute gene family. The first genes that she knocked out encoded two closely related members of the most highly conserved group of Argonautes, named ALG-1 and ALG-2 (see the black branch of the tree, Figure 8).

When Alla knocked out these genes by RNAi, she observed a striking phenotype. But in order to explain the significance of her findings, I have to digress for a moment and tell you about some previous work that set the stage for Alla's discovery.

Figure 9. Alla Grishok

This previous work goes back to 1993 when, after several years of trying, Victor Ambros' lab succeeded in cloning the *lin-4* gene [11]. One of the reasons *lin-4* had been so hard to clone was that the gene was tiny and did not encode a protein. Instead, the *lin-4* gene appeared to encode two RNA products: an ~70 nucleotide-long RNA capable of forming a double-stranded RNA molecule with a hairpin-like structure, and a single-stranded 22 nucleotide RNA that appeared to be derived from this longer RNA (Figure 10). This short RNA was capable of binding directly to sites in the transcript of the *lin-14* gene, a gene that is negatively regulated by *lin-4* during the normal course of worm development.

Even before we identified RDE-1, we were interested in the possibility of a relationship between the RNAi pathway and the *lin-4* pathway. Indeed, Hiroaki had raised the concern that RNAi-defective mutants could be hard to recover as viable strains since they might also cause disruption of the *lin-4* pathway. Making all of these possibilities even more exciting – while we were conducting genetic screens for RNAi deficient strains, beautiful work was published by Hamilton and Baulcombe [12], linking small RNAs of ~21 nucleotides to viral gene silencing in plants, and by Gary Ruvkun's lab, identifying a second *lin-4*-like worm gene, *let-7* [13]. Whereas *lin-4* was a worm-specific gene, it turned out that the *let-7* gene had homologs in every animal, including humans. Remarkably, every single nucleotide in the twenty-one nucleotide mature *let-7* RNA products from the worm and human were identical to each other. The conservation of *let-7* initiated a gold rush to find small RNA encoding genes, now referred to as micro-RNA genes, in the genomes of numerous organisms. But, despite all of the excitement, the relationship between RNA interference and microRNAs had not really been made yet. As Phil Sharp said at one meeting, "It looks like a horse and smells like a horse", but there was no molecular or genetic evidence that these pathways were linked.

While this exciting work was going on in worms and plants, biochemists were making rapid progress in reconstituting elements of the silencing pathway in *Drosophila* cell extracts. David Bartel's group along with Phil Zamore, Tom Tuschl and Phil Sharp at MIT, and Greg Hannon's group at Cold Spring Harbor, spearheaded these efforts [14, 15]. They showed that activities present in *Drosophila* cells could process double-stranded RNA into tiny RNAs approximately 21 nucleotides long. Tom Tuschl and colleagues were the first to show that these small RNAs could silence gene expression in vertebrate cells [16]. Thus genetic studies in worms had identified small RNAs as silencing agents beginning in 1993, experimental studies of virus infections in plants identified small RNAs accumulating in infected plants, biochemical studies in fly extracts identified small RNAs in extracts, and finally experimental studies identified silencing activity in cellular assays with vertebrate cells. But were these small RNA molecules only similar in size, or did their similarity extend to mechanism? The answer to that key question was still unknown.

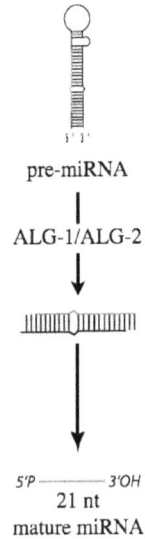

pre-miRNA

ALG-1/ALG-2

5'P ——————— 3'OH
21 nt

mature miRNA

Figure 10.

Alla's work provided an answer. When Alla knocked out *alg-1* and *alg-2*, she observed a phenotype that was very similar to that observed when you knock out *let-7*. To confirm this connection we collaborated with Gary Ruvkun and Amy Pasquinelli, who had recently developed probes for following the processing of the *lin-4* and *let-7* precursor RNAs into their mature 21 nucleotide RNAs. In wild-type animals, the precursor forms are barely detectable. However, we found that, after inactivation of *alg-1* and *-2*, this precursor accumulates to high levels while the product, the mature twenty-one nucleotide RNA, is greatly diminished [17] (Figure 11).

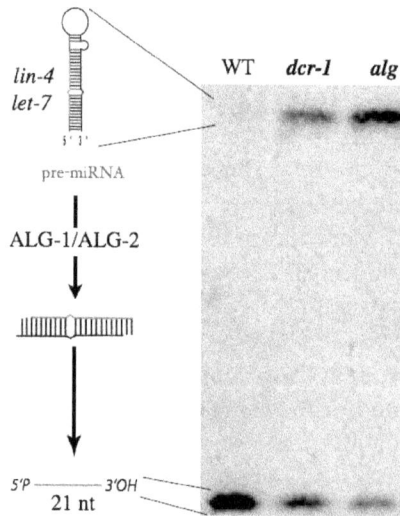

Figure 11.

We also looked at the involvement of Dicer in this process. Dicer was identified by Greg Hannon's lab as a nuclease required for processing long double-stranded RNA into approximately 21-nucleotide fragments in *Drosophila* cells. We were able to show, as did several other groups [18, 19], that when you knock out Dicer you also see defects in the processing of these microRNAs (Figure 11).

With these findings, the first link was established between RNA interference and a natural developmental mechanism for regulating gene expression. This was extremely exciting, and we envisioned a model (Figure 12), in which the RNAi and microRNA pathways utilized different members of the RDE-1 family and converged on Dicer. Downstream of Dicer these pathways appeared to diverge again, through the action of unknown effectors that direct different types of silencing, including mRNA destruction, transcriptional silencing and inhibition of translation. And yet, we still had not identified the RDE-1 family member involved in transposon and transgene silencing.

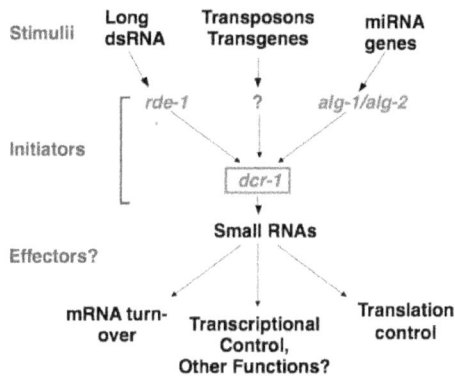

Figure 12.

At that time we thought of the RDE-1 family members (also known as Argonaute proteins) as initiators of the silencing pathways. Genetic studies had placed RDE-1 at an upstream step in the pathway and, as I showed you, ALG-1 and -2 are required for processing the microRNA precursors. However, there was mounting evidence that these proteins might also function downstream in the silencing step. Definitive support for this idea came from Greg Hannon's group through a collaboration with Ji-Joon Song and Leemor Joshua-Tor at Cold Spring Harbor [20]. They showed that Argonaute proteins have structural similarity to an enzyme domain that can cut RNA, and they presented a model for how Argonaute proteins can bind the ends of the short RNAs and utilize the sequence information to find and destroy target mRNAs in the cell. These studies demonstrated that Argonaute proteins represent the long sought "slicer" activity (or the cop) that lies at the heart of the RNA-induced silencing pathway.

We were surprised to learn that RDE-1 was probably the slicer enzyme because our genetic studies had placed RDE-1 activity at an upstream step in the pathway. However, we realized that this observation could be explained if Argonautes function more than once during RNAi in *C. elegans*. For example (Figure 13), we imagined that RDE-1 could function along with small RNAs derived from processing of the trigger dsRNA in an initial round of target mRNA cleavage. The cleaved target mRNA could then serve as a template for an RNA-dependent RNA polymerase that produces new siRNAs that could, in turn, interact with other Argonautes to mediate efficient silencing of the gene.

Experimental tests of this model were recently published [21]. Surprisingly, we found that, rather than a single additional gene, multiple RDE-1 homologs function together to mediate silencing at the downstream step in the pathway. It was necessary to construct a strain containing six different Argonaute mutants in order to see a strong defect in RNAi. All of these functionally related genes reside within the expanded (red) family of Argonaute genes depicted in Figure 8. These downstream Argonautes are limiting for RNAi. When they are overexpressed RNAi is enhanced, and when they are inactive RNAi is decreased. These observations suggest that these genes have been amplified in order to mediate efficient gene silencing.

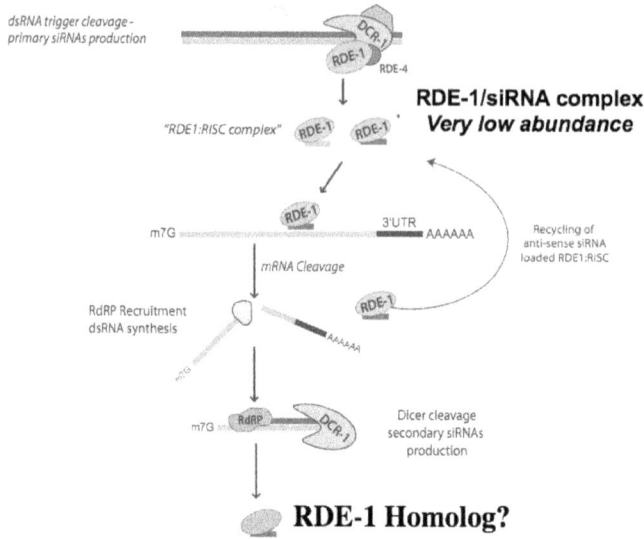

Figure 13.

The mechanism of silencing mediated by these downstream Argonautes remains unknown. It could be through mRNA destruction, but comparison of members of this group of Argonautes to RDE-1 and other members of the family suggest that these downstream Argonautes are not likely to have an intact RNA-cleaving nuclease domain. Our studies of these proteins indicate that they also function in endogenous silencing pathways (Figure 14), including pathways likely to have a role in silencing transposons, transgenes and other genes at the chromatin level [21].

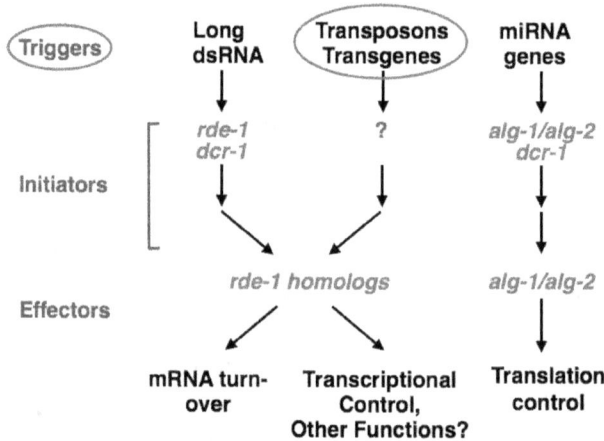

Figure 14.

The last concept I want to discuss relates to the question of how RNAi can interact with chromatin to silence genes, and the potential importance of this mechanism for gene regulation during both development and evolution. As indicated earlier in my talk, many of the genes involved in RNAi are also required for the silencing of transgenes in the germline. For example, the gene *mut-7* was identified in our screens for genes essential for RNAi [6], but had also been identified in earlier studies as a gene required for transposon [22] and transgene silencing. While RNAi appeared to have a post-transcriptional effect, several studies suggested that transgene silencing involves regulation at the level of the DNA (or more precisely, the chromatin). For example, some of the genes required for transgene silencing in *C. elegans* were related to genes of the polycomb group that interact with chromatin to direct gene silencing in other organisms [23]. Beautiful direct evidence for a link between RNAi and chromatin silencing has more recently come from work in the fission yeast *S. pombe*, where a complex containing an Argonaute protein and known chromatin interacting factors has been shown to interact directly with silenced genes in the nucleus [24]. To explain how RNAi could regulate DNA directly, I have to tell you a little bit about the physiological nature of DNA inside cells. Your DNA isn't just lying around by itself. The unit of packaging for DNA is a protein structure called the nucleosome. The DNA wraps around the nucleosome twice, and the nucleosomes are in turn wrapped up and packaged into even thicker fibers. Chromosomes are composed of these protein/DNA fibers, also referred to as chromatin. Partly, what's achieved by packaging the DNA into chromatin is a silencing effect. Structural studies of the nucleosome core suggest that short protein tails stick out past the DNA in such a way that they are readily accessible for modification [25]. The modification of these tails, and the resulting regulatory effects on gene expression, is turning out to be a fascinating subject – one that I'm sure this committee will need to consider in the future. Interestingly, mechanisms are now emerging that explain how small RNAs can guide the modification of these chromatin tails [24]. I will illustrate these mechanism with a model that could not only explain chromatin-based silencing mediated by RNAi but also provide a mechanism for the RNA-mediated evolution concept mentioned earlier (Figure 15).

A Chromatin-RNA Feedback Loop

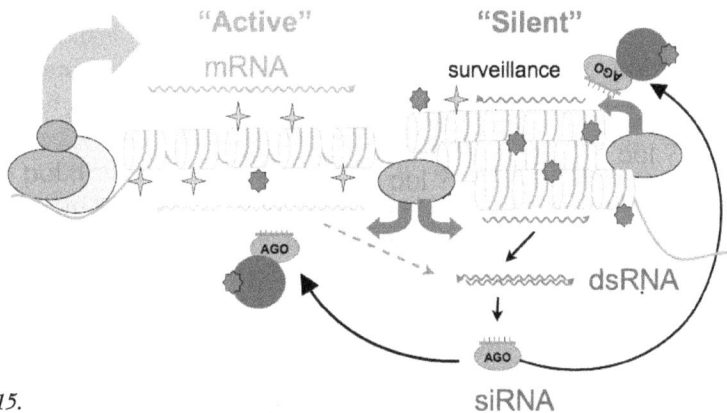

Figure 15.

In this model the DNA (green line) is shown wrapped around the nucleosomes, which are in turn packaged into the higher-order chromatin structures. Modifications to the nucleosome tails that confer an active conformation are shown as four-pointed stars, while silencing marks are shown as multi-pointed red stars. In the active conformation the regulatory region of the gene, called the promoter, is free of nucleosomes and is shown bound by the RNA-polymerase complex, (the complex that produces messenger RNAs and the subject of this year's Chemistry prize). In the "silent" region a different kind of polymerase activity is recruited. Instead of producing mRNA, this hypothetical polymerase produces transcripts that enter an RNAi-like silencing pathway. The silencing RNAs could arise by virtue of Dicer-mediated processing of double-stranded RNA. For example double-stranded RNA could form as the result of bi-directional transcription within the region, or through recruitment of an RNA-dependent RNA polymerase that recognizes some feature of the surveillance RNA. Alternatively, it is also possible that short-interfering RNAs are made directly by transcription from nucleosomal DNA and are loaded onto Argonaute proteins without going through a dsRNA intermediate. Whatever the mechanism for generating the small RNAs, the resulting Argonaute/small RNA complexes could then interact through sequence-specific interactions with nascent surveillance transcripts, or directly with the DNA, to guide chromatin-modifying enzymes back to the locus to reinforce silencing. These silencing complexes could also function *in trans* to silence other genes with related sequence, such as repeated members of a transposon family.

The concept of transcription occurring not simply to express the gene, but also to regulate it, is extremely powerful. Silencing marks present at low levels within "active regions" could modulate gene expression by specifying the production of intermediate levels of silencing RNAs that in turn specify an intermediate level of gene expression. According to this model, the DNA is like the hardware in a computer and the RNA/chromatin interactions are like the software. The RNA, through interactions with the chromatin, determines not only what regions of the DNA are active, but also the level of activity. When the DNA is replicated and the chromatin is disassembled, the RNA can help reinstall the silencing marks, essentially programming the resulting daughter cells to adopt gene expression patterns like the mother cell. As Weismann pointed out, asymmetric cell division could segregate this regulatory potential such that the daughter cells become different at one or many different loci [2]. This mechanism could help explain how a somatic cell nucleus can be reprogrammed to undergo embryonic development after transfer into an egg. Mechanisms like this could also help explain how cells are able to maintain their gene expression programs for decades during an organism's life span.

However, and here is where evolution comes in, this RNA/chromatin feedback mechanism could also function within the germline. Chromatin marks in the germline could specify a level of surveillance-RNA expression that keeps some genes off entirely, and modulates others such that, when they are activated during somatic development, their level of expression is proportional to the amount of silencing RNA produced at the locus. The feedback loop is self-sustaining but is likely to be subject to natural variations in levels. Upward or downward variations that occur naturally could be selected and transmitted in the germline from one generation to the next. This kind of reversible change in levels of gene expression could play an important role in helping organisms adjust to changes and variations in their environments.

We know from experiments in *C. elegans* that silencing induced by RNAi can be transmitted for multiple generations [26], and that chromatin-modifying factors appear to play a role in this inheritance mechanism [27]. Given the existence of these phenomena, it is tempting to speculate that all genes might continuously sample, through natural variation, different levels of heritable small RNA/chromatin interactions. Variation of this type could have a major impact on fitness and evolution, providing a rapid mechanism for evolutionary change mediated through RNA-chromatin interactions, *without any underlying changes in the DNA sequence.* I will end by saying we simply don't know yet how important small RNAs will turn out to be during development and evolution. I encourage you all to think about the possibilities, to learn more about biology and RNAi, and if you get inspired and excited, please join the adventure and help explore the many unknowns that are still waiting to be addressed.

REFERENCES

1. Watson, J.D. and Crick, F.H., Molecular structure of nucleic acids; a structure for deoxyribose nucleic acid. *Nature,* **1953.** 171(4356): p. 737-8.
2. Weismann, A., *The Germ-Plasm: A Theory of Heredity.* **1893,** New York: Charles Scribner's Sons. vxiii, 477.
3 Mayr, E., *The Growth of Biological Thought.* **1982,** Cambridge, MA: The Belknap Press of Harvard University Press, 974.
4. Fire, A., Xu, S., Montgomery, M.K., Kostas, S.A., Driver, S.E., and Mello, C.C., Potent and specific genetic interference by double-stranded RNA in *Caenorhabditis elegans. Nature,* **1998.** 391(6669): p. 806-11.
5. Timmons, L. and Fire, A., Specific interference by ingested dsRNA. *Nature,* **1998.** 395(6705): p. 854.
6. Tabara, H., Sarkissian, M., Kelly, W.G., Fleenor, J., Grishok, A., Timmons, L., Fire, A., and Mello, C.C., The *rde-1* gene, RNA interference, and transposon silencing in *C. elegans. Cell,* **1999.** 99(2): p. 123-32.
7. Wilson, J.E., Connell, J.E., and Macdonald, P.M., Aubergine enhances oskar translation in the *Drosophila* ovary. *Development,* **1996.** 122(5): p. 1631-9.
8. Schmidt, A., Palumbo, G., Bozzetti, M.P., Tritto, P., Pimpinelli, S., and Schafer, U., Genetic and molecular characterization of sting, a gene involved in crystal formation and meiotic drive in the male germ line of *Drosophila melanogaster. Genetics,* **1999.** 151(2): p. 749-60.
9. Bohmert, K., Camus, I., Bellini, C., Bouchez, D., Caboche, M., and Benning, C., AGO1 defines a novel locus of *Arabidopsis* controlling leaf development. *The EMBO journal,* **1998.** 17(1): p. 170-80.
10. Catalanotto, C., Azzalin, G., Macino, G., and Cogoni, C., Gene silencing in worms and fungi. *Nature,* **2000.** 404(6775): p. 245.

11. Lee, R.C., Feinbaum, R.L., and Ambros, V., The *C. elegans* heterochronic gene *lin-4* encodes small RNAs with antisense complementarity to *lin-14*. *Cell*, **1993**. 75(5): p. 843-54.

12. Hamilton, A.J. and Baulcombe, D.C., A species of small antisense RNA in posttranscriptional gene silencing in plants. *Science*, **1999**. 286(5441): p. 950-2.

13. Reinhart, B.J., Slack, F.J., Basson, M., Pasquinelli, A.E., Bettinger, J.C., Rougvie, A.E., Horvitz, H.R., and Ruvkun, G., The 21-nucleotide *let-7* RNA regulates developmental timing in *Caenorhabditis elegans*. *Nature*, **2000**. 403(6772): p. 901-6.

14. Zamore, P.D., Tuschl, T., Sharp, P.A., and Bartel, D.P., RNAi: double-stranded RNA directs the ATP-dependent cleavage of mRNA at 21 to 23 nucleotide intervals. *Cell*, **2000**. 101(1): p. 25-33.

15. Hammond, S.M., Bernstein, E., Beach, D., and Hannon, G.J., An RNA-directed nuclease mediates post-transcriptional gene silencing in *Drosophila* cells. *Nature*, **2000**. 404(6775): p. 293-6.

16. Elbashir, S.M., Harborth, J., Lendeckel, W., Yalcin, A., Weber, K., and Tuschl, T., Duplexes of 21-nucleotide RNAs mediate RNA interference in cultured mammalian cells. *Nature*, **2001**. 411(6836): p. 494-8.

17. Grishok, A., Pasquinelli, A.E., Conte, D., Li, N., Parrish, S., Ha, I., Baillie, D.L., Fire, A., Ruvkun, G., and Mello, C.C., Genes and mechanisms related to RNA interference regulate expression of the small temporal RNAs that control *C. elegans* developmental timing. *Cell*, **2001**. 106(1): p. 23-34.

18. Hutvagner, G., McLachlan, J., Pasquinelli, A.E., Balint, E., Tuschl, T., and Zamore, P.D., A cellular function for the RNA-interference enzyme Dicer in the maturation of the *let-7* small temporal RNA. *Science*, **2001**. 293(5531): p. 834-8.

19. Ketting, R.F., Fischer, S.E., Bernstein, E., Sijen, T., Hannon, G.J., and Plasterk, R.H., Dicer functions in RNA interference and in synthesis of small RNA involved in developmental timing in *C. elegans*. *Genes Dev*, **2001**. 15(20): p. 2654-9.

20. Song, J.J., Smith, S.K., Hannon, G.J., and Joshua-Tor, L., Crystal structure of Argonaute and its implications for RISC slicer activity. *Science*, **2004**. 305(5689): p. 1434-7.

21. Yigit, E., Batista, P.J., Bei, Y., Pang, K.M., Chen, C.C., Tolia, N.H., Joshua-Tor, L., Mitani, S., Simard, M.J., and Mello, C.C., Analysis of the *C. elegans* Argonaute family reveals that distinct Argonautes act sequentially during RNAi. *Cell*, **2006**. 127(4): p. 747-57.

22. Ketting, R.F., Haverkamp, T.H., van Luenen, H.G., and Plasterk, R.H., MUT-7 of *C. elegans*, required for transposon silencing and RNA interference, is a homolog of Werner syndrome helicase and RNaseD. *Cell*, **1999**. 99(2): p. 133-41.

23. Kelly, W.G. and Fire, A., Chromatin silencing and the maintenance of a functional germline in *Caenorhabditis elegans*. *Development*, **1998**. 125(13): p. 2451-6.

24. Verdel, A., Jia, S., Gerber, S., Sugiyama, T., Gygi, S., Grewal, S.I., and Moazed, D., RNAi-mediated targeting of heterochromatin by the RITS complex. *Science*, **2004**. 303(5658): p. 672-6.

25. Richmond, T.J. and Davey, C.A., The structure of DNA in the nucleosome core. *Nature*, **2003**. 423(6936): p. 145-50.

26. Grishok, A., Tabara, H., and Mello, C.C., Genetic requirements for inheritance of RNAi in *C. elegans*. *Science*, **2000**. 287(5462): p. 2494-7.

27. Vastenhouw, N.L., Brunschwig, K., Okihara, K.L., Muller, F., Tijsterman, M., and Plasterk, R.H., Gene expression: long-term gene silencing by RNAi. *Nature*, **2006**. 442(7105): p. 882.

Portrait photo of Dr. Mello by photographer John Mottern.

Physiology or Medicine 2007

**Mario R. Capecchi, Sir Martin J. Evans and
Oliver Smithies**

*"for their discoveries of principles for introducing specific gene
modifications in mice by the use of embryonic stem cells"*

THE NOBEL PRIZE IN PHYSIOLOGY OR MEDICINE

Speech by Professor Christer Betsholtz of the Nobel Assembly at Karolinska Institutet.
Translation of the Swedish text.

Your Majesties, Your Royal Highnesses, Ladies and Gentlemen,

This year's Nobel Prize in Physiology or Medicine rewards discoveries that have given us new and powerful methods for studying and understanding the role of our genes. These genes carry their information as DNA code, a kind of script which in humans had been fully read for the first time in 2001. But reading genetic script is one thing – understanding its significance is another.

To study the role of a gene, we need to be able to change it in a specific way and then observe what happens or doesn't happen. This approach is empirical and resembles that of a child as it learns the meaning of words. Small children perform their word experiments by inserting or leaving out words in different contexts and, based on the reactions of people around it, guessing the meaning of the words. As an example: When a child tries out an exciting new word it has heard, and the reaction of its parents is "Shame on you, don't ever say that!" the child naturally draws the conclusion that this word is very important and also presumably usable.

Doing equivalent experiments with genetic language is of course somewhat more complicated, though similar in principle. We have more than 22,000 genes, distributed among 3 billion DNA "letters". Making a specific, or targeted, genetic modification can thus be compared to correcting an error in a text document 30 times larger than the Swedish National Encyclopaedia. Nowadays, given computers and word processing software, this is of course no major problem. We merely specify what text we want to remove and what we want to insert in its place, then we let the computer find the right place in the document and make the substitution.

This is a good analogy to what Mario Capecchi and Oliver Smithies discovered in their path-breaking experiments during the first half of the 1980s. Independently of each other, they found that DNA molecules which resemble parts of normal genes, yet differ from them in crucial respects, can be inserted in the right place in the genome.

Their discoveries made it possible to carry out targeted gene modifications in individual cells in a culture, but an important problem remained to be solved. Every cell in our bodies carries our entire genome, so if we want to understand the function of a given gene in its full context – in real life – the same genetic change must be introduced into all the cells of the body. If a targeted gene modification in *one* cell is comparable to finding a needle in a haystack, then the challenge here is to find a needle in hundreds of billions of haystacks.

Martin Evans solved this problem through his discovery of embryonic stem cells – cells from early embryos that can be cultured, grown and genetically modified in a test tube. Like a fertilised egg cell, these embryonic stem cells can give rise to all the cells of the body, thereby passing their genes, including modified ones, onward to future generations.

But making hereditary, targeted changes in genes is something we are neither able, willing nor allowed to do in humans. Instead we do this in mice, with which we share most of our genes. Here it is important that we apply accepted standards and ethical principles for animal experiments, and that in doing so we weigh the expected benefit to humanity against the number of animals used and any suffering that may be caused. The usefulness of mice with targeted gene modifications can hardly be exaggerated. "Knockout mice", in which the functions of individual genes have been knocked out, have already shed light on the role of several thousands of our genes and provided us with important new knowledge that, among other things, is being used today in the development of new drugs for treating virtually all important human diseases. The 2007 Nobel Prize in Physiology or Medicine is indeed a real knockout.

Professors Capecchi, Evans and Smithies,

In the early 80s your ideas about how mice with precisely tailored genetic changes could be obtained were met with scepticism. In the beginning of the 90s, reported successful examples of gene-targeted mice were still considered anecdotal. Today, information about the physiological functions of *all* mammalian genes is within reach. Few discoveries have had greater impact on contemporary biomedical sciences than yours. On behalf of the Nobel Assembly at Karolinska Institutet it is my privilege and pleasure to express our warmest congratulations and our deepest admiration as I now ask you to step forward to receive the Nobel Prize from the hands of His Majesty the King.

MARIO R. CAPECCHI

THE MAKING OF A SCIENTIST II

Preface

In 1996, as a Kyoto Prize laureate, I was asked to write an autobiographical sketch of my early upbringing. Through this exercise, shared by all of the laureates, the hope was to uncover potential influences or experiences that may have been key to fostering the creative spirit within us. In my own case, what I saw was that, despite the complete absence of an early nurturing environment, the intrinsic drive to make a difference in our world is not easily quenched and that given an opportunity, early handicaps can be overcome and dreams achieved. This was intended as a message of hope for those who have struggled early in their lives. As I have previously noted, our ability to identify the genetic and environmental factors that contribute to talents such as creativity are too complex for us to currently predict. In the absence of such wisdom our only recourse is to provide all children with the opportunities to pursue their passions and dreams. Our understanding of human development is too meager to allow us to predict the next Beethoven, Modigliani, or Martin Luther King.

The content of the autobiographical sketch was based on my own memories, on conversations with my aunt and uncle, who raised me once I arrived in the United States, and on conversations with my mother. Because of the added exposure resulting from the winning of the Nobel Prize, I have received letters from people who knew me in Italy during those formative early years. In addition members of the press have taken an interest in my story and have sought independent corroboration. An amazing and wonderful surprise is that they have discovered a half-sister of whom I was completely unaware. She is two years younger than I, and was given up for adoption before she was one year old. Most recently I had the opportunity to meet my half-sister. She was a very nice person, as a sister should be. I am grateful for all of these new sources of information and revelation. Where appropriate, I will weave the new information into this retelling of my story.

Autobiographical Sketch

I was born in Verona, Italy on October 6, 1937. Fascism, Nazism, and Communism were raging through the country. My mother, Lucy Ramberg, was a poet; my father, Luciano Capecchi, an officer in the Italian Air Force.

Figure 1. A photograph of my mother, Lucy Ramberg, at age 19.

This was a time of extremes, turmoil and juxtapositions of opposites. They had a passionate love affair, and my mother wisely chose not to marry him. This took a great deal of courage on her part. It embittered my father.

I have only a few pictures of my mother. She was a beautiful woman with a passion for languages and a flair for the dramatic (see Figure 1). This picture was taken when she was 19. She grew up, with her two brothers, in a villa in Florence, Italy. There were magnificent gardens, a nanny, gardeners, cooks, house cleaners, and private tutors for languages, literature, history, and the sciences. She was fluent in half a dozen languages. Her father, Walter Ramberg, was an archeologist specializing in Greek antiquities, born and trained in Germany. Her mother was a painter born and raised in Oregon, USA. In her late teens, my grandmother, Lucy Dodd, packed up her steamer trunks and sailed with her mother from Oregon to Florence, Italy, where they settled. My grandmother was determined to become a painter. This occurred near the end of the 19th century, a time when young women were not expected to set off on their own with strong ambitions of developing their own careers.

My grandmother became a very gifted painter. Let me share with you a couple of her paintings, which also illustrate the young lives of her children. These paintings are very large, approximately seven feet by five feet. The first painting (Figure 2) is the center panel of a triptych depicting my mother and her two brothers Walter and Edward (both of whom became physicists) surrounded by olive trees at the villa in Florence. The influence of the French impressionist painters is evident. The second painting (Figure 3) is of my mother, age 8, and her younger brother Edward, age 6, having a tea party, again at the villa in Florence. Their father, the German archeologist, was killed as a young man in World War I. My grandmother finished raising her

Figure 2. A painting done by my grandmother, Lucy Dodd Ramberg, of her three children, left to right, Edward, Lucy, and Walter. It was painted at their villa in Florence, Italy in 1913.

Figure 3. A painting by Lucy Dodd Ramberg of my mother, Lucy, and uncle Edward having tea at the villa in Florence, Italy (1913).

Figure 4. A photograph of the chalet where my mother and I lived in Wolfgrübben just north of Bolzano, Italy. In the foreground is my mother, Lucy.

three children on her own by painting, mostly portraits, and by converting the family villa into a finishing school for young women, primarily from the United States.

My mother's love and passion was poetry. She published in German. She received her university training at the Sorbonne in Paris and was a lecturer at that university in literature and languages. At that time, she joined with a group of poets, known as the Bohemians, who were prominent for their open opposition to Fascism and Nazism. In 1937, my mother moved to the Tyrol, the Italian Alps. Figure 4 shows the chalet north of Bolzano, in Wolfgrübben, with my mother in the foreground. We lived in this chalet until I was 3½ years old. In the spring of 1941, German officers came to our chalet and arrested my mother. This is one of my earliest memories. My mother had taught me to speak both Italian and German, and I was quite aware of what was happening. I sensed that I would not see my mother again for many years, if ever. She was incarcerated as a political prisoner in Germany.

I have believed that her place of incarceration was Dachau. This was based on conversations with my uncle Edward, my mother's younger brother. During World War II, my uncle lived in the United States. Throughout these war years, he made many attempts to locate where my mother was being held. The most reliable information indicated that the location was near Munich. Dachau is located near Munich and was built to hold political prisoners. My mother survived her captivity, but after the war, despite my prodding, she refused to talk about her war experiences.

Reporters from the Associated Press (AP) have found records that my mother was indeed a prisoner during the war in Germany. In fact, they have found records of German interest in my mother's political activities preceding 1939. In that year, they had her arrested by the Italian authorities and jailed in Perugia and subsequently released. However, the AP reporters did not find records indicating that my mother was incarcerated in Dachau. Though Germans were noted for their meticulous record keeping, it would be difficult now to evaluate the accuracy of the existing war records, particularly for cases where data is missing. It is clear, however, that exactly where in Germany my mother was held has not yet been determined. Regardless of which prison camp was involved, her experiences were undoubtedly more horrific than mine. She had aged beyond recognition during those five years of internment. Following her release, though she lived until she was 82 years old, she never psychologically recovered from her wartime experiences.

My mother had anticipated her arrest by German authorities. Prior to their arrival, she had sold most of her possessions and gave the proceeds to an Italian peasant family in the Tyrol so that they could take care of me. I lived on their farm for one year. It was a very simple life. They grew their own wheat, harvested it, and took it to the miller to be ground. From the flour they made bread which they took to the baker to be baked. During this time, I spent most of my time with the women of the farm. In the late fall, the grapes were harvested by hand and put into enormous wooden vats. The children, including me, stripped, jumped into the vats and mashed the grapes with our feet. We became squealing masses of purple energy. I still remember the pungent odor and taste of the fresh grapes. Most recently, members of the Dolomiten Press have located this farm and I had the opportunity to visit it. It is still owned by the same family that occupied it when I was there. The old farm house has been taken down and a new one erected. However, the pictures of the old farm house, as well as the surrounding land are remarkably consistent with my memories.

World War II was now fully under way. The American and British forces had landed in Southern Italy and were proceeding northward. Bombings of northern Italian cities were a daily occurrence. As constant reminders of the war, curfews and blackouts were in effect every night; no lights were permitted. In the night we could hear the drone of presumed American and British reconnaissance planes which we nicknamed "Pepe." One hot afternoon, American planes swooped down from the sky and began machine gunning the peasants in the fields. A senseless exercise. A bullet grazed my leg, fortunately not breaking any bones. I still have the scar, which, many years later my daughter proudly had me display to her third-grade class in Utah.

For reasons that have never been clear to me, my mother's money ran out after one year and, at age 4½, I set off on my own. I headed south, sometimes living in the streets, sometimes joining gangs of other homeless children, sometimes living in orphanages, and most of the time being hungry. My recollections of those four years are vivid but not continuous, rather like a series of snapshots. Some of them are brutal beyond description, others more palatable.

There are records in the archives of Ritten, a region of the Southern Alps of Italy, that I left Bozen to go to Reggio Emilia on July 18, 1942. AP reporters exploring this history have suggested that my father came to the farm, picked me up, and that we went together to Reggio Emilia where he was living. I have no memory of his coming to the farm, nor of having travelled with him to Reggio Emilia. I have recently received a letter from a man who remembers me as the youngest member of his street gang operating in Bolzano, which is on the way to Reggio Emilia.

I did end up in Reggio Emilia, which is approximately 160 miles south of Bolzano. I knew that my father lived in Reggio Emilia and I have previously noted that I had lived with him a couple of times from 1942-1946, for a total period of approximately three weeks. The question has been raised why I didn't live with him for a much longer period. The reason was that he was extremely abusive. Amidst all of the horrors of war, perhaps the most difficult for me to accept as a child was having a father who was brutal to me.

Recently, I have also received a very nice letter from the priest in Reggio Emilia who ran the orphanage in which I was eventually placed. I remember him because he was one of the very few men I encountered in Reggio Emilia who showed compassion for the children and took an interest in me. I am surprised, but pleased, that after all these years he still remembers me among the thousands of children he was responsible for over the years. Further, I believe I was at that orphanage for only several months, the first time in the fall of 1945, after which I ran away, followed by a second period, in the same orphanage, in the spring of 1946. But his memory is genuine, for he recounts incidents consistent with my memories that could only have been known through our common experience.

In the spring of 1945, Munich was liberated by the American troops. My mother had survived her captivity and set out to find me. In October 1946, she succeeded. As an example of her flair for the dramatic, she found me on my ninth birthday, and I am sure that this was by design. I did not recognize her. In five years she had aged a lifetime. I was in a hospital when she found me. All of the children in this hospital were there for the same reasons: malnutrition, typhoid, or both. The prospects for most of those children ever leaving that hospital were slim because they had no nourishing food. Our daily diet consisted of a bowl of chicory coffee and a small crust of old bread. I had been in that hospital in Reggio Emilia for what seemed like a year. Scores of beds lined the rooms and corridors of the hospital, one bed touching the next. There were no sheets or blankets. It was easier to clean without them. Our symptoms were monotonously the same. In the morning we awoke fairly lucid. The nurse, Sister Maria, would take our temperature. She promised me that if I could go through one day without a high fever, I could leave the hospital. She knew that without any clothes I was not likely to run away. By late morning, the high, burning fever would return and we would pass into oblivion. Consistent with the diagnosis of typhoid, many years later I received a typhoid/paratyphoid shot, went into shock, and passed out.

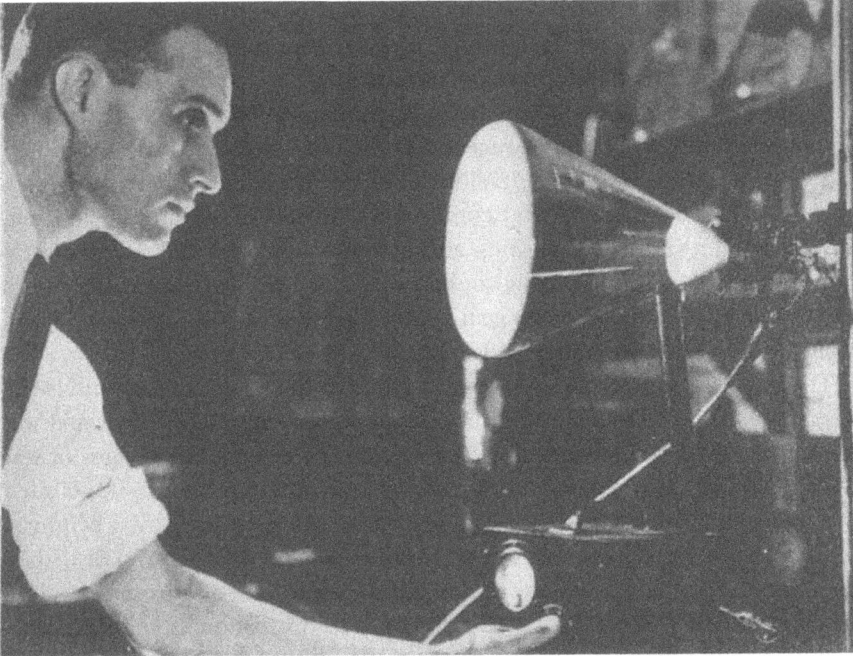

Figure 5. A photograph of my uncle Edward Ramberg working in his laboratory at RCA Princeton, New Jersey.

The same day that my mother arrived at the hospital, she bought me a full set of new clothes, a Tyrolean outfit complete with a small cap with a feather in it. I still have the hat. We went to Rome to process papers, where I had my first bath in six years, and then on to Naples. My mother's younger brother, Edward, had sent her money to buy two boat tickets to America. I was expecting to see roads paved with gold in America. As it turned out, I found much more: opportunities.

On arriving in America, my mother and I lived with my uncle and aunt, Edward and Sarah Ramberg. Edward, my mother's younger brother was a brilliant physicist. He was a Ph.D. student in quantum mechanics with Arnold Sommerfeld and translated one of Sommerfeld's major texts into English. Among Edward's many contributions was his discovery of how to focus electrons, knowledge which he used in helping to build the first electron microscope at RCA. Edward's books on electron optics have been published in many languages. During my visit to Japan to celebrate the Kyoto Prize, several Japanese physicists approached me to express how grateful they were for my uncle's texts from which they learned electron optics. Another achievement, of which he was less proud was being a principal contributor to the development of both black and white and color television. While I grew up in his home, television was not allowed. Figure 5 shows a photograph of my uncle working in his laboratory.

My aunt and uncle were Quakers and they did not support violence as solutions to political problems anywhere in the world. During World War

II, my uncle did alternative service rather than bear arms. He worked in a mental institution in New Hampshire, cleared swamps in the south, and was a guinea pig for the development of vaccines against tropical diseases. After the war he settled in a commune in Pennsylvania, called Bryn Gweled, which he helped found. People of all races and religious affiliations were welcomed in this community. It was a marvelous place for children: it contained thick woods for exploration and had communal activities of all kinds—painting, dance, theater, sports, electronics, and many sessions devoted to the discussion of the major religious philosophies of the world. Every week there were communal work parties, putting in roads, phone lines, and electrical lines, building a community center and so on.

The contrast between living primarily alone in the streets of Italy and living in an intensely cooperative and supportive community in Pennsylvania was enormous. Time was needed for healing and for erasing the images of war from my mind. I remember that for many years after coming to the United States I would go to sleep tossing and turning with such force that by morning the sheets were torn and the bed frame broken. This activity disturbed my aunt and uncle to the extent that Sarah would take me from one child psychologist or psychiatrist, to another. These professionals were not very helpful, but the support of the community was. The nightly activity eventually subsided. There may be lessons to be learned from such experiences for the treatment of the children from Darfur, the Congo, and now Kenya.

Sarah and Edward took on the challenge of converting me into a productive human being. This, I am sure, was a very formidable task. I had received little or no formal education or training for living in a social environment. Quakers do not believe in frills, but rather in a life of service. My aunt and uncle taught me by example. I was given few material goods, but every opportunity to develop my mind and soul. What I made of myself would be entirely up to me. The day after I arrived in America, I went to school. I started in the third grade in the Southampton public school system. Sarah also took on the task of teaching me to read, starting from the very beginning.

The first task was to learn English. I had a marvelous third grade teacher. She was patient and encouraging. The class was studying Holland, so I started participation in class functions by painting a huge mural on butcher block paper with tulips, windmills, children ice skating, children in Dutch costumes, and ships. It was a collage of activities and colors. This did not require verbal communication.

I was a good, but not serious, student in grade school and high school. Academics came easily to me. I attended an outstanding high school, George School, a Quaker school north of Philadelphia. The teachers were superb, challenging, enthusiastic, competent, and caring. They enjoyed teaching. The campus was also magnificent, particularly in the spring when the cherry and dogwood trees were bursting with blossoms. An emphasis on Quaker beliefs permeated all of the academic and sports programs. A favorite period for many, including me, was Quaker meeting, a time set aside for silent meditation, and taking stock of where we were going. My wife and I sent our daughter to George School for her own last two years in high school so that she might also benefit from the personal virtues it promotes, and we think she has.

Sports were very important to me at George School, and physical activity has remained an important activity for me to this day. I played varsity football, soccer, and baseball, and wrestled. I was particularly proficient at wrestling. I enjoyed the drama of a single opponent, as well as the physical and psychological challenges of the sport. After George School, I went to Antioch, a small liberal arts college in Ohio.

At Antioch College I became a serious student, converting to academics all of the energy I had previously devoted to sports. Coming from George School, I carried the charge of making this a better, more equitable world for all people. Most of the problems appeared to be political, so I started out at Antioch majoring in political science. However, I soon became disillusioned with political science since there appeared to be little science to this discipline, so I switched to the physical sciences—physics and chemistry. I found great pleasure in the simplicity and elegance of mathematics and classical physics. I took almost every mathematics, physics, and chemistry course offered at Antioch, including Boolean algebra and topology, electrodynamics, and physical chemistry.

Although I found physics and mathematics intellectually satisfying, it was becoming apparent that what I was learning came from the past. The newest physics that was taught at Antioch was quantum mechanics, a revolution that had occurred in the 1920's and earlier. Also, many frontiers of experimental physics, particularly experimental particle physics, were requiring the use of larger and larger accelerators, which involved bigger and bigger teams of scientists and support groups to execute the experiments. I was looking for a science in which the individual investigator had a more intimate, hands-on involvement with the experiments. Fortunately, Antioch had an outstanding work-study program; one quarter we studied on campus, the next was spent working on jobs related to our fields of interest. The jobs, in my case laboratory jobs, were maintained all over the country, and every three months we packed up our bags and set off for a new city and a new work experience. So one quarter off I went to Boston and the Massachusetts Institute of Technology (MIT).

There I encountered molecular biology as the field was being born (late 1950's). This was a new breed of science and scientist. Everything was new. There were no limitations. Enthusiasm permeated this field. Devotees from physics, chemistry, genetics, and biology joined its ranks. The common premises were that the most complex biological phenomena could, with persistence, be understood in molecular terms and that biological phenomena observed in simple organisms, such as viruses and bacteria, were mirrored in more complex ones. Implicit corollaries to this premise were that whatever was learned in one organism was likely to be directly relevant to others and that similar approaches could be used to study biological phenomena in many organisms. Genetics, along with molecular biology, became the princi-

pal means for dissecting complex biological phenomena into workable subunits. Soon all organisms came under the scrutiny of these approaches.

I became a product of the molecular biology revolution. The next generation. As an Antioch college undergraduate, I worked several quarters in Alex
Rich's laboratory at MIT. He was an x-ray crystallographer, with very broad
interests in molecular biology. While at MIT I was also fortunate to be influenced by Salvador Luria, Cyrus Leventhal and Boris Magasanik, through
courses, seminars, and personal discussions. At that time Sheldon Penman
and Jim Darnell were also working in Alex Rich's laboratory. When placed in
the same room, these two were particularly boisterous, providing comic relief
to the fast moving era.

After Antioch, I set off for what I perceived as the "Mecca" of molecular biology, Harvard University. I had interviewed with Professor James D.
Watson, of "Watson and Crick" fame, and asked him where should I do my
graduate studies. His reply was curt and to the point: "Here. You would be
fucking crazy to go anywhere else." The simplicity of the message was very
persuasive.

James D. Watson had a profound influence on my career (see Figure 6).
He was my mentor. He did not teach me how to do molecular biology; because of my Antioch job experiences, I had already become a proficient ex-
perimenter. Jim instead taught me the process of science—how to extract the
questions in a field that are critical to it and at the same time approachable
through current technology. As an individual, he personified molecular biol-

Figure 6. A photograph of James D. Watson.

ogy, and, as his students, we were its eager practitioners. His bravado encouraged self-confidence in those around him. His stark honesty made our quest for truth uncompromising. His sense of justice encouraged compassion. He taught us not to bother with small questions, for such pursuits were likely to produce small answers. At a critical time, when I was contemplating leaving Harvard as a faculty member and going to Utah, he, being familiar with my self-sufficiency, counseled me that I could do good science anywhere. The move turned out to be a good decision. In Utah I had the luxury to pursue long-term projects that were not readily possible at Harvard, which, in too many cases had become a bastion of short-term gratification.

Doing science in Jim's laboratory was exhilarating. As a graduate student, I was provided with what appeared to be limitless resources. I could not be kept out of the laboratory. Ninety-hour weeks were common. The lab was filled with talented students, each working on his or her own set of projects. Represented was a mixture of genetics, molecular biology, and biochemistry. We were cracking the genetic code, determining how proteins were synthesized, and isolating and characterizing the enzymatic machinery required for transcription. At this time, Walter Gilbert was also working in Jim's laboratory. He was then a member of the physics department, but had also been bitten by the molecular biology bug. Jim and Wally complemented each other brilliantly, because they approached science from very different perspectives. Jim was intuitive, biological; Wally quantitative, with a physicist's perspective. They were both very competitive. As students, we received the benefit of both, but also their scrutiny. They were merciless, but fair. You had to have a tough hide, but you learned rigor, both with respect to your science and your presentations. Once you made it through Jim's laboratory, the rest of the world seemed a piece of cake. It was excellent training. Despite the toughness, which at times was hard, Jim was extremely supportive. He also made sure that you, the student, received full credit for your work. Despite the fact that Jim was responsible for all of the resources needed to run his laboratory, if you did all of the work for a given paper, then you were the sole author on that paper. Among all of the laboratory heads in the world, I believe that Jim Watson was among very few in implementing this policy.

The summer before I started graduate school, Marshal Nirenberg had announced that polyU directs the synthesis of polyphenylalanine in a cell free protein synthesizing extract. That paper was a bombshell! I decided I would generate a cell-free extract capable of synthesizing real, functional proteins. Jim's laboratory had started working on the RNA bacteriophage, R17. Its genome also served as messenger RNA to direct the synthesis of its viral proteins. That would be my message. The cell-free protein synthesizing extract worked beautifully. Authentic viral coat protein and replicase were shown to be synthesized in the extract[1]. Further, the coat protein was functional, it bound to a specific sequence of the R17 genome, thereby modulating the synthesis of the replicase. To this day, the high affinity of the viral coat protein for this RNA sequence is exploited as a general reporter system to track RNA trafficking within living cells and neuronal axons. In collaboration with

Figure 7. A photograph of Karl G. Lark.

Gary Gussin, also a graduate student in Jim's laboratory, this system was used to determine the molecular mechanism of genetic suppression of nonsense mutations[2]. In collaboration with Jerry Adams, another graduate student in Jim's laboratory, the system was also used to determine that initiation of the synthesis of all proteins in bacteria proceeded through the use of formyl-methionine-tRNA[3,4]. A similar mechanism is involved in the initiation of protein synthesis in all eukaryotic organisms. Finally, I used the same *in vitro* system to show that termination of protein synthesis unexpectedly utilized protein factors, rather than tRNA, to accomplish this end[5,6]. Jim Watson would later offer the very complimentary comment "that Capecchi accomplished more as a graduate student than most scientists accomplish in a lifetime." It was, indeed, a productive time, but it wasn't work; it was sheer joy.

While a graduate student in Jim's laboratory, I was invited to become a junior fellow of the Society of Fellows at Harvard. Being a junior fellow was very special. The society's membership, junior and senior fellows, represented a broad spectrum of disciplines; all the members were talented, and most of them were much more verbal than I. Social discourse centered around meals, prepared by an exquisite French chef and ending with fine brandy and Cuban cigars. Frequent guests at these dinners were the likes of Leonard Bernstein. Surreal maybe, but also very special.

From Jim's laboratory, I joined the faculty in the Department of Biochemistry at Harvard Medical School, across the river in Boston. During my four years at Harvard Medical School I quickly rose through the ranks, but then, I unexpectedly decided to go to Utah. I was looking for something different. There were excellent scientists in the department I was in at Harvard Medical School, but the department was not built with synergy in mind. Each research group was an island onto itself. At that time, they were also unwilling to hire additional young faculty and thereby provide the department with a more youthful, energetic character. At the University of Utah, I would be joining a newly formed department that was being assembled by a very talented scientist and administrator, Karl G. Lark (Figure 7). He had excellent taste in scientists and a vision of assembling a faculty that would enjoy working together and striving together for excellence. I could be a participant in the growth of that department and help shape its character. Furthermore, the University's administration, led then by President David P. Gardner, was in synchrony with this vision and a strong supporter. Gordon had already attracted Baldomero (Toto) Olivera, Martin Rechsteiner, Sandy Parkinson, and Larry Okun to Utah. After I arrived at Utah, we were able to bring to Utah such outstanding scientists as Ray Gesteland, John Roth, and Mary Beckerle. Utah also provided wide open space, an entirely new canvas upon which to create a new career (see Figures 8). These are views from one of the homes in Utah which I have shared with my wife, Laurie Fraser, and daughter, Misha. The air is clean, and I can look for long distances. The elements of nature are all around us. What a place to begin a new life!

Figure 8. Views from one of our homes in Utah and a photograph of my wife, Laurie Fraser, and daughter, Misha, just after she was born. Misha is now graduating from the University of California, Santa Cruz as an arts major.

REFERENCES

1. Capecchi, M. R. (1966). Cell-free protein synthesis programmed with R17 RNA: Identification of two phage proteins. *J. Mol. Bol.* 21:173–193.
2. Capecchi, M. R. and Gussin, G. N. (1965). Suppression *in vitro*: Identification of a serine-tRNA as a "Nonsense Suppressor." *Science* 149:417–422.
3. Adams, J. M. and Capecchi, M. R. (1966). N-formylmethionine-tRNA as the initiator of protein syntheses. *Proc. Natl. Acad. Sci. USA* 55:147–155.
4. Capecchi, M. R. (1966). Initiation of *E. coli* proteins. *Proc. Natl. Acad. Sci. USA* 55:1517–1524.
5. Capecchi, M. R. (1967). Polypeptide chain termination *in vitro*: Isolation of a release factor. *Proc. Natl. Acad. Sci. USA* 58:1144–1151.
6. Capecchi, M. R. and Klein, H. A. (1970). Release factors mediating termination of complete proteins. *Nature* 26:1029–1033.

GENE TARGETING 1977–PRESENT

Nobel Lecture, December 7, 2007

by

MARIO R. CAPECCHI

University of Utah, School of Medicine, 15 North 2030 East, Salt Lake City, UT 84112-5331, USA.

EARLY EXPERIMENTS

My entry into what was going to become the field of gene targeting started in 1977. The size of my laboratory in Utah, devoted to this project, was modest: myself and two competent technicians—my wife Laurie Fraser, and Susan Tamowski. I was experimenting with the use of extremely small glass needles to inject DNA directly into nuclei of living cells. In the laboratory adjacent to ours, Dr. Larry Okun, a neuroscientist, was recording intracellular electrical potentials in cultured neurons from chick dorsal root ganglia. His apparatus for penetrating the cells non-destructively to measure these electrical potentials appeared to be ideal for conversion into a "microsyringe" to allow pumping of defined quantities of macromolecules, including DNA, into mammalian cells in culture. Larry (Figure 1) graciously helped me enormously with the process of conversion. I should further add that Larry Okun, has been over many years, too many to count on one's fingers and toes, my favorite person to discuss science, politics, and trivia. But his rigorous insight into science, in particular, has been of immeasurable help to me throughout my tenure at the University of Utah. Having enticed me and my wife[1] to come to Utah from Boston in the first place, by organizing an unbelievably beautiful 10 day backpacking trip in the nearby Wind River Mountains of Wyoming, along a series of mountain lakes bursting with trout every evening, he owed us quite a bit, and he delivered. Once assembled, the injection apparatus (Figure 2) was quite effective, allowing me to do 1000 nuclear injections per hour of well-defined volumes of solution (in the range of femtoliters) containing chosen macromolecules.

In 1977, Wigler and Axel showed that cultured mammalian cells deficient in the enzyme, thymidine kinase, Tk$^-$, could be restored to Tk$^+$ status by the introduction of functional copies of the herpes virus *thymidine kinase gene* (*HSV-tk*)[2]. Although an important advance for the field of somatic cell genetics, their protocol—the use of calcium phosphate co-precipitation to introduce the DNA into the cultured cells by phagocytosis—was not very efficient. With their method, stable incorporation of functional copies of HSV-tk occurred in approximately one cell per million cells exposed to the DNA calcium phosphate co-precipitate[2]. It seemed that the low efficiency might be

Figure 1. A photograph of Lawrence M. Okun.

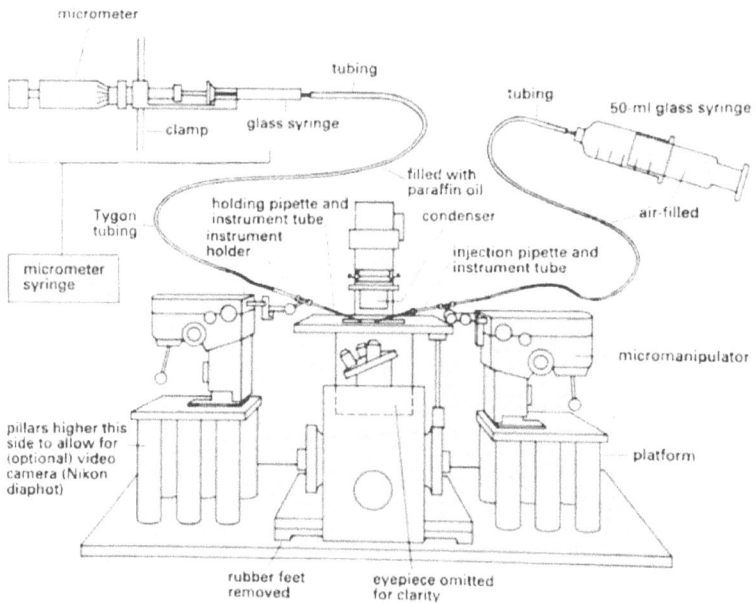

Figure 2. A schematic of the apparatus I used to inject DNA into nuclei of mammalian cells in culture. Micromanipulators are used to guide the needle, tip diameter 0.1 μm, containing the DNA solution into nuclei of living cells while being viewed through a light microscope.

a problem of delivery. Most of the DNA taken up by the cells did not appear to be delivered to the nucleus, where it could function, but instead was destined for lysosomes, where it was degraded. I sought to determine whether I could introduce functional copies of the *HSV-tk* gene directly into nuclei of cultured TK⁻ cells using the microinjection apparatus described above. This procedure proved to be extremely efficient; one cell in three that received the DNA stably passed functional copies of the *HSV-tk* gene onto its daughter cells[3]. An immediate outcome of these experiments was that the high efficiency of DNA transfer that we observed by microinjection made it practical for investigators to use the same methodology to generate transgenic mice containing random insertions of exogenous DNA. This was accomplished by injection of the desired DNA into nuclei of one-celled mouse zygotes with the resulting embryos allowed to come to term after transfer to the uterus of foster mothers[4–8]. The generation of transgenic mice, in which chosen exogenous pieces of DNA have been randomly inserted within the mouse genome, has become a cottage industry.

However, I found that the efficient transfer of functional *HSV-tk* genes into the host cell genome required that the injected *HSV-tk* genes be linked to an additional short viral DNA sequence[3]. It seemed plausible to me that highly evolved viral genomes which, as part of their life cycle, resided in the host cell genome, might contain bits of DNA sequence that enhanced their ability to establish themselves within the host cell genome. I searched the genomes of the lytic simian virus, SV40, and the ASV retroviral provirus for the presence of such sequences and found them[3]. When linked to the injected *HSV-tk* gene, these sequences increased the frequency with which TK⁺ cells were generated by a factor of 100 over that produced by HSV-tk DNA injected alone. I showed that this enhancement did not result from independent replication of the injected *HSV-tk* DNA as an extra-chromosomal plasmid, but rather that the efficiency-enhancing sequences were either increasing the frequency with which the exogenous DNA was inserting itself into the host genome or increasing the probability that the *HSV-tk* gene, once integrated into the host genome, was being expressed in the recipient cells. The latter turned out to be correct. These experiments were completed before the idea of gene-expression enhancers had emerged and contributed to the definition of these special DNA sequences[9]. Further, the emerging idea of enhancers profoundly influenced our contributions to the development of gene targeting vectors. Specifically, it alerted us to the importance of using appropriate enhancers to mediate expression of newly introduced selectable genes (used to select for successfully altered recipient cells), regardless of the inherent expression characteristics of the host chromosomal sites into which we were targeting those genes[10].

HOMOLOGOUS RECOMBINATION

Although the ability to introduce exogenous DNA randomly into the host cell genome with very high efficiency by microinjection was itself extremely

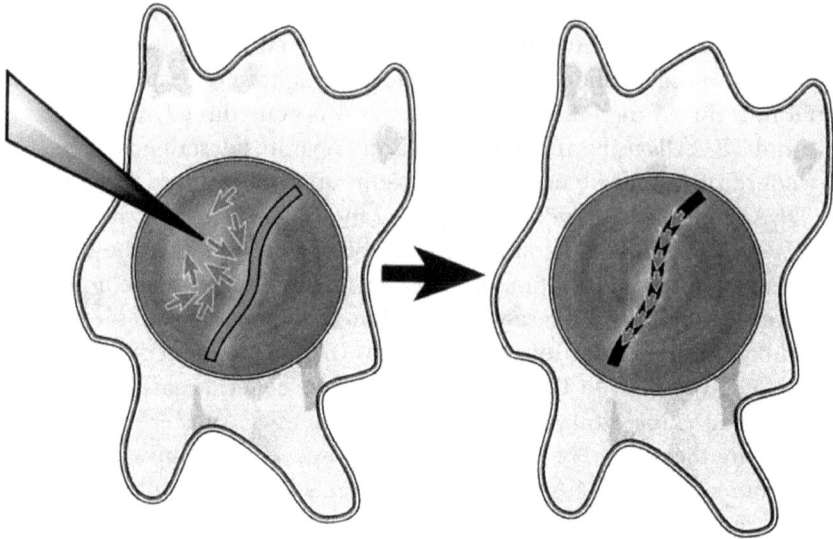

Figure 3. Formation of highly ordered head-to-tail DNA concatemers following introduction of multiple copies of the same DNA sequence into mammalian cell nuclei.

useful, the observation that I found most fascinating from these early DNA-injection experiments was that, when multiple copies of the HSV-tk plasmid were injected into a given cell, though many of them became randomly inserted into the host cell's genome, they would all be found at a single locus, as a highly ordered head-to-tail concatemer (see Figure 3)[11]. This was the key observation and stimulus for the targeting project that followed.

It was clear, however, that this project would now progress more rapidly with the efforts of additional investigators. Fortunately, two very gifted post-doctoral fellows, Drs. Kim Folger Bruce and Eric Wong chose to join my group at this time (see Figure 4). It seemed that the highly-ordered concatemers of exogenous genes found at the insertion sites could not arise by a random mechanism, but were likely generated either by replication of the injected DNA before insertion (for example by a rolling circle-type mechanism of DNA replication) or by homologous recombination between the co-injected HSV-tk plasmids. We proved that they were generated by homologous recombination[11]. This conclusion was very significant because it demonstrated that mammalian somatic cells contain an efficient enzymatic machinery for mediating homologous recombination. The high efficiency of this machinery became evident from the observation that when more than 100 HSV-tk plasmid molecules were injected per cell, they were all incorporated into a single, ordered, head-to-tail concatemer[11]. These experiments were also the first demonstration of homologous recombination between co-introduced DNA molecules in cultured mammalian cells. From these results it was immediately apparent to me that if we could harness this efficient machinery to mediate homologous recombination between a newly introduced DNA molecule of our choice and the same DNA sequence in the recipient

Figure 4. Photographs of Kim Folger Bruce and Eric Wong who worked in my laboratory from 1981–1985 and 1983–1986 respectively.

cell's genome, we would have the ability to mutate any endogenous cellular gene in cultured cells, in any chosen way. It was thus these experiments that provided us the incentive for vigorous pursuit of gene targeting in mammalian cells. Interestingly, these experiments were done prior to our hearing that gene targeting could be readily achieved in yeast[12]. The results derived from the analysis of mechanisms of gene targeting in yeast did, however, influence our thinking during subsequent development of gene targeting in mammalian cells[13–15].

The next step in our quest for achieving mammalian gene targeting required our becoming more familiar with the homologous recombination machinery in mammalian cells, for example its substrate preferences and what were the most common reaction products resulting from homologous recombination. At this time Dr. Kirk Thomas also joined my research group as a postdoctoral fellow and became a critical contributor to our research (Figure 5). By examining homologous recombination between co-injected DNA molecules, we learned, among other things, that linear DNA molecules, rather than circular or super-coiled molecules were a preferred substrate for homologous recombination, that the efficiency of homologous recombination was cell-cycle dependent, showing a peak of activity in early S phase; and that, although both reciprocal and non-reciprocal exchanges occurred, there was a distinct bias towards the latter[16–18]. These results contributed substantially to our choice of experimental design for the next stage of our quest— the detection of homologous recombination between newly introduced, exogenous DNA molecules and their endogenous chromosomal homologs in recipient cells.

Figure 5. Photograph of Kirk R. Thomas who was in my laboratory from 1983 to 2002 first as a postdoctoral fellow and then as a Senior Scientist of the Howard Hughes Medical Insitute.

In 1980, I submitted a grant application to the National Institute of Health proposing to test the feasibility of such gene targeting in mammalian cells. These experiments were emphatically discouraged by the reviewers on the grounds that there might be only a vanishing small probability that the newly introduced DNA would ever find its matching sequence within the host cell genome, a prerequisite for homologous recombination. Despite the rejection, I decided to put all of our effort into continuing this line of research. This was a big gamble on our part. Aware that the frequency of gene targeting to homologous sites was likely to be low and that the far more common competitive reaction would be random insertion of the targeting vector into non-homologous sites of the host cell genome, we proposed to use selection to eliminate cells not containing the desired homologous recombination events. One first test of gene targeting (Figure 6) used, as the chromosomal target, DNA sequences that we had previously randomly inserted into the host cell genome. Thus, the first step in this scheme required generating cell lines containing random insertions of a defective *neomycin-resistance gene* (*neo^r*) that contained either a small deletion or a point mutation in that gene. In the second step, the targeting-vector DNA also contained a defective *neo^r* gene, with a mutation that differed from the one present at the host cell target site. Homologous recombination, between the two defective *neo^r* genes, one in the targeting vector and the second residing in the host cell chromosome, could generate a functional *neo^r* from the two defective parts and render the cells resistant to the drug G418, which is lethal to cells without a

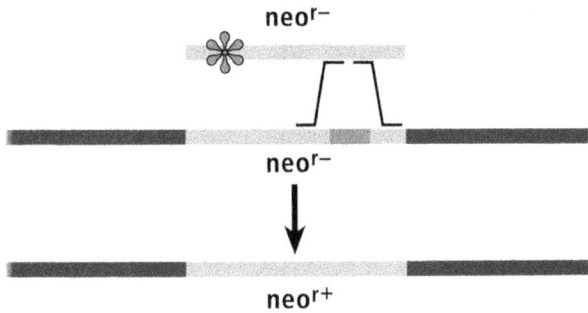

Figure 6. Regeneration of a functional neo^r gene by gene targeting. The recipient cell contains a defective neo^r with a deletion mutation (del). The targeting vector contains a 5'-point mutation (*). With a frequency of approximately 1 in 1000 cells receiving an injection of the targeting vector containing the point mutation, the chromosomal copy of neo^r is corrected with the information supplied by the targeting vector.

functional *neo^r* gene. Thus, successful gene targeting events would yield cells capable of growth in medium containing G418.

For the first step we generated recipient cell lines containing a single copy of the defective *neo^r* gene, cell lines containing multiple copies of the defective gene as a head-to-tail concatemer and, by inhibiting concatemer formation, even cell lines containing multiple defective *neo^r* genes as single copies inserted in separate chromosomes. These different recipient cell lines allowed us to evaluate how the number and location of the target sites within a recipient cell's genome influenced the targeting frequency. By 1984, we had good evidence that gene targeting in cultured mammalian cells indeed occurs[19]. At this time, I submitted another grant application to the same National Institute of Health study section that had rejected our earlier proposed gene targeting experiments. Their response was "We are glad that you didn't follow our advice."

To our delight, correction of the defective chromosomally-located *neo^r* genes by homologous recombination with our microinjected gene-targeting vector occurred at an absolute frequency of 1 per 1,000 injected cells[20]. This frequency was many orders of magnitude greater than the reversion frequency of the individual *neo^r* mutations by themselves. Furthermore, the frequency was not only higher than we expected, particularly considering that the extent of DNA sequence homology between the targeting vector and the target locus was less than 1,000 base pairs, but the relatively high targeting frequency made it practical for us to examine a number of parameters influencing that frequency[20]. An important lesson learned from testing the different recipient cell lines was that all of the chromosomal target positions analyzed seemed equally accessible to the homologous recombination machinery, indicating that a large fraction of the mouse genome was likely to be manipulable through gene targeting[20].

At this time, Oliver Smithies and his colleagues reported their classic experiment of targeting modification of the β-globin locus in cultured mammalian cells[21]. This elegant experiment demonstrated that it was feasible to disrupt an endogenous gene in cultured mammalian cells. Oliver and I pursued gene targeting independently. We had separate visions in mind and different approaches to its implementation. Through the years we have been extremely fortunate in our ability to share expertise and reagents, as well as enjoying each other's fellowship. That is not to say we were not competitive. Science is very competitive and a high premium is placed on being first. Equally important, however, science is also a very communal enterprise in which all are dependent on past and concurrent contributions by many, many other investigators for advances and inspiration. Where would either Oliver's lab or mine have gotten without the ability to generate viable mouse chimeras, initially starting with mouse morulas, and then extending the technology to injected cells from the innercell mass, EC cells, and ES cells into the pre-implantation embryos? The contributions and progression of this technology by Mintz, Gardner, Stevens, Martin, and Evans are apparent[22–28], providing just some examples of the many investigators whose efforts have been essential to our eventual ability to do gene targeting in living mice.

Having established that gene targeting could be achieved in cultured mammalian cells and having determined some of the parameters that influenced its frequency, we were ready to extend gene targeting to the living mouse. The low frequency of gene targeting, relative to random integration of the targeting vector into the recipient cell genome, made it impractical to attempt gene targeting directly in one-celled mouse zygotes. Instead, it seemed our best option was to carry out the gene targeting in populations of cultured embryo-derived stem (ES) cells, from which the relatively rare targeted recombinants could then be selected and purified. These purified cells, when subsequently introduced into preimplantation embryos and allowed to mature in a foster mother, would be expected to contribute to the formation of all tissues of the mouse, including the germ line. Fortunately, at a Gordon Conference in the summer of 1984, when we were ready to initiate these experiments, I heard that Martin Evans had isolated from mouse embryos ES cells capable of contributing in just this way to the formation of the germ line and to do so at a reasonable frequency[27,28]. Martin's ES cells appeared to be much more promising in their potential to contribute to the embryonic germline than were the previously described embryonal carcinoma (EC) cells[24,29].

GENE TARGETING IN ES CELLS

In the winter of 1985, my wife, Laurie Fraser, and I spent a week in Martin Evans' laboratory learning how to derive and culture mouse ES cells, as well as how to generate mouse chimeras from these cells by their addition to recipient preimplantation embryos. Also instrumental in our learning these techniques were Dr. Elizabeth Robertson and Alan Bradley, a postdoctoral fellow and graduate student in Evans' laboratory, respectively. This is an excellent example of how science progresses from the collective sharing of expertise and resources. We have always been grateful for Martin's generosity.

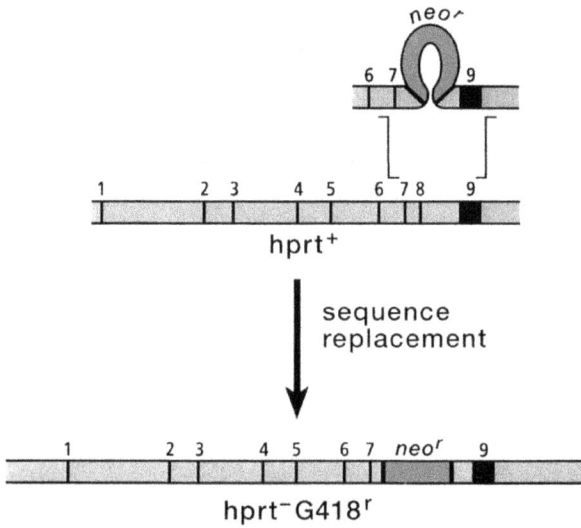

Figure 7. Disruption of *hprt* gene by gene targeting in mouse ES cells. The targeting vector contains genomic sequences from the mouse *hprt* gene disrupted in the eighth exon by *neo*ʳ. After homologous pairing between the vector and the cognate sequences in the endogenous hprt gene of the ES cell genome, a homologous recombination event replaces the ES cell genomic sequences with vector sequences containing the *neo*ʳ gene. The resulting cells are able to grow in medium containing the drugs G418, which kills cells without an inserted functional *neo*ʳ gene, and 6-TG, which kills cells with an undisrupted functional *hprt* gene.

In the beginning of 1986, our effort thus shifted to doing gene targeting experiments in mouse ES cells. We decided also to switch to electroporation as a means of introducing our targeting vectors into ES cells. Although microinjection was orders of magnitude more efficient than electroporation as a means for generating cell lines with targeted mutations, injections are done one cell at a time. With electroporation, we could introduce the gene targeting vector into 10^7 cells in a single experiment, easily producing large numbers of cells containing targeted mutations, even at the lower efficiency. In addition, it was apparent to us that, as a technology, electroporation would be more readily adopted by other laboratories, relative to microinjection, thereby making gene targeting more user friendly to more scientists.

To rigorously determine the quantitative efficiency of gene targeting in ES cells as well as to evaluate the parameters that affect the gene targeting frequency, we chose as our target locus the *hypoxanthine phosphoribosyl transferase* (*hprt*) gene. There were two reasons for this choice. First, since *hprt* is located on the X chromosome, and the ES cell lines that we were using were derived from a male mouse, only a single *hprt* locus had to be disrupted in the recipient cells to yield *hprt*⁻ cell lines. Second, a good protocol for selecting cells with a disrupted *hprt* gene existed, based on the drug 6-thioguanine (6TG), which kills cells with a functional *hprt* gene[30]. The strategy we used was to generate a gene-targeting vector that contained an *hprt* genomic sequence that was disrupted by insertion of *neo*ʳ in one of the gene's exons

Figure 8. A photograph of Chuxia Deng who worked in my laboratory as a graduate student from 1987 to 1992.

(Figure 7). The exon we chose, exon 8, encodes the active catalytic site for this enzyme. Homologous recombination between this targeting vector and the endogenous *hprt* locus would generate hprt⁻ ES cells resistant to growth in medium containing both 6TG (killing cells with untargeted hprt⁺ loci) and G418 (killing cells lacking the inserted *neo*ʳ gene, as described above). All cell lines generated from cells selected in this way, had lost *hprt* function as a result of the targeted disruptions of the *hprt* locus[10]. Thus, the *hprt* locus provided Chuxia Deng (Figure 8), then a graduate student in our laboratory, an ideal locus for further study of the parameters that influenced the targeting efficiency[31–33].

Because we foresaw that the *neo*ʳ gene would probably be used as a positive selectable gene for the disruption of many genes in ES cells, it was essential that its expression be mediated by an enhancer that would function in ES cells, regardless of the expression status of the target locus. Here our previous experience with enhancers proved of value. We knew from those experiments that the activities of promoter-enhancer configurations are very cell-type specific. To encourage strong *neo*ʳ expression in ES cells, we chose to drive its expression with a mutated polyoma virus enhancer that had been selected for strong expression in mouse embryonal carcinoma cells, which we presumed to be similar to mouse ES cells[10,34]. Subsequently, the strategy described above namely using a *neo*[r] driven by an enhancer that allows strong expression in ES cells independent of chromosomal locations, has become the standard for the disruption of most genes in ES cells.

The experiments described above showed that mouse ES cells were a good recipient host for gene targeting. In addition the drug-selection protocols required to identify ES cell lines containing the desired gene-targeting event did not appear to alter their pluripotent potential[10]. I believe that this paper was pivotal in the development of the field, encouraging other investigators to begin to use gene targeting in mice as a means for determining the function of chosen mammalian genes in the living animal.

The ratio of homologous, i.e. targeted insertions to random insertions at non-homologous sites in ES cells is approximately 1 to 1,000[10]. Because the disruption of most genes does not produce a cellular phenotype, that is selectable in cell culture, investigators seeking to disrupt a gene of choice would need to undertake tedious DNA screens through many cell colonies to identify the rare ones containing the desired targeting event.

To address this problem we reported in 1988, a general strategy to enrich cells in which a homologous targeting event has occurred[35]. This enrichment procedure, known as positive-negative selection, was derived from an observation in experiments done in our laboratory, namely that linear DNA molecules, when inserted at random sites in the recipient cell's genome, most frequently retain their ends, while sequences inserted at the target site, by homologous recombination, lose non-homologous ends from the original vector (see Figure 9). Further, contrary to expectations from studies of homologous recombinations in yeast, we showed that even blocking both ends of the homology arms of a targeting vector with non-homologous DNA sequences does not reduce the targeting frequency in mammalian cells[35]. This approach, correspondingly has two components. One component is a positive selectable gene, *neo*[r] used, as described above, to select for recipient cells that have incorporated the targeting vector anywhere in their genomes (that is, at the target site by homologous recombination or at a random site by non-homologous recombination). The second component is a negative selectable gene, HSV-tk, located at the end of the linearized targeting vector and used to select against cells containing random insertions of the target vector (medium containing the drugs, *gangcyclovir* or FIAU, kills cells expressing the *HSV-tk* gene but not cells expressing the endogenous mammalian thymidine kinase gene). Thus the positive selection enriches for recipient cells that have incorporated the targeting vector somewhere in their genome, whereas the negative selection eliminates those that have incorporated it at non-homologous sites. The net effect is enrichment for cells in which the desired homologous targeting event has occurred. The strength of this enrichment procedure is that it is independent of the function of the gene that is being disrupted and succeeds whether or not the gene is expressed in ES cells. The validity of the procedure was shown by using it to enrich for ES

A Gene targeting

x⁻neoᴿHSV–tk⁻(G418ʳ,FIAUʳ)

B Random integration

x⁺neoᴿHSV–tk⁺(G418ʳ,FIAUˢ)

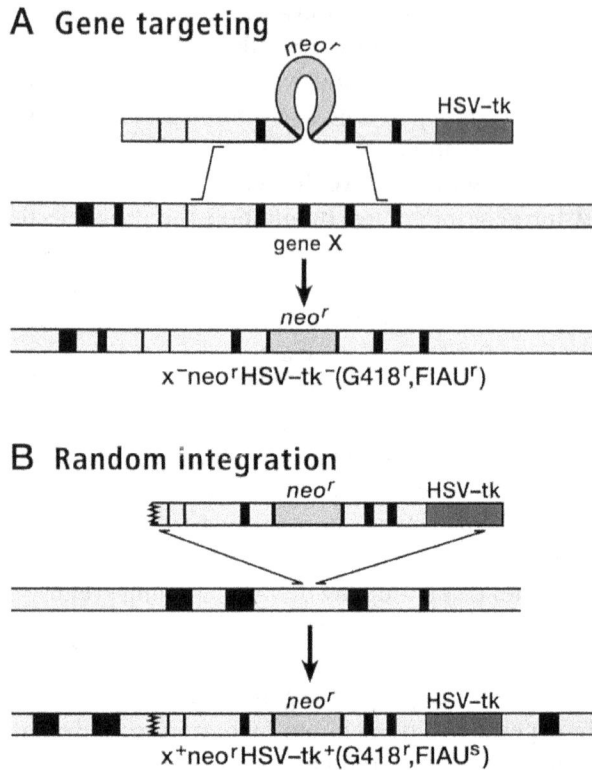

Figure 9. The positive-negative selection procedure used to enrich for ES cells containing a targeted disruption of gene X.a. The linear replacement-type targeting vector contains an insertion of neoʳ in an exon of gene X and a linked HSV-tk gene at one end. It is shown pairing with a chromosomal copy of gene X. Homologous recombination between the targeting vector and the cognate chromosomal gene results in the disruption of one genomic copy of gene X and the losss of the vector's HSV-tk gene. Cells in which this event has occurred will be X$^{+/-}$, neoʳ, HSV-tk⁻ and will grow in medium containing G418 and FIAU. The former requires the presence of a functional neoʳ gene and the latter kills cells containing a functional HSV-tk gene. Integration of the targeting vector at a random site of the ES cell genome by non-homologous recombination. Because non-homologous insertion of exogenous DNA into the chromosome occurs through the ends of the linearized DNA, HSV-tk will remain linked to neoʳ. Cells derived from this type of recombination event will be X$^{+/+}$ neoʳ⁺ and HSVtk⁺ and therefore resistant to growth in G418 but killed by presence of FlAU. Cells that have not received a targeting vector, will be X^{++} neoʳ⁻ and HSVtk⁻ and will be killed by the presence of G418. As a consequence this procedure specifically enriches for cells in which a gene targeting event has occurred.

cells containing targeted mutations in the *int2* gene, now known as *Fgf3*[35]. These experiments were carried out by Suzi Mansour, a talented postdoctoral fellow in our laboratory (Figure 10) and Kirk R. Thomas (Figure 4). Positive-negative selection has become the most frequently used procedure to enrich for cells containing gene targeting events. Using positive-negative selection we have found that the targeting frequency varies from gene to gene. With genes that exhibit a high targeting frequency, a high percentage of clones obtained after positive-negative selection contain the targeting event. The worst

Figure 10. A photograph of Suzanne Mansour. She worked in my laboratory as a post-doctoral fellow from 1987 to 1992.

cases have been ones in which one in a hundred selected clones contains the desired targeting event. If the targeted gene is one expressed in ES cells, then the targeting frequency at that locus is likely to be high.

EXTENSIONS AND MORE RECENT DEVELOPMENTS

The use of gene targeting to evaluate the functions of genes in the mouse is now routine. It is being used in hundreds of laboratories all over the world. Well over 11,000 genes have been disrupted in the mouse using the described procedures. This is quite surprising considering that these disruptions have been done in individual laboratories in the absence of coordinated programs. Now however, there are a number of funded national and international efforts to disrupt every gene in the mouse by gene targeting[36]. In addition, hundreds of human diseases have been modeled in the mouse by the use of gene targeting. These models allow study of the pathology of the diseases in much more detail than is possible in humans. In addition, the models provide a vehicle for subsequent development and evaluation of new therapeutic modalities including drugs.

To date, gene targeting has been used primarily to disrupt genes, producing so called "knockout mice." However gene targeting can be used to alter the sequences of a chosen genetic locus in the mouse in any conceivable manner, thus providing a very general means for "editing" the mouse genome. It can be used to generate gain-of-function mutations or partial loss-of-function mutations. Gene targeting can also be used to restrict the loss of function of a chosen gene to particular tissues, yielding so-called conditional mutations. This is most commonly achieved by combining exogenous (non-mammalian) site-specific recombination systems, such as those derived from bacteriophages or yeast (i.e. Cre/loxP or Flp/FRT respectively), with gene targeting, to mediate excision of a gene only where the appropriate recombinase is produced[37–40]. By control of where Cre- or Flp-recombinase is expressed, for example in the liver, a gene, flanked by loxP or FRT recognition sequences, respectively, can be excised in the desired tissue (e.g. liver). Temporal control of gene function has also been achieved by making the production of the functional recombinase dependent upon the administration of small molecules or even on physical stimuli, such as light[41–44]. Such conditional mutagenesis has been very effective for more accurate modeling of human cancers, which are often restricted to particular tissues and even to specific cells within those tissues, as well as being initiated post birth[45–48]. In human cancers, the interactions between the host tissues and the malignant cells are often critical to its initiation and progression[49,50]. Thus, inclusion of these interactions in the mouse model also becomes critical if the mouse model is to accurately recapitulate the human malignancy.

Gene targeting is an evolving technology and we can anticipate further extensions to its repertoire. To date it has been used primarily to perturb the function of one gene at a time. We can anticipate development of efficient multiplexing systems that will allow simultaneous conditional or non-conditional modulation of multiple genes. We can also anticipate improvements in exogenous reporter genes with parallel improvements in their detection, particularly with respect to capture times, resolution and sensitivity. Such improvements will undoubtedly be necessary if this technology is to make significant inroads in addressing truly complex biological questions, such as the molecular mechanisms underlying higher cognitive functions in mammals.

I have tried to take the reader through a brief, personal journey of my life, my development as a scientist, and our laboratory's development of gene targeting. In the process, I have tried to give credit to those who have helped me along the way to reach our goals. What I have failed to communicate is the enormity of the scientific community and how many scientists actually have helped in untold, countless ways. That list would be in the hundreds and thousands. As a scientist I have been fortunate to have visited many, many laboratories all over the world and to have talked with other scientists about their work and aspirations. It is through these conversations that one's vision broadens and an appreciation of the complexity and beauty of the biological world is reinforced. However the people that have been most influential are the members of my immediate family, Laurie Fraser and Misha Capecchi, my wife and daughter, respectively. Their support has kept me going, their sage advice has kept me from falling down too frequently, their love has made it all worthwhile.

The Nobel Prize has greatly rewarded a major segment of my life and, as a kind of demarcation invites some reflection. I hope that our contributions, among other developments, will be used by many to reduce suffering, improve our health and extend the productivity and fulfillment of our lives. Equally important, I hope that the new biological insights will yield a better understanding of ourselves as human beings and of our relationship to our environment, so that we may become better stewards of a fragile Earth. We live in a closed system and we have to gain the knowledge that will enable us to live in harmony with it. Neither we nor our planet can any longer afford the ravages of wars. Nor can the planet survive needless consumption. We must learn to distribute our resources more equitably among all peoples. As a scientist, I naturally find myself thinking about the future. As a people, we must learn to become more responsible for the consequences of our activities over much longer periods of time so that future generations may also enjoy this splendid world. It is my hope that science can combine with ethics to permit this.

REFERENCES

1. The reference here is to my first wife, Nancy McReynolds. She worked with me at Harvard and then helped me set up my laboratory in Utah. In addition to being my companion, Nancy was a critical contributor to my work before gene targeting. She is now an outstanding geriatric nurse and we are still good friends.

2. Wigler, M., Silverstein, S., Lee, L., Pellicer, A., Cheng, Y., and Axel, R. (1977). Transfer of purified herpes virus thymidine kinase gene to cultured mouse cells. *Cell* 11:223–232.

3. Capecchi, M. R. (1980). High efficiency transformation by direct microinjection of DNA into cultured mammalian cells. *Cell* 22:479–488.

4. Gordon, J. W., Scangos, G. A., Plotkin, D. J., Barbosa, J. A. and Ruddle, F. H. (1980). Genetic transformation of mouse embryos by microinjection of purified DNA. *Proc. Natl. Acad. Sci. USA* 77:7380–7384.

5. Constantini, F. and Lacy, E. (1981). Introduction of a rabbit β-globin gene into the mouse germ line. *Nature* 294:92–94.

6. Brinster, R. L., Chen, H. Y., Trumbauer, M. E., Senear, A. W., Warren, R. and Palmiter, R. D. (1981). Somatic expression of herpes thymidine kinase in mice following injection of a fusion gene into eggs. *Cell* 27:223–231.

7. Wagner, E. F., Stewart, T. A., and Mintz, B. (1981) The human β-globin gene and a functional thymidine kinase gene in developing mice. *Proc. Natl. Acad. Sci. USA* 78:5016–5020.

8. Wagner, E. F., Hoppe, P. C., Jollick, J. D., Scholl, D. R., Hodinka, R. L., and Gault, J. B. (1981) Microinjection of a rabbit β-globin gene in zygotes and its subsequent expression in adult mice and their offspring. *Proc. Natl. Acad. Sci. USA* 78:6376–6380.

9. Levinson, B., Khoury, B. G., VandeWoude, G., and Gruss, P. (1982) Activation of SV40 genome by 72-base pair tandem repeats of Moloney sarcoma virus. *Nature* 295:568–572.

10. Thomas, K. R. and Capecchi, M. R. (1987). Site-directed mutagenesis by gene targeting in mouse embryo-derived stem cells. *Cell* 51:503–512.

11. Folger, K. R., Wong, E. A., Wahl, G. and Capecchi, M. R. (1982). Patterns of integration of DNA microinjected into cultured mammalian cells: Evidence for homologous recombination between injected plasmid DNA molecules. *Mol. Cell. Biol.* 2:1372–1387.

12. Hinnen, A., Hicks, J. B., and Fink, G. R. (1978) Transformation of yeast. *Proc. Natl. Acad. Sci. USA* 75:1929–1933.

13. Petes, T. D. (1980) Unequal meiotic recombination within tandem arrays of yeast ribosomal DNA genes. *Cell* 19(3):765–774.

14. Orr-Weaver, T. L., Szostak, J. W., and Rothstein, R. J. (1981) Yeast transformation: a model system for the study of recombination. *Proc. Natl. Acad. Sci. USA* 78:6354–6358.

15. Szostak, J. W., Orr-Weaver, T. L., Rothstein, R. J., and Stahl, F. W. (1983) The double-strand repair model of recombination. *Cell* 33:25–35.

16. Wong, E. A. and Capecchi, M. R. (1986). Analysis of homologous recombination in cultured mammalian cells in a transient expression and a stable transformation assay. *Somat. Cell Mol. Genet.* 12:63–72.

17. Folger, K. R., Thomas, K. R. and Capecchi, M. R. (1985). Nonreciprocal exchanges of information between DNA duplexes coinjected into mammalian cell nuclei. *Mol. Cell. Biol.* 5:59–69.

18. Wong, E. A. and Capecchi, M. R. (1987). Homologous recombination between coinjected DNA sequences peaks in early to mid-S phase. *Mol. Cell. Biol.* 7:2294–2295.

19. Folger, K. R., Thomas, K. R. and Capecchi, M. R. (1984). Analysis of homologous recombination in cultured mammalian cells. *Cold Spring Harbor Symp. Quant. Biol.* 49:123–138.

20. Thomas, K. R., Folger, K. R. and Capecchi, M. R. (1986). High frequency targeting of genes to specific sites in the mammalian genome. *Cell* 44:419–428.

21. Smithies, O., Gregg, R. G., Koralewski, M. A. and Kurcherlapati, R. S. (1985). Insertion of DNA sequences into the human chromosomal β-globin locus by homologous recombination. *Nature* 317:230–234.

22. Mintz, B. (1965). Genetic mosaicism in adult mice of quadriparental lineage. *Science* 148:1252–1233.

23. Gardner, R. L. (1968). Mouse chimeras obtained by the injection of cells into the blastocyst. *Nature* 220:596–597.

24. Stevens, L. C. (1967). The biology of teratomas. *Adv. Morphog.* 6:1–31.

25. Papaioannou, V. E., McBurney, M., Gardner, R. L. and Evans, M. J. (1975). The fate of teratocarcinoma cells injected into early mouse embryos *Nature* 258:70–73.

26. Martin, G. R. (1981). Isolation of a pluripotent cell line from early mouse embryos cultured in medium conditioned by teratocarcinoma stem cells. *Proc. Natl. Acad. Sci. USA* 78:7634–7638.

27. Evans, M. J. and Kaufman, M. H. (1981). Establishment in culture of pluripotential cells from mouse embryos. *Nature* 292:154–56.

28. Bradley, A., Evans, M. J., Kaufman, M. H. and Robertson, E. J. (1984). Formation of germ line chimeras from embryo-derived teratocarcinoma cell lines. *Nature* 309:255–256.

29. Kleinsmith, L. J. and Pierce, G .B. (1964). Multipotentiality of single embryonal carcinoma cells. *Cancer Res.* 24:1544–51.

30. Sharp, J. D., Capecchi, N. E. and Capecchi, M. R. (1973). Altered enzymes in drug-resistant variants of mammalian tissue culture cells. *Proc. Natl. Acad. Sci. USA* 70:3145–49.

31. Thomas, K. R., Deng, C. and Capecchi, M. R. (1992). High-fidelity gene targeting in embryonic stem cells by using sequence replacement vectors. *Mol. Cell Biol.* 12:2919–23.

32. Deng, C. and Capecchi, M. R. (1992). Re-examination of gene targeting frequency as a function of extent of homology between the target vector and the target locus. *Mol. Cell Biol.* 12:3365–71.

33. Deng, C., Thomas, K. R. and Capecchi, M. R. (1993). Location of crossovers during gene targeting with insertion and replacement vectors. *Mol. Cell Biol.* 13:2134–2146.

34. Linney, E. and Donerly, S. (1983). Fragments from F9 PyEC mutants increase expression of heterologous genes in transfected F9 cells. *Cell* 35:693–699.

35. Mansour, S. L., Thomas, K. R. and Capecchi, M. R. (1988). Disruption of the proto-oncogene int2 in mouse embryo-derived stem cells: A general strategy for targeting mutations to nonselectable genes. *Nature* 336:348–352.

36. The International Mouse Knockout Consortium. A mouse for all reasons (2007). *Nature Rev. Genetics* 2:743–55.

37. Lewandoski, M. (2002). Conditional control of gene expression in the mouse. *Nature Rev. Genetics* 2:743–755.

38. Branda, C. S. and Dymecki (2004). Talking about a revolution: the impact of site-specific recombinases on genetic analysis in mice. *Dev. Cell* 6:7–28.

39. Glaser, S. Anastassiadis and Steward, A. F. (2005). Current issues in mouse germline engineering. *Nature Genet.* 37:1187–1193.

40. Schmid-Supprian, M. and Rajewsky, K. (2007). Vagaries of conditional gene targeting. *Nature Immunol.* 8:665–668.

41. Gossen, M. and Bujard, H. (2002). Studying gene function in eukaryotes by conditional gene inactivation. *Ann. Rev. Genet.* 36:153–173.

42. Berens, C. and Hillen, W. (2003). Gene regulation by tetracyclines. *Eur. J. Biochem.* 270:3109–21.

43. India, A. K., Warot, X., Brocard, J., Borpert, J. M., Xiao, J. H., Chambon, P. and Metzger, D. (1999). Temporally-controlled site-specific mutagenesis in the basal layer of the epidermis: comparison of the recombinase activity of the tamoxifen-inducible Cre-ERᵀ and Cre-ERᵀ² recombinases. *Nuc. Acid Res.* 27:4324–27.

44. Hayashi, S. and McMahon, A. P. (2002). Efficient recombination in diverse tissues by a tamoxifen-inducible form of Cre: A tool for temporally regulated gene activation/inactivation in the mouse. *Dev. Biol.* 244:305–18.

45. Jonkers, J. and Berns, A. (2002). Conditional mouse models of sporadic cancer. *Nature Rev. Cancer* 2:251–65.
46. Frese, K. K. and Tuveson, D. A. (2007). Maximizing mouse cancer models. *Nature Rev. Cancer* 7:645–58.
47. Keller, C., Arenkiel, B. R., Coffin, C. M., El-Bardeesy, N., DePinho, R. A., and Capecchi, M. R. (2004). Alveolar rhabdomyosarcomas in conditional Pax3:Fkhr mice: cooperativity of Ink4a/ARF and Trp53 loss of function. *Genes Dev.* 8:2614–26.
48. Haldar, M., Hancock, J. D., Coffin, C. M., Lessnick, S. A., and Capecchi, M. R. (2007). A conditional mouse model of synovial sarcoma: insights into a myogenic origin. *Cancer Cell* 11:375–88.
49. Hanahan, D. and Weinberg, R. A. (2000). The hallmarks of cancer. *Cell* 100:57–70.
50. Weinberg, R. A. (2007). The biology of cancer. Garland Science Press. Taylor and Frances Group, pp. 1–796.

Portrait photo of Mario R. Capecchi by photographer Ulla Montan.

Martin Evans.

MARTIN J. EVANS

I was born on the first day of January 1941 in the front bedroom of my grandparents' house in Rodborough near Stroud in Gloucestershire where my mother had come to escape the bombing in London. "A fine strapping lad" was the news my grandfather received as Dr Mold came to the top of the stairs. Later my father's factory was bombed out and evacuated from East London to Hertford where they continued to manufacture military transport vehicles. My parents found a tiny rural cottage to live in near Wareside. It was here that I have my first memories. We had mains water but nothing else. I remember the grand acquisition of a paraffin stove for easier cooking and also after the war helping my father to install some 12 volt battery operated lights and a generator. The switches were from motor vehicles and the wires were ex army telephone cable. In the few hours he was not at work my father ran the local communication signals for the Home Guard. Starting with nothing eventually they had their own telephone network covering Hertfordshire and run from our living room. I remember a slit trench in front of the house as an emergency shelter and seeing the local fields being tilled by prisoners of war. It was here that I remember what I have described as my first experiment – covert mixing of sand cement and water because I couldn't understand how adding water could make it become solid. It didn't work (as any good experimental first attempt!) because I added too much water; but it is a vivid memory.

After the war we moved to live in Raynes Park and it was there that, on what should have been my first day at school, I suffered a burst appendix and had it not been for the first antimicrobial drugs (M&B 693) would not have survived. I remember vividly seeing the low winter sun, red through the darkened glass of the ambulance, and hearing its shrill bell. I was awakened by a beautiful nurse with a sip of juice – of exilir! Maybe that's why I have since always loved nurses.

I contracted mumps in hospital and thereafter throughout the next 18 months suffered most available childhood infections. I hardly got to school and spent long periods in bed with various books and toys including meccano. When I was seven we moved to Orpington and in my new school I was kept in for extra lessons to learn "joined-up" writing instead of playing football. I still think that my poor handwriting and lack of soccer skills date from that period.

When I try to think back to my favourite toys, books and activities I do see now that perhaps I did naturally become a scientist. I made pressed flower collections at about the age of eight, later I learned about wild orchids and became adept at finding all the species in Kent. I loved old science books – e.g. "Moving things for Lively Youngsters". I was bought a job – lot of very early wireless construction materials from a house sale including a crystal and catswhisker and beautifully-made hexagonal, plug-in, air filled coils bearing the inscription "What are the Wild Waves saying?"; my father and I were able to make a working crystal set from these. We had an old model steam engine "Puffing Billy" and I still remember the smell of the methylated spirit burner. One Christmas I was fortunate enough to be given the electric experimental set for which I had yearned and in which the main item was an induction coil to generate high voltage shocks for all. I build a tesla coil and other static electric gadgets. I kept aquaria. My father had a mechanical workshop and trained me in operating the metalwork lathe. I used it to make cannons and first of all filled them with matchheads for powder. Then one of my greatest amateur passions began with a chemistry set of the sort which is nowadays (on account of Health and Safety!) unobtainable and I slowly learnt quite a bit of basic chemistry. It was possible to buy supplementary chemicals from helpful chemist shops and I even had a can of metallic sodium. The weak cannon filling matcheads were replaced by a variety of improved explosives and I spent a long time trying to perfect a rocket – most of my attempts failed to lift off and several underwent spectacular explosive failures! I was, however, careful with these explosive mixtures but probably my most dangerous experiment was when I attempted to make a large batch of urea formaldehyde resin in the fireplace of my bedroom; the reaction suddenly took off and overheated, nearly suffocating me. In this and other aspects – when only 10 or 12 I was able to spend all day alone miles from home collecting fresh water fish and specimens, with no-one worrying – my personal explorations seem to have been remarkably fostered and unfettered compared with today's children who are often constrained by our safety-conscious society. I was a boy who dreamt long and hard and could spend all day wondering how to join two cans to make a cylinder.

By the time, therefore, in the middle school of St Dunstan's College we started Chemistry and Physics lessons I was ready, I was keen. I even remember the sudden shock of hearing some of the technical words I had only up to that time read being actually pronounced.

In the prep school I had been taught biology by the Reverend Toller a wonderful man who used his classes to give us foundation in plant and animal science and cosmology. I still remember the start of one term of lessons when he said "Go out tonight and look up at the stars – look carefully they are not all the same colour". In the middle school we didn't do Biology before the sixth form but in my year we had the opportunity to start by dropping Maths after the O level exam in the fourth year. Although my main subjects were Chemistry and Physics I chose to do Biology probably just because it had been denied. This meant that because I had dropped the Maths the only choice I had for advanced level in the sixth form was Chemistry with Botany and Zoology.

I won a major scholarship to Christ's college in Cambridge. The Cambridge Natural Science Tripos has the great advantage of allowing breadth and choice. I embarked on Zoology, Botany and Chemistry with the clear intention of following specialisation in Chemistry. After the first term, however, I found that the Zoology course which was a dry systematics was not for me, whereas the Botany in which plant physiology and function illuminated the adaptive radiation and speciation lit my fire. I took a drastic change in my planned courses, dropped Zoology started Biochemistry and doubled the Botany. Chemistry my erstwhile love was becoming less of an option for although my Maths and Physics could just manage to mid degree level I was probably going to struggle too much at Part 2. Biochemistry part 2 was the option I had never anticipated but suited me perfectly. In the days I was there it was taught magnificently at the highest level. Enzyme function was taught by the world expert Malcolm Dixon by then seemingly a quiet, mumbling old man. (Younger then than I am now!) One day I arrived late for the 9am lecture and had to sit in the front row. I heard what he was saying fully for the first time and he was giving us the most exquisite analysis of a paper published that week in *Nature*! Ever after I made sure of a front seat! Molecular Genetics was just starting, Sanger had sequenced Insulin, the principles of protein synthesis were being established and the genetic triplet code was being solved. Jacob and Monod won the Nobel Prize and Monod gave an illuminating series of lectures which together with a seminar series organised by Sidney Brenner in his College rooms in Kings, inspired me in the new concepts of control of genetic readout through messenger RNAs.

Sadly, I was never able to sit my final exams for which I had been working so avidly as I succumbed to Glandular Fever. This effectively scuppered my chances of acquiring the postgraduate position and grant I wanted and so I took a post as a research assistant with Dr Elizabeth Deuchar at University College in London – a most fortunate choice. She was an excellent supervisor who encouraged but didn't direct, a policy I have myself followed in supervising graduate students. For some it works most excellently and I am justly proud of many of my students. It was just what I needed. It did result in my taking a long time to complete the PhD and in a lot of non realised innovation. I experimented, I developed techniques, I learnt skills – viewed coldly I probably wasted a lot of time but I was not under the pressure that so many are today.

My ambition was to isolate developmentally controlled m-RNA but at that time none of the cloning tools or probes upon which we now rely were available. All I could look at were double reciprocally labelled (14C and 3H) profiles of polyribosomes and messenger RNA from dissected blastula and gastrula ectoderm by Sucrose Density Gradient centrifugation and RNA by agarose electrophoresis. In modern terms I was looking at animal cap development in culture before induction and after commitment to either a neural or an epidermal ectoderm.

At that time I saw two impediments to further progress: the difficulty of getting enough material for biochemical analysis, and the lack of any foreseeable genetics. In discussion with Robin Weiss and Ann Burgess I cast around for a more tractable system and Robin suggested looking at mouse teratocarcinomas. In 1966 Leroy Stevens and Barry Pierce both published reviews of their formative studies. Stevens had developed a strain of mice with a high incidence of spontaneous testicular teratomas (129 Sv) and demonstrated that the tumours depended upon "Pluripotent embryonic cells (that) appear to give rise to both rapidly differentiating cells and others which, like themselves, remain undifferentiated." Pierce had, importantly, demonstrated the clonality of these multiply differentiating tumours. Stevens generously sent me stocks of mice and tumours and Robin, who was at that time setting the foundations of his pivotal studies on retroviral tumour viruses, together with Pavel Veseley, who was visiting from Prague, taught me tissue culture.

When I was in Cambridge I had met Judith and we were now married with two small boys. I am blessed with a wonderful, interesting, intelligent, independent and supportive wife. By then I had become an assistant lecturer and was teaching an intensive course in molecular developmental biology. I have always found during my academic career that investment in home and family life together with the imperatives of teaching provide the necessary counterbalance to the inevitable slow progress of research. One is doing a good job, at the University, at home and can therefore take the extensive rough with the occasional smooth of research.

The science progressed towards every goal and is related elsewhere. Memorable highlights include my trips to Oxford visiting Richard Gardner to make chimaeras. On one occasion I remember our workshop had rigged up an old black wax-oven with a huge battery connected by curly red wires as a temperature-controlled transport vessel and I carried this object, which had all the appearance of a cartoon bomb, home on the crowded tube – with no notice taken, or complaints, by fellow passengers. This was at the height of bomb attacks by the IRA!

When I was approaching the salary bar from lecturer to senior lecturer at UCL I was wondering about my future and I applied for a post at the Genetics department in Cambridge. I remember at the interview under extensive cross examination by the whole department that I told them that I would be aiming to use mouse teratocarcinoma stem cells as a vector to mouse genetics. I don't think they believed that it would become possible. After many months and when I had virtually forgotten about my application I received a phone call assuming I would take the job, which (after a little hesitation) I did. I later learnt that I was not their first choice – I don't know who got away! Cambridge was a difficult move for me but eventually good. It was a good time to move my children. The boys had not yet started secondary school and we were very worried by the available schools where we lived in London. Clare was just four and therefore needed to start primary. Judith had been working in family planning in London but there were virtually no positions available in Cambridge so when the opportunity came she moved into general practice nursing. This was a newly developing field and over the years she was instrumental in helping to set up local groups and training. It was a very proud moment for us all when she was awarded an MBE for services to practice nursing.

It was at Cambridge that I met Matt Kaufman who provided the vital trick of using delayed blastocysts for our isolation of ES cells. This was a very productive collaboration of the best sort where we were able to combine our complementary approaches and expertise and I was later sorry to see him go when he left to take the prestigious chair of Anatomy in Edinburgh.

I had an excellent laboratory including Robin Lovell-Badge, David Latchman, Liz Robertson and Allan Bradley among others. Later, when they had all left, I remember a colleague saying, " … you'll never be able to rebuild it now … ". It was a challenge, but slowly and in particular with Bill Colledge and Mark Carlton we started to be able to use the genetic promise of the ES cells. Because we mutated, trapped and targeted somewhat eclectically we were drawn into a number of fascinating fields of biology and medicine. I am glad to have seen our work with cystic fibrosis and breast cancer moving from mouse models to potential therapeutic application but I do feel that the major contributions are in fundamental understanding of biology.

I have always been interested – indeed waylaid – by the leading edges of technology even during my PhD years when I pioneered (but did not publish) agarose gel electrophoresis for RNA fractionation. Also, much later, I was instrumental in showing that Green Fluorescent Protein and RNAi could be made to work in mammalian cells.

I have spent a long time with personal computers starting with the Olivetti programmer 101 which was a programmable calculator with (as far as I remember) a memory of eight numbers in an acoustic delay loop and the ability to store its programme on a magnetic card. I was able to program this to replace the use of the University's IBM mainframe to normalise dual label results with only two separate entries of the data compared with preparation of a suite of punched cards. I later progressed through machine language programming to the use of Basic on a Hewlett-Packard calculator – an amazing advance. Hooked, I used an Apple II with an attached CP/M card as a word processor and later was responsible for Cambridge University using the Torch box addition to the BBC microcomputer as the standard word processing system. This came about from a fairly drunken presentation where on the basis of an order for just two units I persuaded the manufacturers to give the University a main dealer discount! I subsequently became enamoured with (and wasted a lot of time on) UNIX on the desktop.

When I took ES cells to Oliver Smithies he introduced me to PCR and his own amazing thermal cycling system. I needed to source the Taq Polymerase and make my own thermal cycler. It was then that I made acquaintance with Dr Peter Dean who has since become a lifelong friend. He was just in the process of setting up his own molecular biologicals supply company and I was able to find him a source of Taq Polymerase and together in discussion we invented a much neater thermal cycler which, by using a low heat capacity block was able to be heated rapidly by a quartz halogen lamp and cooled by a fan. We established a company and went into manufacture and it was, for some years, the best thermally performing machine on the market. This was an adventure into commerce. I was also a co-founder of a biotechnology company with much larger aspirations (Animal Biotechnology Cambridge Ltd) and here learnt a lot about business management.

At a time when funding was at a very low ebb I was visited by Michael Morgan of the Wellcome Trust and in conversation asked him how, if ever, will we be able to retain the best postdocs in the UK? Would it be possible to set up an institute of developmental biology? To my amazement he just said, "Send the Welcome Trust an outline proposal". Brigid Hogan and I had discussed the need for an institute of mammalian developmental biology and so together with her we put together a proposal. Although I didn't know it at the time and Michael was quite unable through confidentiality to tell me there was a potential institute in discussion and the one component missing was mammalian. Brigid in London kept getting rumours that John Gurdon was planning something and eventually these rumours became so insistent that, although it appeared that there was nothing whatsoever happening in Cambridge, I phoned him. There was a long pause and John then said – "you'd better come to talk with me". It transpired that he together with Ron Laskey had prepared a proposal which, when we cautiously swapped them, proved to be closely comparable to ours. This was the start of the Welcome/ CRC Institute for Cancer and Developmental Biology which became such a success and which was such a joy to be part of.

In 1999, I moved from leading a personal research group in a scientifically buzzing Institute to heading a large and newly formulated School of Biosciences in Cardiff University. This was a huge change of role but one which I found I was rather better prepared for than I might have anticipated. All my experience both in Cambridge Departmental and University committees as well as my wrangles with commercial management had prepared me well. I was also pleased to be able to help facilitate scientific careers on a wider scale than hithertofore and to help lead and develop a newly re-energised Welsh University into the 21st century. It was hard but extremely rewarding.

A Lasker Prize, Knighthood and now the Nobel Prize brings recognition and quiet, unexpected satisfaction. I have always been a bit outside the mainstream and I have been outspoken and opinionated with a wide interest in science but I'm still at heart that small thoughtful enquiring boy.

EMBRYONIC STEM CELLS:
THE MOUSE SOURCE – VEHICLE FOR
MAMMALIAN GENETICS AND BEYOND

Nobel Lecture, December 7, 2007

by

Martin J. Evans

Cardiff School of Biosciences, Cardiff University, Museum Avenue, Cardiff CF10 3US, Wales, UK.

In a developmental system there may be a complexity of environment, a progressive developmental time course and a multiplicity of cells and interacting components. Isolation of such systems into culture allows both simplification and experimental access. Tissue-culture of disaggregated cells, in particular, allows for isolation and purification of cell type by cloning and detailed manipulation of culture conditions. It also gives an entrée to a genetic analysis via somatic cell genetics. It was for such reasons that I sought an *in vitro* developmental system. Genetic knowledge for any culturable system from higher organisms was at that time sparse but best for chick and mouse.

I had been looking for messenger RNA changes during development mainly in early Xenopus embryos. By the end of the 60's it was becoming clear to me that for such molecular studies not only was a larger scale manipulable system needed it but also one with a better genetic potential. Excellent organ culture systems of the early and mid development of chick were available but these were difficult to deconstruct into the tissue-culture level. Some genetics was available but the chick karyotype was a problem for somatic cell genetic approaches. Mammalian embryos were, by contrast, extremely inaccessible but *in vitro* culture of the early preimplantation stages had been developed (for a comprehensive review see (Cockroft, 1997)). Mammalian long-term tissue culture was better established and mouse genetics was at least comparable to that of the chick. Somatic cell genetic approaches were more available.

Robin Weiss drew my attention to two important reviews of work with mouse teratocarcinomas published in 1967 which pointed the way to an opportunity to develop a tissue culture system for studies of cellular differentiation (Pierce, 1967; Stevens, 1967). Stevens reviewed his work in which he had established inbred strains of mice with a high incidence of spontaneous testicular teratocarcinomas, had shown that these tumours were transplantable and demonstrated their origin from primordial germ cells in the foetal testis. He also showed that they could be experimentally induced by ectopic transplantation both of geminal ridges containing these primordial germ cells and of early embryos i.e. transplantation of sources of pluripotential cells. Prophetically Stevens and Little in their paper of 1954 (Stevens and Little, 1954) set the field by saying of their transplantable teratocarcinomas *"Pluripotential embryonic cells appear to give rise to both rapidly differentiating cells and others which like themselves, remain undifferentiated"*; this is the definition for an Embryonic Stem Cell.

Pierce and his colleagues provided a long series of experimental studies including demonstrating growth of cells from the tumours in tissue culture but the single most important demonstration was that of Kleinsmith and Pierce (1964) who showed that transplantation of a single cell *in vivo* could result in a fully differentiating teratocarcinoma; unequivocally establishing the presence of the pluripotential tumour stem cells. These cells were named following the human nomenclature as Embryonal Carcinoma cells.

In May 1969, Leroy Stevens very generously sent me stocks of 129 inbred mice some carrying transplantable teratocarcinomas (induced as described in (Stevens, 1970)). From tumours passaged from this stock I was able to establish clonal tissue cultures which retained their full pluripotency as demonstrated by their ability to differentiate as a tumour *in vivo* (Evans, 1972). One significant feature of this isolation and cloning was that irradiated chick embryo fibroblasts were used as a feeder layer. It was noted that when the use of this irradiated feeder layer was discontinued the cultures spontaneously generated differentiated cell types (E cells) as well as maintaining the stem cell line (C cells). Retrospectively we can see that these E cells do indeed arise by differentiation from the stem cells and that they provide a balanced, mixed population where the pluripotentcy of the stem cells is maintained by the feeder effect of the associated differentiated products. At the time, however, the processes of differentiation were unclear and it was not possible to see extensive differentiation *in vitro*. I published a detailed discussion of the situation in 1975 (Evans, 1975) which shows the difficulty of interpretation just as we were beginning to see differentiation *in vitro*. It is also of retrospective interest that it was here where I first proposed that pluripotential embryonic cells should be able to be cultured directly from normal embryos something which was not to be achieved for another six years; *" I should like to suggest that it may be quite feasible to obtain cultures of pluripotent cells directly from the embryo now that experience has been obtained handling such cells, and that the earlier results of Cole, Edwards & Paul (Cole et al., 1966) with cultures from rabbit blastocysts should not necessarily inhibit further efforts in this direction."*

During this time, moreover, studies started to indicate the very close relationship of these cultured Embryonal Carcinoma cells to their primordial germ cell and normal embryo counterparts.

Repeated innoculation of syngeneic mice with irradiated teratocarcinoma cell cultures results in antisera reacting with the cell surface of the EC cells (reviewed by Jacob, 1977). Unaltered teratomas are still, however, produced in these hyperimmunised mice by inoculation with live EC cells. The same specific cell surface antigens are present upon the cells of early mouse embryos and germ cells (Artzt *et al.*, 1973). Although originally these antigens were thought to be cell surface proteins associated with the wild-type t-locus this became disproved by careful genetic studies (Erickson and Lewis, 1980) and it subsequently became apparent from studies with human monoclonal autoantibody sera that they were cell surface carbohydrate moieties. (Childs *et al.*, 1983; Gooi *et al.*, 1981; Kapadia *et al.*, 1981). The branching of these carbohydrate chains differs on the ES cells and their differentiated progeny as also seen in development of the early mouse embryo. Studies with these and the usefully discriminatory Forssman antigen demonstrated that the EC cells had a similar cell surface phenotype only to the pluripotential cells of the early pre and postimplantation mouse embryo (Evans *et al.*, 1979; Stinnakre *et al.*, 1981). In addition high resolution two dimensional electrophoretic analyis of nascent protein synthesis suggested that the EC cells were very similar to early embryo cell types but in particular matched 5 day ectoderm (Lovell-Badge and Evans, 1980).

Perhaps the most dramatic indication of the similarity of EC cells to the early embryo was their ability to become reincorporated into a mouse blastocyst and develop into a healthy mouse with tissue contributions from the EC cells. The first indications of this were published by Brinster (Brinster, 1974). My experiments together with Richard Gardner and his colleagues (Papaioannou *et al.*, 1978; Papaioannou *et al.*, 1975) showed that very extensive chimaerism across most tissues of the mouse was achievable using tissue culture EC cells but that some of these animals showed later-origin tumours of differentiated cell types. (Some EC cell stocks (apparently those which through progression in cell culture differentiated more poorly) gave rise to animals bearing early-origin undifferentiated tumours.) None of these mice were able to transmit the teratoma-origin genome through their gametes, most probably because the cells used were aneuploid. In any case it became apparent that for effective germline transmission both euploidy and excellent and uncompromised cellular differentiation would be needed.

Progress in understanding the differentiation of the cells *in vitro* (which can be very extensive and equivalent to that seen in a teratoma) gave rise to one of the important conceptual breakthroughs – the realization that the differentiation of the EC cells was not abnormal, disorganized, random or stochastic but followed the normal pathways of early embryonic development. We noticed that in every situation where the EC cells were allowed to differentiate the first differentiated cells to appear were primary embryonic endoderm (Evans and Martin, 1975; Martin and Evans, 1975b). We had been investigating the relationship of the C-cells and E-cells in the culture and used very careful recloning of isolated single cells on feeder layers. Homogeneous cultures of the C-cells (the embryonal carcinoma cells) were able to be maintained by passage on feeder layers but when the feeders were removed the cells clumped up and some became detached as small colonies which formed embryoid bodies (Martin and Evans, 1975a). EC cell clumps in suspension formed simple embryoid bodies and when these were allowed to develop further they became more complex cystic bodies in suspension or if allowed to attach to the

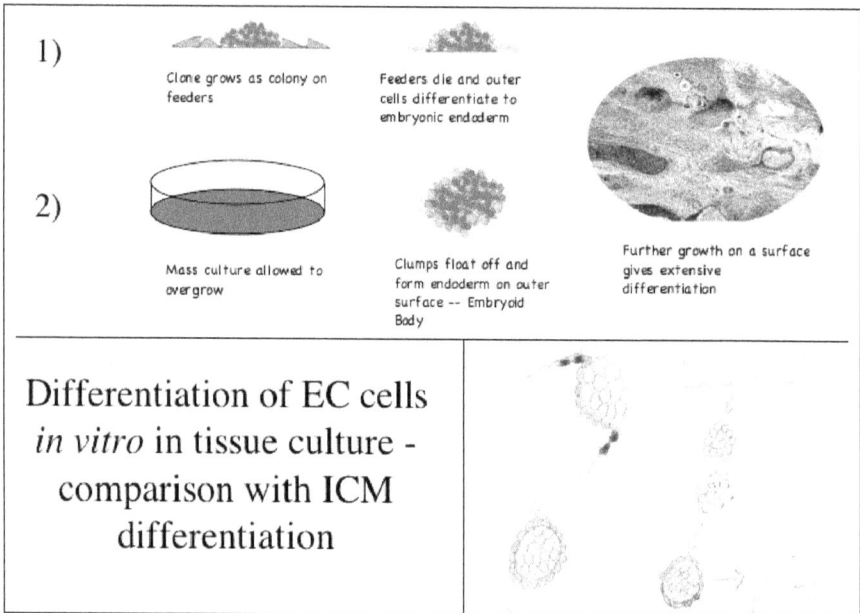

1) Clone grows as colony on feeders

Feeders die and outer cells differentiate to embryonic endoderm

2) Mass culture allowed to overgrow

Clumps float off and form endoderm on outer surface -- Embryoid Body

Further growth on a surface gives extensive differentiation

Differentiation of EC cells *in vitro* in tissue culture - comparison with ICM differentiation

Figure 1.

tissue culture surface spread out and developed into a mixture of cells and tissues which on section proved equivalent in complexity and organisation to the wide diversity of tissues seen in a teratoma (Martin and Evans, 1975c). Similar extensive differentiation was seen if an individual colony arising from a clone on feeders was allowed to continue to grow after the feeders died out (Evans and Martin, 1975). It is clear that the initiation of this differentiation is the same process as that seen when cells on the blastocoelic surface of the inner cell mass (ICM) differentiate into primary endoderm; an isolated ICM becomes surrounded by a rind of endoderm (Rossant, 1975).

In about 1980, I had been trying again to isolate cells from ICM's and had generated numbers of endodermal cultures. I devoted some time to consider why it had not proved possible to isolate cells equivalent to EC cells directly from early mouse embryos and this is written up in a review published in 1981 (Evans, 1981a). The main points of note are:

1) EC cells from culture form teratocarcinomas upon transplantation *in vivo.*

2) Teratocarcinomas containing EC cells were able to be made by ectopic transplantation of embryos from the 2-cell stage through to the dissected embryonic ectoderm from embryos of 7.5 days of development.

3) The cell surface phenotype and the spectrum of protein synthesis suggested that the closest match to EC cells was later then the 3.5 day ICM and earlier than the 6.5 day ectoderm.

4) EC cells in culture enter into differentiation as though they are ICM cells.

5) EC cells could cooperate with ICM of a blastocyst in the development of a chimaeric mouse.

I considered that there might be three classes of reason why EC cells had not been grown directly from explanted embryos or dissected embryo tissues.

1) There might be only very small numbers of founder cells available and that therefore success *in vitro* would depend upon the highest efficiency of cloning. By that time I had been slowly improving the cloning efficiency of passaged EC cells (both mouse EC cells and Human teratocarcinoma derived cells) and using this as a test for optimising the media and conditions arrived at a mix known around the lab at the time as "Martin's Magic Medium" or MMM. The feeder layer used was also optimised by the same test. Retrospectively an optimised medium and procedure is entirely necessary but numerous variants are possible.

2) The timing might be more critical than *in vivo* where processes of onward development or even regression could take place more readily. Retrospectively, we now know that cultures of ES cells have been satisfactorily established from cleaving embryos through to late 4.5 day so this was not the main problem.

3) It was known that the amount of differentiation of teratocarcinomas tended to diminish with tumour passage. EC cell lines diminished in their readiness to differentiate *in vitro* with tissue culture passage. This raised the possibility that adaptation to tumour and to tissue culture growth involved selection of cell lines which were slower to trigger differentiation and that maybe native cells directly from the embryo would differentiate so readily that the stem cell line was immediately lost. Thus conditions most conducive to maintenance of the undifferentiated stem cell state would be needed. In addition to the media supporting the best cloning efficiency, this meant using optimised feeder cell layers and using repeated dissagregation and passage so as not to allow the cells to form local concentrations. I said "*embryonal carcinoma lines which differentiate in vitro are difficult to maintain in an undifferentiated state, even with the help of feeder layers. It is very likely that even these lines have already been highly selected for the ability to be maintained in tissue culture and concomitantly for less ready differentiation. Their genuine embryonic counterpart may differentiate and lose its pluripotency and rapid growth characteristics all too readily under culture conditions*" (Evans, 1981a). Retrospectively this was probably the most cogent reason. Freshly isolated ES cell lines can differentiate precipitately if not prevented.

Collaboration with Matt Kaufman brought, critically, expertise and experience with early mouse embryo manipulation. He had been exploring the developmental potential of parthenogenetic embryos in particular haploid embryos and had discovered that such haploid embryos could be persuaded to develop to an early postimplantation stage (Kaufman, 1978). These embryos tend to have a reduced cell number at the blastocyst stage and in order to allow a compensatory increase before their implantation Kaufman had utilised implantational delay. We therefore sought to use such delayed blastocysts as a source of haploid cells in culture. In the first place we used diploid delayed blastocysts from strain 129 mice and upon explantation I was able to see outgrowth of instantly recognisable EC-like cells. These were able to be picked and maintained in passage tissue culture and had all the expected properties of the sought after primarily isolated pluripotent cells (Evans and Kaufman, 1981). Most importantly they were euploid XY cells and with careful culture maintained a stable karyotype. Interestingly the XX cells from female embryos were also isolated but had a less stable karyotype presumably because of the long term chromosomal imbalance without X-inactivation (Robertson *et al.*, 1983).

We viewed these cells as normal derivatives from the embryo and confidently expected that they would prove useful vectors to the mouse germ line. Martin later in the year, using a different method, reported the establishment of similar cultures directly from embryos but these did not retain a normal karyotype (Martin, 1981). She provided the important nomenclature of Embryonic Stem Cells.

Together with Liz Robertson and Allan Bradley we were soon able to show that progeny of the ES cells were able to form functional germ cells (both sperm and ova) in chimaeric mice. Interestingly male ES cells were often able to transform the sexual differentiation of a female host blastocyst and result in a male chimaeric mouse where, as only the ES derived cells carring a Y chromosome were able to make sperm, 100% of the germline transmission was from the tissue culture derived cells (Bradley *et al.*, 1984).

Transgenesis and mutagenesis was clearly the next step and I chose to use retroviral vectors which have the advantage of cleanly integrated transgenesis and that any mutation caused by this integration is clearly marked by the foreign DNA. It should be remembered that at this time the genetic maps were rudimentary, there was little gene and virtually no genomic sequence data. Thus clean transgene integration associated with mutation was a route to gene discovery.

Using this technique we were able to demonstrate the transmission of sequences introduced by retroviral vectors *in vitro* into the mouse germline (Evans *et al.*, 1985; Robertson *et al.*, 1986) and used several methods to recover newly induced mutation of endogenous loci.

The way was now clear to an experimental genetics for mice. Transgenes could be introduced in culture and the structure verified before introduction into the germline. New mutations could be tested both *in vitro* and *in vivo*. It was around this time that the possibilities of using homologous recombination gene targeting that had been developed by both Oliver Smithies and Mario Capecchi to specifically alter endogenous loci became available and subsequently this has been the most important method for the experimental genetics. These techniques depend upon the availability of cloned sequence for the target gene and the advances in knowledge of mammalian and in particular mouse genomic sequences has been pivotal. Possibly about one quarter of available loci have already been targeted, and indeed complete coverage of specifically induced mutation in the mouse is now planned (Austin *et al.*, 2004). This is all dependent upon the technology of using mouse ES cells as a vector to the mouse germline. We have described studies on numbers of induced and specifically targeted mutations. I shall here, however, mention only some examples of our experiments using retroviral vectors and one example of gene targeting using homologous recombination.

In the first place it would be useful to be able to select a specific mutation in culture. The most feasible candidate was Hprt which being X-linked is present as only a single copy in XY cells and in which mutation is selectable because in its absence the cells are resistant to otherwise lethal incorporation of 6-thoguanine. Two independent mutations were recovered from ES cell cultures superinfected with retroviral vectors and transferred to the mouse germline (Kuehn *et al.*, 1987). Retrospectively one of the two turned out to be not the expected clean proviral insertion but example of retroposition of an endogenous processed message. This is an interesting observation of an unusual event; such elements are commonly found in genomic sequences and may well be the products of retroviral reverse transcriptase. Our proof that the alpha tubulin processed pseudogene was indeed the cause of this Hprt mutation is an interesting example of the of the use of homologous recombination in ES cells. It is particularly clear because it is without complications of associated vector or selection elements (Carlton *et al.*, 1995).

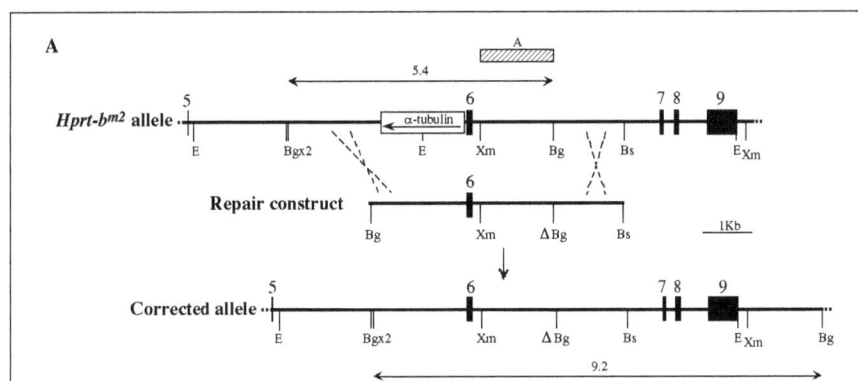

Figure 2. Adapted from (Carlton *et al.*, 1995). We found that the change in the Hprt-b^{m2} allele appeared to be not the expected retroviral vector insertion but an insertion of an α-tubulin processed pseudogene in inverse orientation close to but not disrupting the coding sequence of exon 6. In order to prove that this genomic change was responsible for the mutation we used homologous recombination with a repair construct which was a purified DNA fragment identical with the normal gene sequence across this interval but with a single change to remove a Bgl II restriction site for diagnostic purposes. This construct efficiently targeted the mutant allele and restored function. Southern blot analysis confirmed that only change was removal of the α-tubulin insert. This is a clear example of homologous recombination gene targeting without any complications of selectable markers or associated vectors.

A retroviral vector insertion transmitted through the germline may be screened for phenotypic effect. The absence of homozygous offspring from a heterozygote intercross is indicative of an embryonic lethality. One such example is the insertion 413d which identified a homozygous lethal locus (subsequently renamed nodal). Robertson and her colleagues (Conlon *et al.*, 1991; Robertson *et al.*, 1992) demonstrated that death occurred in the homozygous embryos at an early postimplantation stage but was not a cell autonomous lethality as ES cells homozygous for the insertion could be isolated from blastocysts. Kuehn (Zhou *et al.*, 1993) cloned the locus and showed that it was expressed as a secreted factor controlling axis formation in gastrulation. It is interesting to note that nodal expression may be a key controller of differentiation of ES cells (Takenaga *et al.*, 2007).

Direct physical phenotype may also be observed, for instance (Carlton *et al.*, 1998) described a dominant mutation resulting from the proviral integration which caused a craniofacial dysmorphology resulting from constitutive upregulation of Fg F3 and Fg F4 in the developing skull.

Another very useful technique has been that of gene trapping (Joyner *et al.*, 1992; Skarnes *et al.*, 1992) (reviewed by Evans *et al.*, 1997) where a reporter gene is used to find retroviral vector insertion which falls within a functional locus. Numerous interesting mutations have been recovered in this way and the complex developmental and behavioural consequences of partial inactivation of the histone H3.3a may be quoted as an example (Couldrey *et al.*, 1999).

These types of approaches allow gene function discovery by phenotype but nowadays gene targeting technology allows any designer mutation to be introduced into the mouse germline as a direct experimental approach. In addition to simple mutation, methods have been developed which allow both spatial and temporal control of gene deletion or of function (e.g. see review by Clarke, 2000). All these studies are dependant upon the combination of *in vitro* cell genetic manipulation and selection coupled with true *in vivo* observation of the physiological consequences in the context of the whole animal. This has been made possible by tissue culture of embryonic stem cells.

I have been interested in the relationship between embryonal carcinoma cells, normal embryo cells and embryonic stem cells for many years (Evans, 1981b). It was the close relationship between EC cells and early embryo pluripotential cells as shown by both their cell surface phenotype and by the extensive match of nascent protein synthesis patterns that helped to lead the way to the isolation of Embryonic Stem Cells. Together with Susan Hunter we have been utilising an analysis of global transcriptional patterns to compare Embryonic Stem Cells in culture with normal early mouse embryo pluripotential tissues. These studies show considerable differences between ICM from blastocysts of either 3.5 or 4.5 days of development and ES cells but a remarkable match with ectoderm from 5.5 days of development (Figure 3). This match is all the more remarkable as we are comparing cells isolated directly from the normal, unmanipulated, *in vivo* embryo with ES cells from an established cell line growing in an artificial serum-containing tissue culture medium on a plastic surface. It was always possible that mouse ES cells are effectively an artefact of culture and only become "normalised" by re-incorporation into an embryo and re-entrained into normal development by virtue of the influence from the environment of the host embryo. Alternatively they might represent a normal stem cell population. These present studies suggest that any culture adaptations away from a normal are minimal.

Embryonic Stem Cells have, therefore, delivered a major platform technology for experimental genetic manipulation which is delivering most important theoretical understanding and practical medical benefit. They are also proving greatly instrumental in delivering a second platform technology of

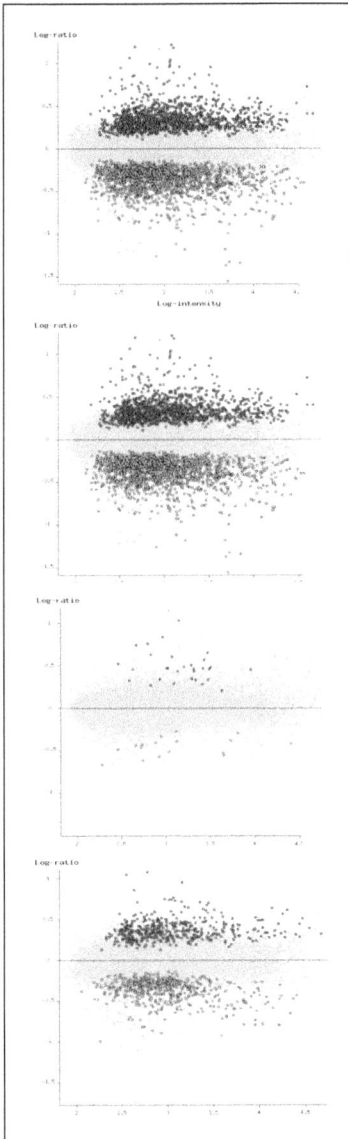

Figure 3. ES cells in culture were compared with tissues isolated from early mouse embryos by microarray transcriptomics. c-DNA was isolated from ICM's or dissected epiblast and compared with c-DNA isolated from ES cells, amplified by two rounds T7 transcription and hybridised to slides printed with the 15k NIA c-DNA probes. Unpublished results Susan Hunter and Martin Evans.

Charts of log ratio vs log intensity
A 3.5 day immunosurgically isolated ICM vs ES cells
B 4.5 day immunosurgically isolated ICM vs ES cells
C 5.5 day dissected epiblast vs ES cells
D 6.5 day dissected primary ectoderm vs ES cells

Global anova analysis; red spots significantly overexpressed by embryo samples, green spots significantly underexpressed by embryo samples. False discovery rate set to <0.01.
Note the exceptional match between ES cells and 5.5 day epiblast.

stem cell based regenerative medicine. One of the original aims of the tissue culture of EC cells was to provide a tractable system for the study of cellular determination and differentiation in vitro. This was achieved but with the mouse cells has not yet been fully exploited. With the advent of human ES cells and the possibilities of using them as a renewable source of tissue-specific precursors for tissue transplant therapies and regenerative medicine (see review by Lerou and Daley, 2005) the importance of understanding and controlling ES cell determination and differentiation in vitro has been highlighted. It is clear that the utility of isolation, maintenance and use of pluripotential stem cells has a long and important future.

REFERENCES

Artzt, K., Dubois, P., Bennett, D., Condamine, H., Babinet, C. and Jacob, F. (1973). Surface antigens common to mouse cleavage embryos and primitive teratocarcinoma cells in culture. *Proc Natl Acad Sci USA* **70**, 2988-92.

Austin, C. P., Battey, J. F., Bradley, A., Bucan, M., Capecchi, M., Collins, F. S., Dove, W. F., Duyk, G., Dymecki, S., Eppig, J. T. *et al.* (2004). The knockout mouse project. *Nat Genet* **36**, 921-4.

Bradley, A., Evans, M. J., Kaufman, M. J. and Robertson, E. J. (1984). Formation of germ-line chimaeras from embryo-derived teratocarcinoma cell lines. *Nature* **309**, 255-256.

Brinster, R. L. (1974). The effect of cells transferred into the mouse blastocyst on subsequent development. *J Exp Med* **140**, 1049-56.

Carlton, M. B., Colledge, W. H. and Evans, M. J. (1995). Generation of a pseudogene during retroviral infection. *Mamm Genome* **6**, 90-5.

Carlton, M. B. L., Colledge, W. H. and Evans, M. J. (1998). A Crouzon-like craniofacial dysmorphology in the mouse is caused by an insertional mutation at the *Fgf3/Fgf4* locus. *Developmental Dynamics* **212**, 242-249.

Childs, R. A., Pennington, J., Uemura, K., Scudder, P., Goodfellow, P. N., Evans, M. J. and Feizi, T. (1983). High-molecular-weight glycoproteins are the major carriers of the carbohydrate differentiation antigens I, i and SSEA-1 of mouse teratocarcinoma cells. *Biochem J* **215**, 491-503.

Clarke, A. R. (2000). Manipulating the germline: its impact on the study of carcinogenesis. *Carcinogenesis* **21**, 435-41.

Cockroft, D. L. (1997). A comparative and historical review of culture methods for vertebrates. *Int J Dev Biol* **41**, 127-37.

Cole, R. J., Edwards, R. G. and Paul, J. (1966). Cytodifferentiation and embryogenesis in cell colonies and tissue cultures derived from ova and blastocysts of the rabbit. *Dev Biol* **13**, 385-407.

Conlon, F. L., Barth, K. S. and Robertson, E. J. (1991). A novel retrovirally induced embryonic lethal mutation in the mouse: assessment of the developmental fate of embryonic stem cells homozygous for the 413.d proviral integration. *Development* **111**, 969-81.

Couldrey, C., Carlton, M. B., Nolan, P. M., Colledge, W. H. and Evans, M. J. (1999). A retroviral gene trap insertion into the histone 3.3A gene causes partial neonatal lethality, stunted growth, neuromuscular deficits and male sub-fertility in transgenic mice. *Hum Mol Genet* **8**, 2489-95.

Erickson, R. P. and Lewis, S. E. (1980). Cell surfaces and embryos: expression of the F9 teratocarcinoma antigen in T-region lethal, other lethal, and normal pre-implantation mouse embryos. *J Reprod Immunol* **2**, 293-304.

Evans, M. (1981a). Origin of mouse embryonal carcinoma cells and the possibility of their direct isolation into tissue culture. *J Reprod Fertil* **62**, 625-31.

Evans, M. J. (1972). The isolation and properties of a clonal tissue culture strain of pluripotent mouse teratoma cells. *J Embryol Exp Morphol* **28**, 163-76.

Evans, M. J. (1975). Studies with Teratoma Cells In Vitro. In *THe Early Development of Mammals (British Society for Developmental Biology Symposium 2)*, (ed. B. Wild), pp. 265-284. Cambridge: Cambridge University Press.

Evans, M. J. (1981b). Are teratocarcinomas formed from normal cells? In *Germ Cell Tumours*, (ed. C. J. Anderson W. G. Jones and A. Milford-Ward): Taylor and Francis.

Evans, M. J., Bradley, A., Kuehn, M. R. and Robertson, E. J. (1985). The ability of EK cells to form chimeras after selection of clones in G418 and some observations in the integration of retroviral vector proviral DNA into EK cells. In *CSH Symposia on Quantitative Biology*, vol. 50.

Evans, M. J., Carlton, M. B. and Russ, A. P. (1997). Gene trapping and functional genomics. *Trends Genet* **13**, 370-4.

Evans, M. J. and Kaufman, M. H. (1981). Establishment in culture of pluripotential cells from mouse embryos. *Nature* **292**, 154-6.

Evans, M. J., Lovell-Badge, R. H., Stern, P. L. and Stinnakre, M. G. (1979). Cell lineages of the mouse and embryonal carcinoma cells: Forssman antigen distribution and patterns of protein synthesis. In *INSERM Symposium 10*, (ed. N. L. Douarin), pp. 115-129: Elsevier.

Evans, M. J. and Martin, G. R. (1975). The differentiation of clonal teratocarcinoma cell culture *in vitro*. In *Roche Symposium on Teratomas and Differentiation*, (ed. D. Solter and M. Sherman): Acad. Press NY.

Gooi, H. C., Feizi, T., Kapadia, A., Knowles, B. B., Solter, D. and Evans, M. J. (1981). Stage-specific embryonic antigen involves alpha 1 goes to 3 fucosylated type 2 blood group chains. *Nature* **292**, 156-8.

Jacob, F. (1977). Mouse teratocarcinoma and embryonic antigens. *Immunol Rev* **33**, 3-32.

Joyner, A. L., Auerbach, A. and Skarnes, W. C. (1992). The gene trap approach in embryonic stem cells: the potential for genetic screens in mice. *Ciba Found Symp* **165**, 277-88; discussion 288-97.

Kapadia, A., Feizi, T. and Evans, M. J. (1981). Changes in the expression and polarization of blood group I and i antigens in post-implantation embryos and teratocarcinomas of mouse associated with cell differentiation. *Exp Cell Res* **131**, 185-95.

Kaufman, M. H. (1978). Chromosome analysis of early postimplantation presumptive haploid parthenogenetic mouse embryos. *J Embryol Exp Morphol* **45**, 85-91.

Kleinsmith, L. J. and Pierce, G. B., Jr. (1964). Multipotentiality of Single Embryonal Carcinoma Cells. *Cancer Res* **24**, 1544-51.

Kuehn, M. R., Bradley, A., Robertson, E. J. and Evans, M. J. (1987). A potential animal model for Lesch-Nyhan syndrome through introduction of HPRT mutations into mice. *Nature* **326**, 295-298.

Lerou, P. H. and Daley, G. Q. (2005). Therapeutic potential of embryonic stem cells. *Blood Rev* **19**, 321-31.

Lovell-Badge, R. H. and Evans, M. J. (1980). Changes in protein synthesis during differentiation of embryonal carcinoma cells, and a comparison with embryo cells. *J Embryol Exp Morphol* **59**, 187-206.

Martin, G. R. (1981). Isolation of a pluripotent cell line from early mouse embryos cultured in medium conditioned by teratocarcinoma stem cells. *Proc Natl Acad Sci USA* **78**, 7634-8.

Martin, G. R. and Evans, M. J. (1975a). Differentiation of clonal lines of teratocarcinoma cells: formation of embryoid bodies in vitro. *Proc Natl Acad Sci USA* **72**, 1441-5.

Martin, G. R. and Evans, M. J. (1975b). The formation of embryoid bodies *in vitro* by homogenous embryonal carcinoma cell cultures derived from isolated single cells. In *Roche Symposium on Teratomas and Differentiation*, (ed. D. Solter and M. Sherman): Acad. Press, NY.

Martin, G. R. and Evans, M. J. (1975c). Multiple differentiation of clonal teratoma stem cells following embryoid body formation *in vitro*. *Cell* **6**, 467-474.

Papaioannou, V. E., Gardner, R. L., McBurney, M. W., Babinet, C. and Evans, M. J. (1978). Participation of cultured teratocarcinoma cells in mouse embryogenesis. *J Embryol Exp Morphol* **44**, 93-104.

Papaioannou, V. E., McBurney, M. W., Gardner, R. L. and Evans, M. J. (1975). Fate of teratocarcinoma cells injected into early mouse embryos. *Nature* **258**, 70-73.

Pierce, G. B. (1967). Teratocarcinoma: model for a developmental concept of cancer. *Curr Top Dev Biol* **2**, 223-46.

Robertson, E., Bradley, A., Kuehn, M. and Evans, M. J. (1986). Germ-line transmission of genes introduced into cultured pluripotential cells by a retroviral vector. *Nature* **323**, 445-448.

Robertson, E. J., Conlon, F. L., Barth, K. S., Costantini, F. and Lee, J. J. (1992). Use of embryonic stem cells to study mutations affecting postimplantation development in the mouse. *Ciba Found Symp* **165**, 237-50; discussion 250-5.

Robertson, E. J., Evans, M. J. and Kaufman, M. J. (1983). X-chromosome instability in pluripotential stem cell lines derived from parthenogenetic embryos. *J. Embryol. Exp. Morph.* **74**, 297-309.

Rossant, J. (1975). Investigation of the determinative state of the mouse inner cell mass. II. The fate of isolated inner cell masses transferred to the oviduct. *J Embryol Exp Morphol* **33**, 991-1001.

Skarnes, W. C., Auerbach, B. A. and Joyner, A. L. (1992). A gene trap approach in mouse embryonic stem cells: the lacZ reported is activated by splicing, reflects endogenous gene expression, and is mutagenic in mice. *Genes Dev* **6**, 903-18.

Stevens, L. C. (1967). The biology of teratomas. *Adv Morphog* **6**, 1-31.

Stevens, L. C. (1970). The development of transplantable teratocarcinomas from intratesticular grafts of pre- and postimplantation mouse embryos. *Dev Biol* **21**, 364-82.

Stevens, L. C. and Little, C. C. (1954). Spontaneous Testicular Teratomas in an inbred Strain of Mice. *Proc Natl Acad Sci USA* **40**, 1080-1087.

Stinnakre, M. G., Evans, M. J., Willison, K. R. and Stern, P. L. (1981). Expression of Forssman antigen in the post-implantation mouse embryo. *J Embryol Exp Morphol* **61**, 117-31.

Takenaga, M., Fukumoto, M. and Hori, Y. (2007). Regulated Nodal signaling promotes differentiation of the definitive endoderm and mesoderm from ES cells. *J Cell Sci* **120**, 2078-90.

Zhou, X., Sasaki, H., Lowe, L., Hogan, B. L. and Kuehn, M. R. (1993). Nodal is a novel TGF-beta-like gene expressed in the mouse node during gastrulation. *Nature* **361**, 543-7.

Portrait photo of Sir Martin J. Evans by photographer Ulla Montan.

OLIVER SMITHIES

My fraternal twin, Roger, and I were born prematurely on June 23rd, 1925, in Halifax, England, an industrial town in the West Riding of Yorkshire, although we lived outside Halifax at 2, Woodhall Crescent on Wakefield Road, a row house rented from the town. My father, William Smithies, was at that time working for his father, Fred Smithies, who paid him erratically. My mother, neé Doris Sykes, was a college graduate and taught English at the Halifax Technical College (where she met and fell in love with my father who was one of her students and younger than she). Not long after our birth, my father found a regularly paying job selling small life insurance policies to local farmers and their families. He was a kind and gentle man with a natural mechanical aptitude that he had inherited or learned from his father. A car was needed for a person selling insurance to scattered customers. So we were unusual in our neighborhood in the 1930s in having one. Not that the car was very special; it was a two cylinder Jowett and was in constant need of repair. I have vivid memories of "helping" my father, when I was about 8 or 9 years old, to select the least worn exhaust valves to use in keeping it running. (The stems of the valves wore badly.)

Our sister, Nancy, was 5 years younger than us, and a welcome addition to the family. She was a beautiful fair-skinned ginger-haired baby, and we 5 year old twins suggested naming her "Buttercup". All three of us were generally healthy and happy, although Nancy would not have survived infected tonsils without the then newly discovered miracle antibiotic drug "Prontosil" – the first of the sulfonamide drugs. I had a similar incident at age 7, but without the Prontosil, and was bedridden for 10 weeks after a bout with "rheumatic fever". This illness left me with what I now know was a trivial mitral valve murmur. However at that time the condition was considered serious, and I was not allowed to take part in sports for the next 7 years. But in the time that I might otherwise have spent in competitive sports I learned to enjoy reading and making things. And sometime before I was 11, I read a comic strip in which an inventor was the major character. This was what I wanted to be – an inventor! (I didn't know the word "scientist".)

Our mother introduced us joyfully to English literature by reading out loud to us, which she did beautifully, while we waited for my father to come home for the midday meal ("dinner"). Kenneth Grahame's "Wind in the Willows" and Lewis Carroll's "Through the Looking Glass" were favorites. And we heard and enjoyed Chaucer's "Canterbury Tales" spoken in middle English. We were often happy when our father was late. A dictionary was a part of our everyday life as children, and continues to this day to be a constant companion in our house.

The location of the house on Wakefield Road was ideal for children. Behind it was a long oak wood that covered several square miles. In the spring the wood was carpeted with blue bells, and in the fall with acorns. At other times it was a place for children, and lovers. It was also a source of the leaf mold that my maternal grandfather, Ben Sykes, and I collected for his garden. He was a highly intelligent but somewhat short tempered man who lost his job as a company manager because he couldn't get on with the son of his employer, who inherited the business when his father died. When I knew him, Grandfather Sykes was working as a paid gardener, which he enjoyed greatly. To keep his mind active, he began learning to speak French at age 70 plus. He enjoyed keeping bees too, and taught our father to love this activity. Later, when father was away in the army, we looked after his bees, and recovered their swarms. Roger kept bees for the rest of his life, and was still harvesting honey from hives that he had in his garden in a London suburb at age 81 shortly before he died.

Across Wakefield Road from our house was a large field from which we twins would help ourselves to rhubarb – illegally, of course. Beyond the rhubarb field was the Calder Valley canal and the Calder river, both heavily polluted when we lived there – but now recovering well. The Calder valley was even better for children than the long wood. It had caves in disused quarries; and our childhood girl friends, Margaret and Joan Smith, had a farm on the side of the valley. Above the valley was the village of Norland on the edge of a wild heather-covered moor. This moor was another of our playgrounds, and was where my father took his bees for them to collect the heather honey.

My father must have enjoyed mathematics, because I have a particularly vivid memory of him introducing me to decimals at an early age, writing with his finger on the condensate covering the wall above the bath that I was taking. I even remember the color of the wall as being blue. The same love of mathematics was deeply ingrained in Dr. G. E. ("Oddy") Brown who later taught me mathematics at Heath Grammar School. He conveyed enough of the logic and principles of mathematics that I didn't need to take any math courses at the University. Indeed, the examiners of my entrance examination to Oxford University commented that my mathematics was "very promising for a person so young." I suspect that they liked the comment I added to my answer to their question "How much does a Spitfire slow down when it fires its 8 machine guns?" Using their data on muzzle velocities, weight of a bullet, rate of firing, mass of aircraft, etc., etc., I calculated that the aircraft would slow down 150 miles per hour. I tried to calculate this again in several ways, but still got the same result. So I added the comment: "I don't believe this result. I think that the correct answer might be around 35 mph."

I have an equally but quite different vivid childhood memory of being shown, by my Smithies grandfather, how to straighten a bent nail. He, like me, couldn't resist picking up anything that he found lying around because "It might come in useful." This trait was well recognized by Jean Stanier, one of Sandy Ogston's graduate students at the same time as me. Odds and ends of discarded equipment and the like would be set aside and labeled NBGBOKFO – "No bloody good, but okay for Oliver." I still make new devices from what most people would call "junk."

My twin Roger and I went to the school in Copley, a village only a 15 minute walk from our Woodhall Crescent house. Our parents decided to let us go to this unpretentious village school rather than send us to a private school, even though the scholastic levels of the village school were less than desirable. It worked out well. Both of us passed the intelligence test used in 1936, as an entrance examination for acceptance of 11 year olds to a higher level of schooling.

Partly in preparation for this change, we moved to 33, Dudwell Lane, Halifax, a semi-detached house that was part of a collection of rather well designed but inexpensive new houses. This house was only a 15 minute walk from Heath Grammar School, the school which Roger and I now attended. Shortly after moving to 33, I met Harry Whiteley, the only son of the works manager of a local company that made precision time clocks for factories. Harry's and my interests matched perfectly, and we became and still are close friends. Harry's father had set up in the attic of their house ("the loft") a lathe, a good drill press and the hand tools needed for making many things. Harry knew how to use them, and the loft became our playground. I had somewhere read about a radio controlled boat, and we decided to make one. For the transmitter we used a spark coil from a T-model Ford. For the receiver we used a home made coherer, the same device as the one that Marconi had used in his first wireless telegraphy receiver. This was radio transmission at its basic minimum – and we never got it to work. But, encouraged by my grandfather's commercially made receiver, which used a crystal in place of the notoriously fickle coherer, we progressed to winding our own coils and made a much more up-to-date crystal set that worked well. This in turn led to a one-vacuum-tube radio, which I incorporated into my gas mask case instead of the gas mask that all British children were required to carry in the early days of World War II. Our best radio was a super-heterodyne of an advanced design and had four tubes. It worked as a "bread board", but disappointingly not when rebuilt as a more finished product.

When I was about 16, one of my father's friends gave me the engine from a motorcycle. Harry and I made it run, and became interested in owning a complete motorcycle. My first was a 1926 Rudge Whitworth which was notable for having rim brakes that did not work when it rained. Harry helped me exchange the front wheel for one with a safer internal expansion brake, and I used the Rudge regularly to travel to and from college. I also tried, but to no avail, to make it run on a gasoline-water mixture to eke out the very limited gasoline ration. Subsequently, by judicious trading, I managed to acquire motorcycles of increasing power, but always old, and they were an enjoyable and adventurous part of my life for several years. The cars that succeeded the motorcycles were equally old, and kept up my skills as a mechanic. Modern cars and laboratory equipment are unfortunately now only repairable by replacing subassemblies, so the current generation has lost this strong incentive to learn how to use simple tools.

With a motorcycle 1949

Heath Grammar School was an Elizabethan free school founded in 1597. When we attended the school, it had a superb staff of dedicated and highly-educated teachers. History was taught by C. O. Mackley who tried, in vain, to persuade me to study history with him in the sixth form. Chemistry was the task of A.D. Phoenix – who kept order with the flick of the rubber hose from a Bunsen burner. H. Birchall, the games master, tried kindly to bring me up to speed in athletics, but it was a hopeless task with a boy beginning to play games at age 14. My first year in the sixth form, at age 16, was spent with a few other pupils in supervised study of physics, chemistry and mathematics at a more advanced level. The first term of my second year in the sixth was spent in unsupervised study in preparation for the Oxford University scholarship exams. I concentrated on physics (I was thinking of studying the subject at the university, although in the end I chose medical school), and was fortunate in being awarded a Brackenbury Scholarship at Balliol College. Consequently, the remaining two terms in the sixth form were a blast in more ways than one. I was allowed to do whatever I wanted to, which was messing around (alone) in the laboratory. I synthesized many substances that caught my fancy, including phenyl isothiocyanate, which my textbook said was one of the worst smelling substances known to mankind. I made nitrocellulose (a constituent of Nobel's smokeless powder), and mercury fulminate (the detonator for his dynamite). Perhaps from some innate cautiousness I did not try to make them explode. Quite the opposite was inadvertently true of the nitrogen tri-iodide that I prepared. I had spilled traces of it which exploded when Mr. Phoenix wiped the bench (he was heard to say in an exasperated and loud voice "Smithies!") My father had a similar reaction when some that

I had put on the top shelf of our living room sideboard exploded with a puff of purple smoke as he walked by; it was extremely sensitive when dry.

I had three remedies for the homesickness that I felt on first going to Oxford. One was to look out of my college room window in the direction of my home in the north of England. Unfortunately I was actually looking south. I never did get the geography of Oxford right because of this error. The second remedy was to read all the Brontë novels again. The three sisters lived in Haworth, only a few miles from Halifax, and their novels were filled with descriptions of the Yorkshire moors that were such a part of my youth. The third remedy was to go down to the porter's lodge and look for a letter from home. Thereby hangs another tale. Balliol College at that time was heated only by open fireplaces in individual rooms. I lived in a room on the second floor reached by a spiral stone staircase. In the cold damp weather typical of autumn in Oxford, water would condense on the walls and trickle down the staircase. My room was narrow with ill-fitting windows at either end, and with stones covering half of its floor. It was heated (somewhat) with a small fireplace in which I could burn my weekly ration of coal – it was war time. On one occasion when I returned from my homesick visit to the porter's lodge, the corridor was full of smoke and my fire was gone. I followed the trail of smoke and found two second year medical students enjoying *my* fire in *their* grate. We immediately became friends. C. G. A. (Geoffrey) Thomas was one of them – which is how I remember the base-pairing rules of DNA – C with G and A with T.

A. G. "Sandy" Ogston, who had interviewed me during my scholarship exam, was the normal tutor for Balliol college's medical students, but his wartime duties prevented him from being my first tutor. David Whitteridge served in his place. Whitteridge was a brilliant scientist but a hard nosed tutor. I remember him saying to the Master of Balliol (A. D. Lindsay) during our end-of-term meeting that "Smithies can't spell". Lindsay's response "Oh, all interesting people can't spell," was encouraging. Whitteridge's comments "Diffuse, undisciplined, and at times inaccurate" written across my term paper were typically scathing, but deserved. His verbal comment to another student who had copied part of his weekly essay from a source that Whitteridge could recognize was equally to the point – "These scissors and paste jobs will do you no good." Oxford tutors could be ferocious, but that is what made their lessons unforgettable.

I studied anatomy and physiology with a little organic chemistry for two years as a medical student. I surprised the "real" anatomists and myself by winning the anatomy prize, I think because of my answer to one of the exam topics set by Professor Le Gros Clark, who was a pioneer in what we now call cell biology (he was also famous for uncovering the Piltdown-man fraud, and for helping Leakey with his pre-human fossils). I almost walked out of the room on reading the question: "Compare the regenerative powers of muscle, bone and nerve." But I suddenly thought of a principle that I thought made their similarities and differences understandable, and so I stayed. Perhaps Le Gros Clark enjoyed reading my answer as much as I enjoyed writing it.

My third year at Oxford was spent in studying for an honors degree in animal physiology (which included biochemistry). By then Sandy Ogston was back from his wartime duties and had resumed teaching and giving lectures on the application of physical chemistry to biological problems. He was best known for his three-point attachment explanation of how an optically active product can be generated from a symmetric precursor. My weekly tutorials with him were always stimulating and led to many memorable incidents. One occurred during the reading needed to prepare for a tutorial essay on carbohydrate metabolism. After learning something about metabolic pathways, I had been struggling to understand the biological "need" to carry out the complex series of reactions that the body uses to extract energy from carbohydrates. I found the answer in volume 1 of *Advances in Enzymology* in a long article written in 1941 by Fritz Lipmann. In this article Lipmann describes the difference between energy-rich and low-energy phosphate bonds, a difference that makes sense out of the complex series of reactions used to metabolize carbohydrate. I read his article in my Balliol college room with a level of excitement that I still remember. I even recollect the look of the glossy paper, the look of the pages, and the color of the cloth binding of the volume – a very similar feeling to that when I was introduced to decimals by my father.

This introduction to the importance of energy-rich phosphate was the cause of my later coming to Sandy's weekly tutorial with a way to generate an energy-rich phosphate bond from a low energy phosphoester bond by a cyclical oxidation and reduction scheme. Because my scheme could produce energy for nothing, I knew that it was wrong – like the Spitfire slowing down 150 mph – but I didn't know why. Together, Sandy and I – but mainly Sandy – realized that the standard free energy of a reaction (at that time used to classify the energy resulting from a reaction) was not a valid way of calculating how much energy the reaction would produce within a cell. One needed to know the actual concentrations of reactants and products in order to calculate this. My first scientific paper (Ogston & Smithies, 1948) was the outcome of this endeavor. Looking back at the paper, I can see Ogston's analytical mind at work – the paper hints at what is now known to be correct – the need to keep the reactants within a large molecular complex if realistic rates of reaction are to be achieved. This paper was the first of about half a dozen hypothesis papers that I have attempted over my scientific life.

My college "fire-stealing" friends were masters of how to study with the minimum of effort. We learned histology together by playing a show-and-tell game on Sundays that taught us to recognize the tissue on a microscope slide after only a second's glance – just as one recognizes a face. Once identified in this brief time, one could then carefully describe from memory what should be there. If the slide was of liver, for example, we would say "I can see the stellate cells of von Kupfer etc. etc." We never did see them, but this technique, passed on to subsequent generations, meant that Balliol students always came first in the histology examinations. Organic chemistry was equally conquerable if one used all one's senses, as illustrated by Geoffrey Thomas' finding that all the compounds which we were likely to be given could be identified by three tests: "taste, smell and appearance". I put his principle to good use in the final practical examination in Biochemistry. On being presented with a clear colorless, slightly viscous liquid that smelled of caramel and tasted acidic, I thought it might be lactic acid. A confirmatory test was positive, and I finished the exam in less than 10 minutes.

Sandy Ogston's fascination with the relevance of physical chemistry to biological systems was infectious, and I decided to drop out of medical school and do research in this field. The fourth and fifth years of my Oxford period were consequently spent in acquiring a sound background in chemistry. Since I already had a first class honours degree in physiology I did not have to worry about how well I would do in the exams. I could therefore pick and choose among the topics that I would study. I had a grand time. My organic chemistry was confined to biological compounds, my inorganic chemistry could emphasize the simple inorganic materials of biological relevance, Na^+, K^+, F^-, Cl^-, etc., rather than rare earths and the like. And I could emphasize those parts of physical chemistry that I enjoyed or were particularly relevant to biological systems. I remember well studying for and writing what I thought was an outstanding twelve-page essay on "The Pauli exclusion principle and the periodic table", which Ronnie Bell, my first tutor in chemistry, had assigned for one of my early tutorials. I only got half way down its first page when Ronnie spotted a weak link in my argument. The rest of the hour's tutorial was spent in teaching me that "You never, ever, write down anything that you do not understand, or cannot justify."

After completing the undergraduate part of the chemistry degree, and now in my sixth year at Oxford, I joined Sandy's lab in the department of biochemistry as a graduate student. It was a happy place. The oldest of us was Rupert Cecil (a veteran bomber pilot and a wing commander in the Royal Air Force). Rupert, in addition to his own research, managed the complex equipment of the laboratory with complete confidence. One of his responsibilities was a Svedberg ultracentrifuge – a large machine built on a concrete pillar and equipped with a powerful electric oil compressor in a pit below the floor. I never cared for the beast, and studiously avoided being sucked into its tentacles. Nevertheless, my thesis topic centered on an artifactual problem that the ultracentrifuge had generated – "the apparent conversion of the globulin fraction of plasma proteins into the albumin fraction." I was to look for some type of disassociation–reassociation reaction by studying the osmotic pressures of mixtures of proteins. I never did get to that part of my problem, but I had a thoroughly enjoyable two years trying. The outcome was a thesis, half of which was devoted to what are now (to me) un-understandable thermodynamic equations. On later re-writing this part of my thesis for publication I discovered a fatal flaw, so my equations never saw the light of day. The other half was devoted to my development of an extremely precise osmometer. The data it produced were so tight that the line through the experimental points had to be interrupted for them to show. This work was published (Smithies 1953), although the resulting paper has the dubious distinction of never being cited by me or by anyone else. Nevertheless, this thesis work re-enforced my natural inclination to pursue experiments to a conclusion with little regard for the time required to reach this end.

The osmometer required a home made water bath with its temperature controlled to within 0.001°C. This I achieved by using a submerged electric light bulb as a controlled heater. Sandy's next graduate student, Barry Blumberg (Nobel laureate in 1976), inherited my bench – and the water bath. He is said to have destroyed it in a fit of rage induced by the repetitious on-off cycle of its light bulb.

When the time came for me to think about post-doctoral work, Sandy urged me to think about going to the USA. I was not enthusiastic – but was persuaded to overcome my prejudices by Sandy and Robert L. ("Buzz") Baldwin. Buzz was a Rhodes scholar from Madison, Wisconsin, working towards his doctorate with Sandy, and he painted a fine picture of life in Wisconsin. So I applied for and was awarded a Commonwealth Fund fellowship to continue my education as a post-doctoral fellow under the guidance of J. W. (Jack) Williams, a learned physical chemist in the Department of Chemistry at the University of Wisconsin. There were other fine physical chemists in Jack's group, including Bob Alberty, Bob Bock, Dick Golderg and Lou Gosting. My stay with them increased my knowledge of physical chemistry greatly, but the work I did was not particularly rewarding; it culminated in another article that rightly received virtually no attention (Smithies, 1954). In contrast, the reward from the kindness and collegiality of these colleagues and of the other friends that I made in Wisconsin was enormous. They completely removed my foolish preconceptions about "Americans".

My regard for Americans was further increased by my meeting and becoming engaged to Lois Kitze, a graduate student working in virology. But she was reluctant to cross the Atlantic, as I had earlier been in the reverse direction. So, because my acceptance of a Commonwealth Fund fellowship precluded my staying in the United States, I looked for work in Canada. I was fortunate in finding David A. Scott, who in 1954, offered me a job in Toronto. "Scottie" was an older man when I met him, and was winding down a distinguished career in science. He was the first person to crystallize insulin as a poorly soluble zinc salt, which is widely used in the commercial preparation of insulin and still forms the basis for a slow release form of the hormone. He was a Fellow of the Royal Society of Canada, and of the Royal Society of England. When I met him, he was working by himself in a small room in the Connaught Medical Research Laboratories, a part of the University of Toronto, and spent his mornings looking for a protein in plasma which he thought might bind insulin. In the afternoons, he played golf. He offered me the opportunity to work on anything I wished "as long as it is related to insulin". After reading the available literature, I chose to look for a precursor to insulin. I never found it. But the difficulties I encountered in trying to find it, and a childhood memory that the starch which my mother used for my father's shirts turned to a jelly when it cooled, led to my invention of starch gel electrophoresis. The high concentration of starch needed to make a strong gel introduced a new variable into electrophoresis – molecular sieving. Finding the best variety of starch and how to process it for making the gels became necessary when my supplier's stock of processed starch was exhausted. Many hours were spent in testing all the raw starches that I could buy, and then in grocery stores finding potatoes from Holland Marsh, New Brunswick, Prince Edward Island and Idaho from which to make the raw starch. None gave as good gels as those made from my first batch. I eventually found out why: my original supplier had purchased starch processed by a second company that had used raw starch imported by a third company from Denmark because of an attack of potato blight in Canada!

The starch gel method proved very effective. With it I discovered previously unknown differences in the plasma proteins of normal healthy persons, which Norma Ford Walker and I showed were inherited (Smithies and Walker, 1955, 1956). Many new opportunities were opened up, and my friends suggested that I would be helped by having a technician. Somewhat reluctantly I agreed, and was joined part time by Otto Hiller, a young immigrant from Germany. He proved to be an excellent choice. We worked together well and soon became friends. Otto had an excellent mechanical sense, and began to make the starch gel equipment that I and other scientists needed for our work. He came along to Wisconsin when I moved there in 1960, but not as my technician. Instead he set up a business to manufacture the plastic equipment and assemble Heathkit® power supplies which were suitable for the gel electrophoresis. He also arranged for a manufacturer in Denmark to produce a starch suitable for making the gels, and then distributed the starch to scientists all over the world.

Otto and I spent many Saturday afternoons in his "shop" doing the same sorts of things that Harry and I had done in the loft. We assembled a Heathkit® digital alarm clock, and found out that it had a design flaw which caused it to lethally "electrocute" its own Intel CMOS integrated circuit. We worked out a remedy after several replacement chips, and had some enjoyable interactions with the Intel engineers who we found had drawn a Mickey Mouse on an unused part of the chip. This led us to try to make our own precision digital clock, and to attempt bread boarding a microcomputer using Intel chips. But our knowledge and bread boarding technique proved inadequate. So Otto bought a mail order kit for an Altair 8800 microcomputer, while my interest in *making* a computer was replaced by *using* a time sharing GE computer located in Milwaukee, 60 miles from Madison. Communication was by teletype, and the computer language was BASIC. The immediacy of a time-sharing computer suited me, and I subsequently enjoyed directing my student, Bob Goodfleish, while he wrote a group of programs to extract amino acid sequences from our Edman sequenator (Smithies *et al.*, 1971). Nearly 10 years later I had the same enjoyment in directing John Devereux during the writing of a group of programs for analyzing nucleic acid sequences. The resulting paper (Devereux *et al.*, 1984) is my most quoted, with > 6000 citations. More recently I have returned to devising new biological uses of computers, thanks to the existence of generic programs (such as Stella®) that a person can use for modeling complex biological systems without the help of a computer scientist (Smithies *et al.*, 2000; Smithies, 2003). The greatest value of devising these computer models comes, I have found, from their forcing one to clarify which elements in a complex system are most critical, rather from their ability to replicate experimental data or make predictions.

The discovery of inherited differences in plasma proteins shifted my interests towards genetics. This shift, and my wife Lois' homesickness for the States, led me to return to Madison in 1960, to join the strong genetics group at the University of Wisconsin. But I continued to collaborate with my Toronto friends to decipher the molecular/genetic basis of the protein differences found in plasma. This work revealed how homologous recombination could affect protein structure (Smithies *et al.*, 1962). It also led me to hypothesize that antibody variability might be achieved by recombination (Smithies, 1967). As a consequence, I had an enjoyable period devoted to protein sequencing with the automatic Edman sequenator.

This protein sequencing period ended with the advent of DNA cloning, which encouraged me to spend a sabbatical year with Fred Blattner on a floor below mine in the Laboratory of Genetics. During this time I learned to handle bacteria, bacteriophages and DNA (and took flying lessons at a small nearby airfield). Fred was deeply involved in developing safe procedures for cloning DNA, which at that time was thought might be environmentally hazardous. One of the safety tests required volunteers, of which Fred and I were two, to drink milk spiked with a large number of bacteria and then determine how many survived passage through the gut. The little packages of fecal material that we had to bring back to the lab were the sources of much merriment. During this period, I was invited to apply for various chairmanships in genetics, biochemistry and immunology. Somewhat selfishly, considering the great contributions that chairpersons can make to the scientific welfare of their faculty and students, I chose to continue my life as a bench scientist. But without this decision I might not have had the time to start the experiments, begun at age 57, which led to my best gene targeting paper, published after I was 60 (Smithies *et al.*, 1985).

In 1978, Lois and I, by mutual and amicable consent, gave up on our less than ideal marriage. And several years later I followed my mother's example by falling for my post-doctoral student, Nobuyo Maeda. However, we were unable to find a way to continue working together in Wisconsin. So, after more than 25 years, I left Madison to accompany Nobuyo to Chapel Hill, North Carolina, where she had been offered an appointment in the Department of Pathology at the University of North Carolina. Nearly 20 years have passed since that move. We have been happy together, and our science has flourished. The academic environment in Chapel Hill is agreeable and collegiate. The weather changes more gently than in Wisconsin (except for occasional hurricanes), and the winters are less harsh than in the Midwest. As a full time research professor at UNC I have been able to spend even more time at the bench; and all my experiments using gene targeting to generate animal models of human genetic diseases have been carried out in the nurturing environment of the University of North Carolina.

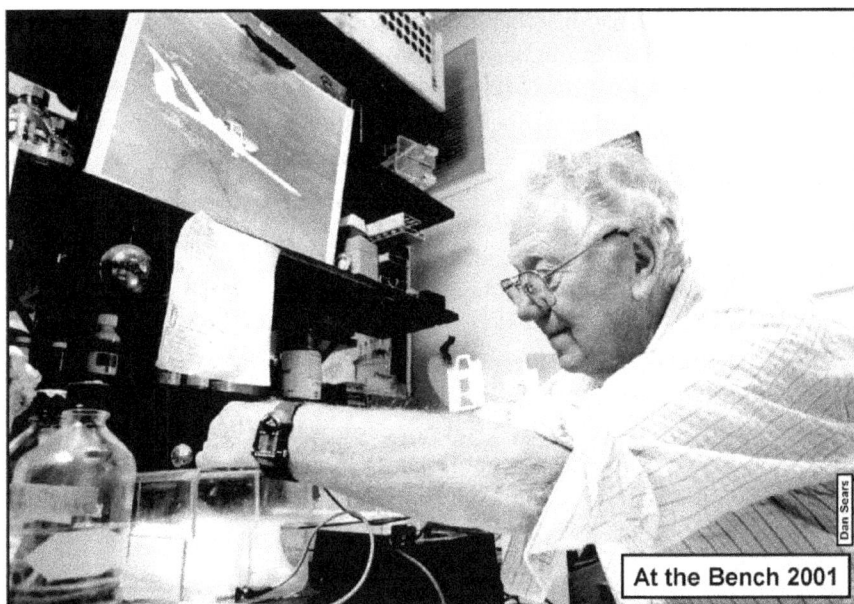

At the Bench 2001

Music has been a part of my non-scientific life, beginning quite early when, as children, Roger and I both sang in the choir at Copley church. We enjoyed the music and also the camaraderie of boys playing pencil games during the sermons. All three of us children were required by our parents to learn to play the piano from 7 until 11, at which time we could choose. Roger chose to learn to play the cello, and he continued playing it and the piano for the rest of his life. Nancy became a professional musician, and taught music in high schools. I stopped music lessons, but continued to sing in the church choir until my voice changed. Later at age 18 during my first year at Oxford I joined the Balliol college choir. In my second year, I auditioned for the Oxford Bach Choir with Sir Hugh Allen – a notoriously brusque conductor, famous for his sharp tongue. He began the audition with a comment and a question "You're from Balliol, I see. This is not your first year, is it?" I agreed. His next question was "Do you know how I know?" I replied "Yes sir, my tie [a Balliol tie] has been washed." The audition never flagged thereafter, even when he asked me to sing my lowest note, only to be interrupted by his secretary saying "Excuse me, Sir Hugh, but this gentleman is a tenor". To which he responded with "Oh, in that case sing your highest note!" followed shortly thereafter with "Stop! Stop!! You'll blow your head off!!!" I sang with his choir for the remainder of my time at Oxford. And I continued to sing tenor with great pleasure with the Symphony Chorus during both my times in Madison, and with the Mendelssohn Choir in Toronto. In Oxford, I learned to play the flute from an ex-army flute teacher. I was not good enough to play in an orchestra, but I happily played for many years with several small groups and with various accompanists.

My interest in flying also began at an early age, before I was 11. I had read all the "Biggles" books by W. E. Johns – fictional accounts of a World

War I fighter pilot. I had also been entranced by the movie serial "Tail Spin Tommy" which played at the Saturday morning "Tuppeny Rush" cinema in Sowerby Bridge, a half hour walk from my home (the admission charge was two pennies). And I had read enough about sailplanes and their instruments to dream of flying them. But World War II broke out when I was 14, and gliding as a sport stopped. It was not until I was 38 that I had my first real encounter with flying. This occurred in 1963, during a visit to Toronto which I had made in order to learn from Gordon Dixon how to sequence proteins. The required experiments did not suit my temperament – so instead I went down to the Toronto Island Airport and spent the next 10 days taking flying lessons. Over the course of the next month, now back in the States, I took enough additional lessons at Morey Airport in Middleton, Wisconsin, to be able to solo (fly by oneself). But I did not continue. Not until the late 1970s, when I was 52, was I able to try again, thanks in part to the stimulus to learn new things that is part of taking a sabbatical year. This time, I took glider lessons from "Jake" Miller and power plane lessons from Field Morey. Field, the son of a Lindberg-era pilot, was and still is a world class flight instructor, and we have had many hours together as student pilot and instructor. And many more as friends, including the time in 1980, when I accompanied him as co-pilot on a record-winning flight for a single engine aircraft across the Atlantic from Goose Bay, Labrador, to Rekjavik, Iceland, and then on to Prestwick, Scotland. We knew it would be difficult because we did not have special fuel tanks. So at the end of the runway at Goose Bay and after being cleared for take off we shut down the engine and topped off the tanks until, after adding several gallons of gasoline, they literally overflowed. After flying for 8½ hours, we landed at Rekjavik with only 3 gallons of fuel left, enough to fly for about another 10 minutes! But we beat the previous record – by 17 minutes. Our record held for nearly 20 years.

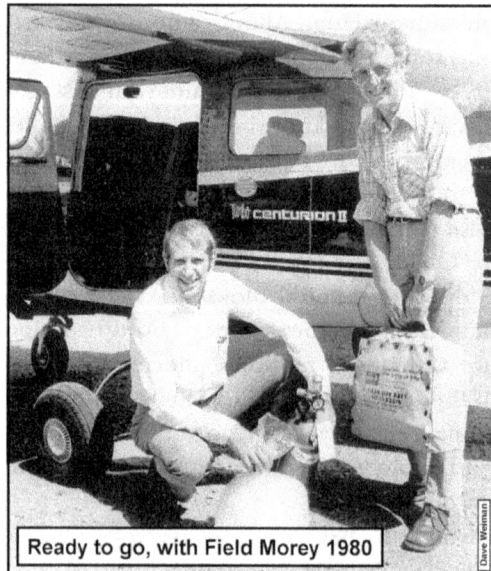

Ready to go, with Field Morey 1980

I learned to fly by instruments with Field, and remember rejoicing with him when "Only one drop dripped" (of sweat from my face). One of my glider students – who, like me, would sweat profusely during instruction – came back from his first solo flight with a big grin on his face, with his hand on the back of his shirt, and with the comment "Look Oliver; it's dry!" Learning to fly is learning to overcome fear with knowledge. This same lesson applies to trying new things in science, and to life in general. I am forever grateful to Field for helping me to learn it, and for giving me the joy of flying airplanes, which still continues after more than 4000 hours of piloting – in all sorts of weather.

Approaches into airports on cloudy days are carried out with the help of two needles on a dial from which indirect evidence the pilot can infer the position of the aircraft; if the needles cross at right angles you can infer that you are on the beam. Our first assay for gene targeting was likewise indirect, being based on finding bacteriophages of a specific type; if we found the bacteriophages we could infer that targeting has occurred. The airplane instrument approach and the gene targeting experiment both have a moment of truth. When the aircraft comes out of the clouds, either the runway is there, or it is not. Likewise, when DNA from a cell colony identified by the indirect bacteriophage assay is tested directly (by a Southern blot), either the gene is altered or it is not. In 1985, at a Gordon Conference during which I first described our success in gene targeting, I told the audience how I was thinking of this airplane analogy while developing the critical Southern blot autoradiograph. On presenting the positive result to the audience I said "And there's the runway!" All the rest of the speakers at that meeting accompanied their critical data slide with the comment "And there's *my* runway!"

REFERENCES

Devereux, J., Haeberli, P., and Smithies, O. (1984). A comprehensive set of sequence analysis programs for the VAX. Nucleic acids research *12*, 387-395.

Ogston, A.G. (1959). The spaces in a uniform random suspension of fibres. Trans. Faraday Soc. *54*, 1754-1757.

Ogston, A. G. and Smithies, O. (1948). Some thermodynamic and kinetic aspects of metabolic phosphorylation. Physiol. Rev. *28*, 283-303.

Smithies, O. (1953). A dynamic osmometer for accurate measurements on small quantities of material: osmotic pressures of isoelectric beta-lactoglobulin solutions. The Biochemical journal *55*, 57-67.

Smithies, O. (1954). The application of four methods for assessing protein homogeneity to crystalline beta-lactoglobulin: an anomaly in phase rule solubility tests. The Biochemical journal *58*, 31-38.

Smithies, O. (1967). Antibody variability. Somatic recombination between the elements of "antibody gene pairs" may explain antibody variability. Science *157*, 267-273.

Smithies, O. (2003). Why the kidney glomerulus does not clog: a gel permeation/diffusion hypothesis of renal function. Proceedings of the National Academy of Sciences of the United States of America *100*, 4108-4113.

Smithies, O., and Walker, N. F. (1955). Genetic control of some serum proteins in normal humans. Nature *176*, 1265-1266.

Smithies, O., and Walker, N. F. (1956). Notation for serum-protein groups and the genes controlling their inheritance. Nature *178*, 694-695.

Smithies, O., Connell, G. E., and Dixon, G. H. (1962). Chromosomal rearrangements and the evolution of haptoglobin genes. Nature *196*, 232-236.

Smithies, O., Gibson, D., Fanning, E. M., Goodfliesh, R. M., Gilman, J. G., and Ballantyne, D. L. (1971). Quantitative procedures for use with the Edman-Begg sequenator. Partial sequences of two unusual immunoglobulin light chains, Rzf and Sac. Biochemistry *10*, 4912-4921.

Smithies, O., Gregg, R. G., Boggs, S. S., Koralewski, M. A., and Kucherlapati, R. S. (1985). Insertion of DNA sequences into the human chromosomal beta-globin locus by homologous recombination. Nature *317*, 230-234.

Smithies, O., Kim, H. S., Takahashi, N., and Edgell, M. H. (2000). Importance of quantitative genetic variations in the etiology of hypertension. Kidney international *58*, 2265-2280.

TURNING PAGES

Nobel Lecture, December 7, 2007

by

OLIVER SMITHIES

Department of Pathology and Laboratory Medicine, University of North Carolina, Chapel Hill, NC 27599-7525, USA.

I am fortunate in having been a bench scientist for almost 60 years, and perhaps somewhat prescient in having kept all my notebooks (of which there are more than 130 since I first began). Together they are a record of my happy life as a scientist. They are also a more or less complete record of the progression and logic of the work that brings me to Stockholm today, and of what I expect to continue when I return to North Carolina. My hope is that in the next 40 minutes or so I can share this progression with you by TURNING PAGES in these notebooks. And I want to talk to a large degree to the people up in the balconies – the students.

The first group of pages documents my CHANCE invention of molecular sieving electrophoresis. My first job was in Toronto, Canada, and I was looking for a precursor for insulin (which I never found!). In the course of this work, I was having trouble in studying insulin with filter paper electrophoresis, as my January 1st, New Year's Day, 1954 page illustrates. [*"Students, note the day!"*]. Insulin stuck to the paper and unrolled like a carpet. – the more protein that I used, the further the carpet unrolled. (Left panel, Figure 1).

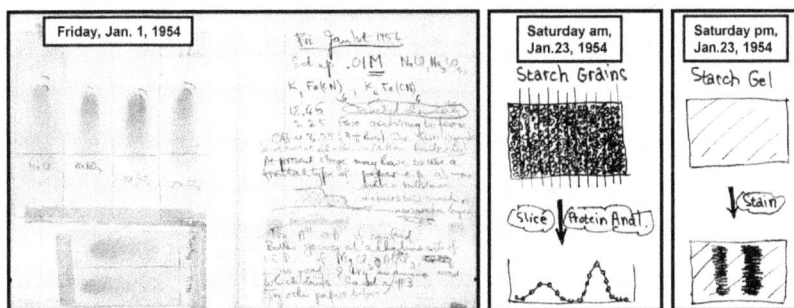

Figure 1.

Then, on January 23rd, 1954 (Middle panel, Figure 1) [*"Notice, students, Saturday morning!"*], I learned of a new method of electrophoresis that used a bed of moist starch *grains* (which do not adsorb proteins) for the electrophoretic medium, instead of moist filter paper (Kunkel and Slater, 1952). But, in order to find the separated proteins when using this method, it was necessary to carry out a protein assay on each of about 40 slices taken from the moist starch bed. I had no technical help, not even a dishwasher, and I couldn't afford the time to do multiple protein assays for each electrophoresis experiment. Happily, however, when I was a boy I sometimes helped my mother with the laundry, and remembered that the boiled starch she used for my father's shirts set into a jelly when it was cold. This memory suggested to me that I could cook the starch grains, make a gel, carry out the electrophoresis, and then just *stain* the gel to find the proteins. (Right panel, Figure 1). As a consequence of raids on them when no one else was around, I knew the whereabouts of the best stockrooms in the Connaught Laboratory where I worked, and so I was able to find some starch and test the gel idea that afternoon. [*"Saturday, still"*] The starch gelled only when its concentration was high, but the result with insulin was, as I recorded in my notebook, "very promising!" I later found out that a high concentration of starch impeded the migration of large proteins more than small proteins. This need to use a high concentration of starch was the chance element in my invention of *molecular sieving* gel electrophoresis (Smithies, 1955). [*"Molecular sieving occurs, students, when you use polyacrylamide gels with proteins and agarose gels with DNA."*]

Three months later, I tried electrophoresing serum – "just for a rough test" – and next day found a total of 11 components. At that time serum was thought to contain only 5 components (albumin, alpha 1, alpha 2, beta and gamma globulins), so I knew I was onto something likely to be important. I stopped looking for the insulin precursor, and began to study serum proteins.

Over the next 7 months I worked the bugs out of the starch gel electrophoresis method using serum from myself and from two of my graduate student friends at the University of Toronto, Gordon H. Dixon and George E. Connell, whom I co-opted to give blood. (Left panel, Figure 2) By the end of October, 1954, I was about ready to publish, when for the first time I ran a sample from a female, Beth Wade (B.W., right panel, Figure 2).

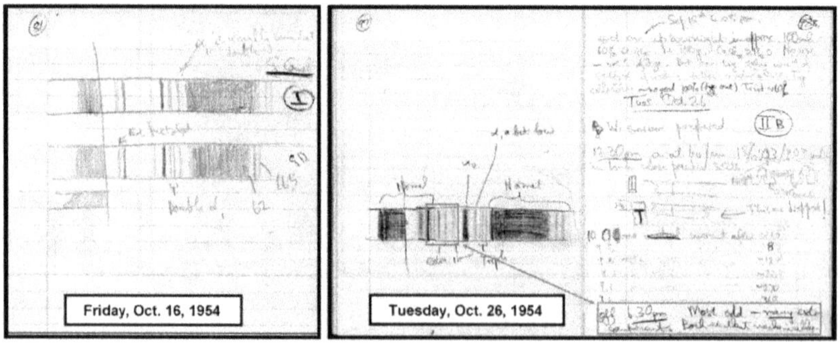

Figure 2.

My notebook entry on that day ("Most odd – *many* extra components") fails to record that I *thought* I'd found a new way of telling males from females! Indeed I called one type M, and the other type F, and found this designation to be correct for several male-female comparisons over the next week or so. But, after a hilarious day when one pair of individuals had the M versus F electrophoretic patterns reversed, the gender distinction proved to be incorrect. In its place, I thought it likely that the differences had a genetic basis. So, I contacted Norma Ford Walker, at the Hospital for Sick Children in Toronto. She was a remarkable lady, "one of the founding members of the institutions of human and medical genetics in North America" (Miller, 2002). And together we showed that the differences in the electrophoretic patterns of individuals were determined by common and completely harmless variations in the gene (Hp) controlling haptoglobin – the chief hemoglobin binding protein in plasma (Smithies and Walker, 1955; 1956).

We identified three common phenotypes (and genotypes): Hp1-1, (Hp^1/Hp^1), Hp2-1 (Hp^2/Hp^1) and Hp2-2 (Hp^2/Hp^2). (Left panel, Figure 3).

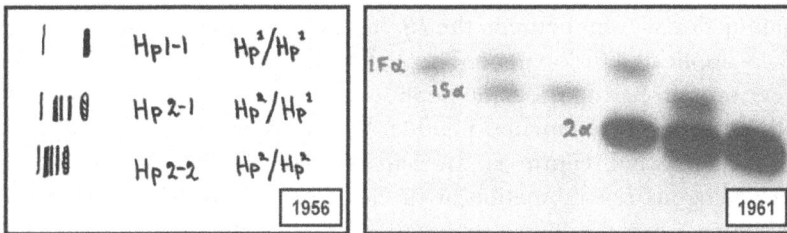

Figure 3.

This finding opened the next chapter in the book of my scientific life – an OPPORTUNITY to study the genetic differences in proteins, starting with the haptoglobins. This I undertook in collaboration with my ex-graduate student friends, Gordon Dixon and George Connell, who had by then come back to the University of Toronto as junior faculty members.

For many years I have advocated and practiced "Saturday morning" experiments, of which you have already had a sample. These experiments have the advantage of not needing to be completely rational, and can be carried out without weighing chemicals, and so forth. [*"But, students, not without proper lab-book notes."*] And I carried out many of these in trying to simplify the complex electrophoretic patterns associated with the products of the Hp^2 gene. One of them included the use of phenol. This was short-lived because phenol dissolved my apparatus! Reducing the protein with beta mercaptoethanol (βME) in the presence of urea, following a suggestion from Gordon, proved to be the key. But not without another hilarious incident that followed my accidental breakage of a bottle of βME over my shoes. I put them on the windowsill for a while. But I didn't have many pairs of shoes, and so I soon began to wear them again. Several days later, during a visit for other reasons to the local police station, I heard two old ladies whispering together. One asked the other, "Do *you* smell it?" Her friend responded, "Yes. Do you think it's a *body?*" My shoes went outside on the windowsill for a while longer.

After learning how to separate haptoglobin into its subunits (alpha and beta), we found that its genetics were more complicated than Norma Ford

Walker and I had thought. Thus when George began purifying haptoglobin from single bottles of donated plasma we found (Right panel, Figure 3) that there are *three* common haptoglobin alleles (Hp^{1F}, Hp^{1S} and Hp^2), not two (published later in Connell *et al.*, 1962). We also noted that the Hp^2 gene, the one which is associated with the complex protein patterns, appeared to produce twice as much alpha subunit as the other two genes (Hp^{1F} and Hp^{1S}). And there were other findings that made us think that the Hp^2 gene was more complicated than the Hp^{1F} and Hp^{1S} genes. For example, when Gordon compared the peptide maps of the hp1Fα, hp1Sα and hp2α haptoglobin subunits, the results were very puzzling, and we had great difficulty in believing them – hp2α appeared to contain all the peptides present in hp1Fα and hp1Sα, plus an extra one. Then, during a get together in Toronto in 1961, I remember saying to Gordon and George, "Let's *believe* our own data." And I suddenly realized that the Hp^2 gene was probably the product of some sort of recombinational event between the Hp^{1F} and Hp^{1S} genes that had generated a partially duplicated fusion gene. The Hp^2 gene would consequently produce a larger protein having the same peptides as a mixture of hp1Fα and hp1Sα together with a novel junction peptide, "J", not present in either hp1Fα or hp1Sα. (Left panel, Figure 4). We had become the first people to detect *non*-homologous recombination at the level of a gene! We called it "non-homologous", because the recombination between the Hp^{1F} and Hp^{1S} genes was within regions that are unrelated in sequence.

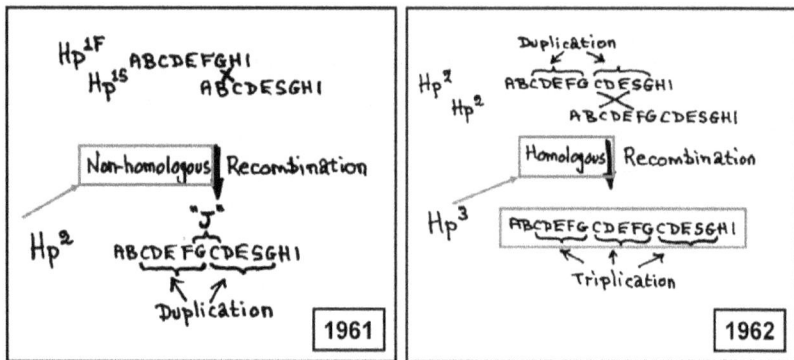

Figure 4.

We decided to present our data and our partial gene duplication hypothesis at the 1961 Second International Conference of Human Genetics in Rome. We also designed an experimental test that George was going to do before we each gave our part of the story at the conference. He would use the ultracentrifuge to determine the sedimentation coefficients of the alpha subunits with the expectation that the hp2α subunit, which our hypothesis said was larger than hp1Fα and hp1Sα, would sediment more rapidly. We met in Rome on the evening before our talks to review George's results, and he broke the bad news – the sedimentation coefficients of the three hpα subunits did *not* differ. What to do? Well, we decided, despite this result, to go ahead with our planned talks, with the understanding that in my part of the presentation I would describe our hypothesis and the experimental test of it that we had carried out. Then I would say "*We don't believe the result,* and I'll go home and invent a new method for determining molecular sizes." The next two pages in my notebooks (Figure 5) show the implementation of that plan (Smithies, 1962). ["*Notice, students, that you shouldn't* always *believe your results!*"]

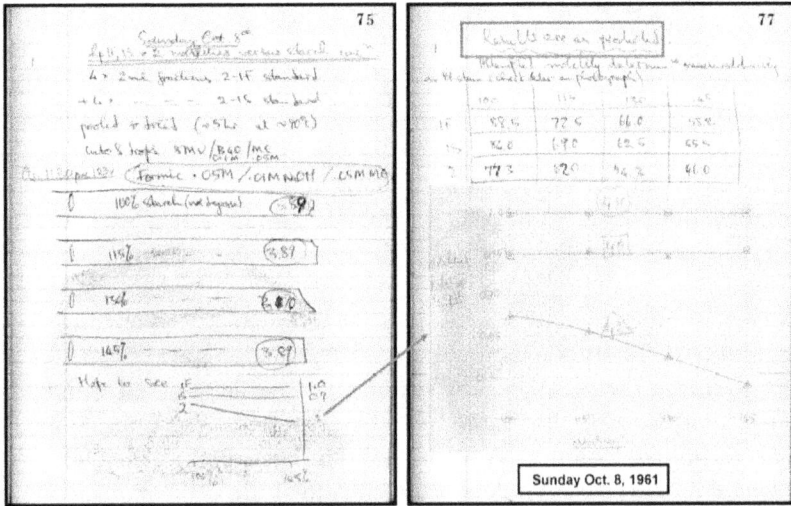

Figure 5.

The new method showed that hp2α *was* bigger than hpIFα and hpISα. (Later, when George got rid of aggregation by adding urea, the ultracentrifuge gave the same result.) Together we published our conclusion that the Hp^2 gene was a partial gene duplication resulting from a non-homologous crossing-over event between the Hp^{IF} and Hp^{IS} genes in a heterozygous individual, Hp^{IF} / Hp^{IS}. (Smithies *et al.*, 1962).

The next part of this chapter in my science concerns the clear distinction between the randomness of *non-homologous* recombination and the predictability of *homologous* recombination. When I told Professor James H. ("Jim") Crow, Chairman of Genetics at the University of Wisconsin, about our results, he referred me to some beautiful classical work involving the genes controlling the development of the eye of the fruit fly, *Drosophilia*. In succession over a period of over 20 years, Tice (1914), Zeleny (1919), Sturtevant (1925) and Bridges (1936) provided evidence that a *unique*, non-homologous recombinational event, which occurred only once, had generated a duplication on the X chromosome of the fruit fly that changed the shape of the eye. They also showed that this duplication enabled unequal but homologous recombinational events that occasionally gave rise to a triplication or to a return to the unduplicated chromosome. We extrapolated this result to the haptoglobin genes, and expected that the same type of event would occur with them – namely that unequal but homologous recombination within the duplicated region of the already larger Hp^2 gene would likewise lead repeatedly to a still larger triplicated gene (Right panel, Figure 4). And we found this larger gene as an uncommon variant (Hp^3, but historically called $Hp^{2)}$) that had arisen independently in all parts of the world where the Hp^2 gene was already in the population. This was my first real understanding of the fundamental difference between the unpredictable nature of non-homologous recombination and the predictability of homologous recombination.

Later, in the late 1970s, I spent a sabbatical period in Fred Blattner's laboratory in the same building as my own laboratory, and learned how to work with DNA and with bacterial and bacteriophage mutants (and, as a concurrent sabbatical activity, learned to fly!). Then, when Fred's Charon bacteriophages were judged to be safe enough for use in cloning human genes, our groups collaborated in isolating and characterizing the two closely related genes that code for the human fetal globins, $^G\gamma$ and $^A\gamma$ (Blattner *et al.*, 1978; Smithies *et al.*, 1978). Subsequently, when we sequenced these two genes, we found clear evidence that DNA had been exchanged between them as a result of another type of homologous recombination, "gene conversion". (Slightom *et al.*, 1980). So, homologous recombination was very much a part of my scientific gestalt. And, not surprisingly, having worked with globin genes, I kept thinking that it ought to be possible to use DNA coding for the *normal* human β globin gene, which was now readily available, to correct the *mutant* human β globin gene that leads to sickle cell anemia, the most frequent disease caused by a single gene in people of African descent. But no one had demonstrated that such an event (now called "gene targeting") was possible with a genome as large as that of humans and other mammals, although it was known to occur in yeast (Hinnen *et al.*, 1978; Szostak and Wu, 1979) with a genome of less than one hundredth the size.

Then in 1982, while teaching a graduate course in molecular genetics at the University of Wisconsin, I came across a beautiful paper that catalyzed me to start writing the next chapter in my book of science – "PLANNING" to use homologous recombination to correct a mutant gene in the human genome. The catalytic paper was published in *Nature* on the first of April, 1982 (Goldfarb *et al.*, 1982). In this paper, the investigators described an elegant gene-rescue procedure to isolate a transforming gene from human T24 bladder carcinoma cells. This gene-rescue procedure depended on using mutant lambda bacteriophages that had a lethal amber chain-termination mutation which could be suppressed if the bacteriophages picked up a copy of *supF* (a mutant tRNA gene able to suppress amber chain-termination mutations). The amber mutant bacteriophages would not grow otherwise. The procedure was complicated, and I had to study the paper carefully in order to use it in teaching. This effort had, however, an unanticipated benefit. During the next 3 weeks I realized that I could use a modified form of Goldfarb's gene-rescue procedure in an assay to determine whether it was possible to place "corrective DNA *in the right place*" in the human genome.

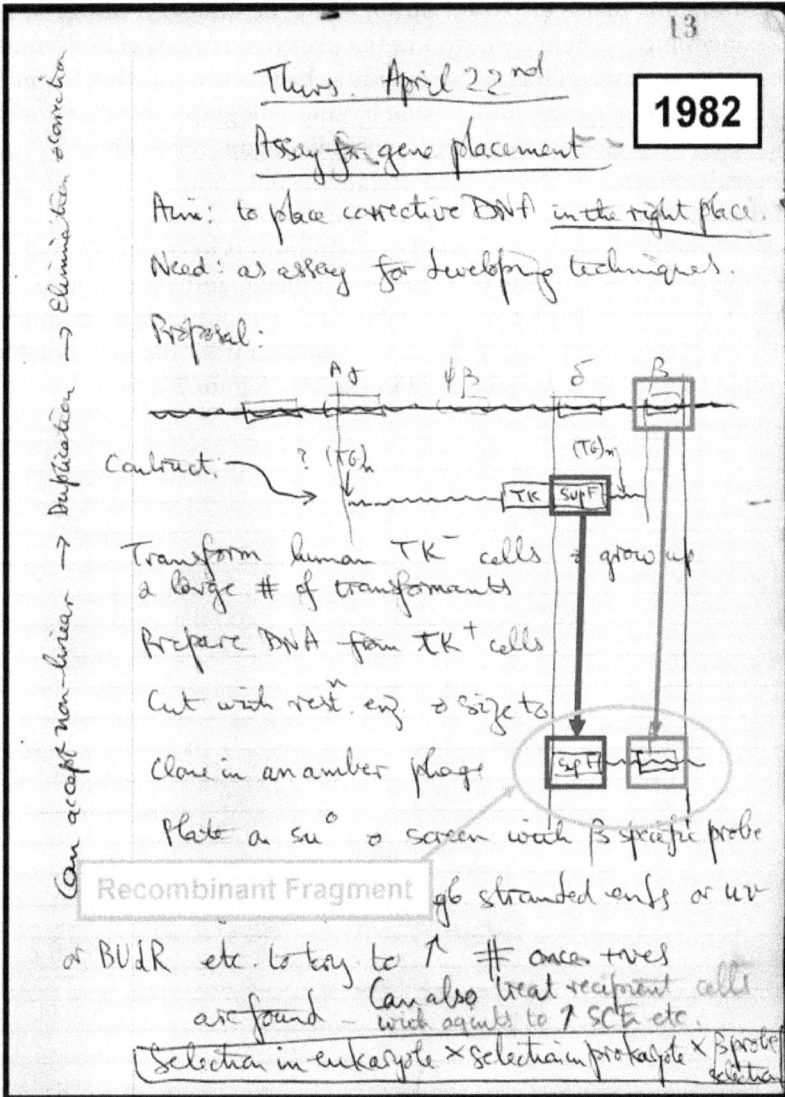

Figure 6.

On April 22nd, 1982, on page 13 of my γ notebook (Figure 6.), I summarized my idea under the heading "Assay for gene placement" (now called "gene targeting"). I proposed to make a DNA construct that included a large fragment of DNA covering the human beta-type globin genes, together with the *supF* gene and the thymidine kinase gene, *TK*. I would then introduce this DNA into human cells that were *TK*, select for transformants that had become TK+, and then use gene rescue to look for a *recombinant fragment* in which the *supF* gene was now next to the β globin gene. This would prove that the incoming DNA had been inserted into the correct place. I was confident that I could detect gene targeting, even if it was *extremely* rare, because I had three levels of selection: selection in the eukaryotic *TK* human cells of transformants that had picked up the *TK* gene and so could grow in a HAT- containing medium; selection in the prokaryotic *E.coli* cells of mutant bacteriophages that could grow because they had picked up DNA fragments containing the s*upF* gene; and selection by autoradiography of bacteriophages that also had β globin sequences. Only homologous recombination could generate the diagnostic recombinant fragment containing *both* the *supF* gene from the incoming DNA *and* the β globin gene from the target locus.

At that time DNA sequencers and DNA synthesizers were not yet available, so making the large targeting construct was difficult, and I had to clone it as a cosmid, which I called Cosos 17. Making this cosmid took me 7 months. Some idea of the complexity of this task is apparent from the next notebook pages that I show but will not attempt to explain (Figure 7.).

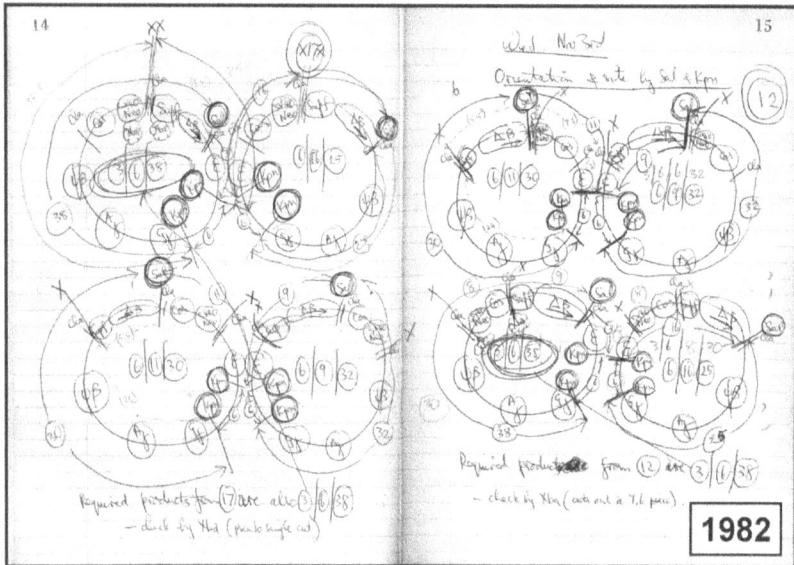

Figure 7.

By the end of 1982, I had sent Cosos 17 to my collaborator Raju Kucherlapati at the University of Illinois. He was to make a calcium phosphate precipitate with this DNA for transfection into another human bladder carcinoma cell line, EJ. Meanwhile, I began work on what turned out to be a scientifically dangerous experiment: I carried out a plasmid by plasmid recombination experiment to test whether the gene-rescue assay would work. The tester plasmid was Δβ17, a small precursor of Cosos 17. The mock target contained the human β globin gene. The good news was that both the recombination and the bacteriophage gene-rescue assay worked (Smithies *et al.*, 1984). The unforeseen bad news was that bacteriophages containing the diagnostic recombinant fragment were now present in the lab.

In May of 1983, Raju sent back to us the first DNA sample, RK41, from a gene-targeting experiment with Cosos 17 and the human EJ bladder carcinoma cells. On June 23rd (my 58th birthday), I started the bacteriophage assay phase of this first real test of the overall scheme. 288 bacteriophages grew; 104 (34%) contained some β globin sequences; but, birthday or not, *none* hybridized to the critical β globin IVS2 probe! (Figure 8). So this first real experiment failed to provide any evidence that homologous recombination had occurred.

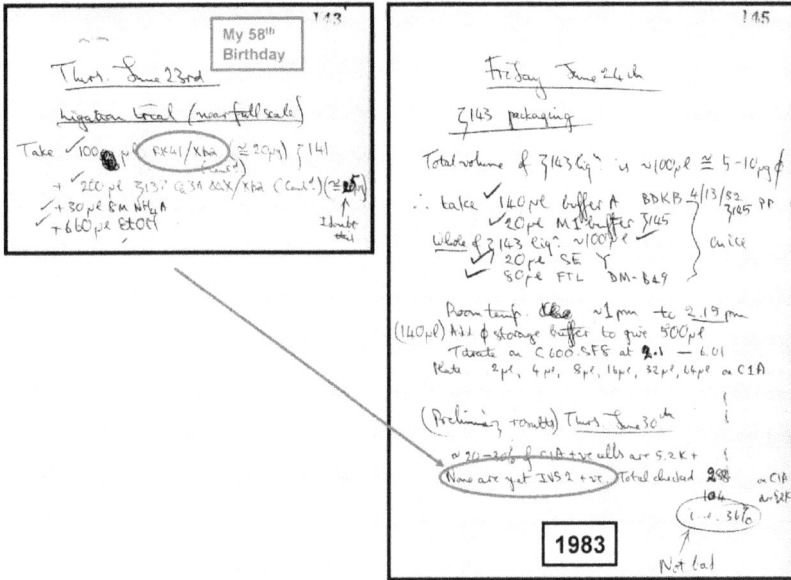

Figure 8.

Over a period of almost a year, my lab and Raju's lab continued experiments with the EJ cells, but without success. These negative results led my graduate student Karen Lyons to suggest that the failure might be because the drug-resistance gene, *Neo^R*, which we were now using instead of *TK*, might not be transcribed when incorporated into the β globin locus of a *bladder*-related cell that does not express β globin. [*"Students, you should keep going when things don't work; but you should also think carefully about what might be wrong."*] Two alternatives were available. We could retain the drug selection, but use cells that *expressed human β globin*; or we could continue to use the EJ bladder carcinoma cells but *without* using drug selection. One of our earlier collaborators, Art Skoultchi, gave us a cell line which he had made that was suitable for the first type of experiment. It was a mouse-human hybrid erythroleukemia cell line (which we called Hu11) that carried a human chromosome 11 and expressed human β globin (Zavodny *et al.*, 1983). Unfortunately the erythroleukemia cells grew in suspension, and could only be transformed by a newly devised procedure – electroporation (Potter *et al.*, 1984) – and no electroporator was then commercially available. So I spent the next few months designing and testing a homemade apparatus, which was constructed inside a baby bathtub from part of a plastic test tube rack and electronic parts from the local Radio Shack store. The final version of the apparatus, illustrated in schematic and real form in Figure 9, does not look impressive – but it worked, and was subsequently used for all the definitive experiments.

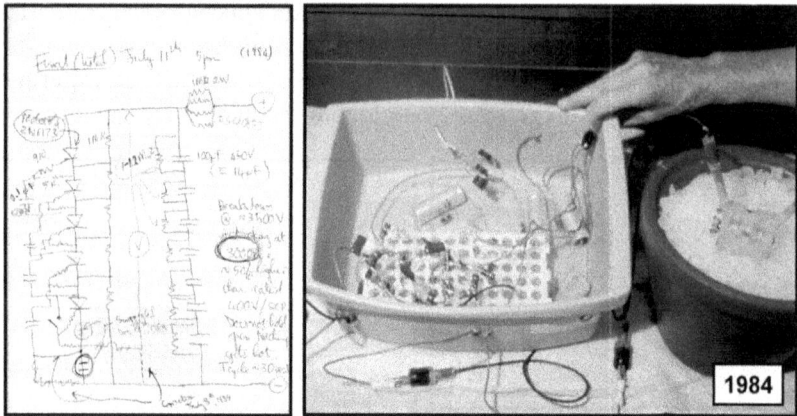

Figure 9.

[*"Students: never make a complex piece of apparatus that can be bought in order to save* money; *but by all means make it to save the* time *that you will have to wait before some manufacturer makes it. "*]

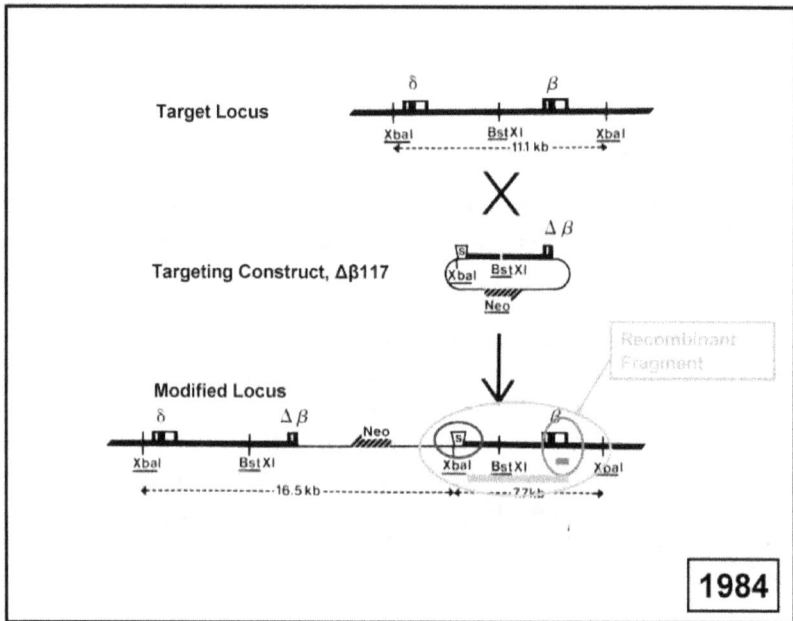

Figure 10.

Meanwhile Raju did an experiment of the second type, using the EJ bladder carcinoma cells without drug selection. This experiment also used a different targeting construct, Δβ117, illustrated in Figure 10. (Smithies *et al.*, 1984). Δβ117 was the recombination tester plasmid Δβ17 which I had modified so that it could be cut (with *Bst* X I) in the region of homology. This type of cut, we had already shown, increases the frequency of homologous recombination in mammalian cells, as it does in yeast (Kucherlapati *et al.*, 1984). Raju treated the bladder carcinoma cells with *Bst*X I-digested Δβ117, grew them up without any drug selection, and then sent us DNA from the cells. My technician, Mike Koralewski tested this DNA with the bacteriophage assay in late August, 1984. He found one IVS2-positive bacteriophage, which I purified and showed had the hoped-for recombinant DNA fragment with *supF* next to β globin IVS2. This was good news.

But we began to have worries. One worry was that this single bacteriophage could have been a contaminant from our recombination tester experiment. (We had had a contamination problem in some earlier gene cloning experiments.) An even more serious worry was that the recombinant fragment present in the bacteriophage might have been formed by recombination in the *bacterial* cells used in the gene rescue assay, rather than in the *mammalian* cells used for the transformation. We were discouraged!

Fortunately, however, I had recently bought an airplane, and had flown it to Florida for a short sailing vacation with my pilot friends. This vacation re-energized me sufficiently that I could face starting the Δβ117 experiments again – with two important changes. First, my postdoctoral fellow, Ron Gregg, who had been trying unsuccessfully to inactivate the *Hprt* gene in human fibroblasts, would electroporate *Bst*X I-digested Δβ117 into the Hu11 cells that express the human β globin gene. Second, after Ron had isolated DNA from drug resistant transfectants, I would digest it with *Xba*I and size separate the restriction enzyme products into two fractions. One fraction would cover the size range 5.5 – 8.5 kb, and another would cover the range 8.5 – 16.5 kb. This fractionation had two purposes. It would reduce the amount of DNA to be packaged into bacteriophages; and, more importantly, it would separate *Xba*I fragments that were 7.7 kb long (the size of the XbaI recombinant fragment) from any fragments that were 11 kb long (the size of the XbaI fragment from the unaltered target locus). If the recombinant fragment was already present in the DNA from the Hu11 cells *before* the DNA had been exposed to bacteria, the 5.5 – 8.5 kb DNA fraction would give IVS-2 positive bacteriophages. If the recombinant fragment was the result of a recombinational event occurring in the bacteria, the 8.5 – 16.5 kb DNA fraction would give IVS-2 positive bacteriophages. In early 1985, this fractionation experiment was completed using size fractionated DNA from a flask containing ~ 1000 drug-resistant colonies. *Two* IVS-2 positive phages were obtained with the 5.5 – 8.5 kb fraction. (Upper panel Figure 11) Now we believed our results.

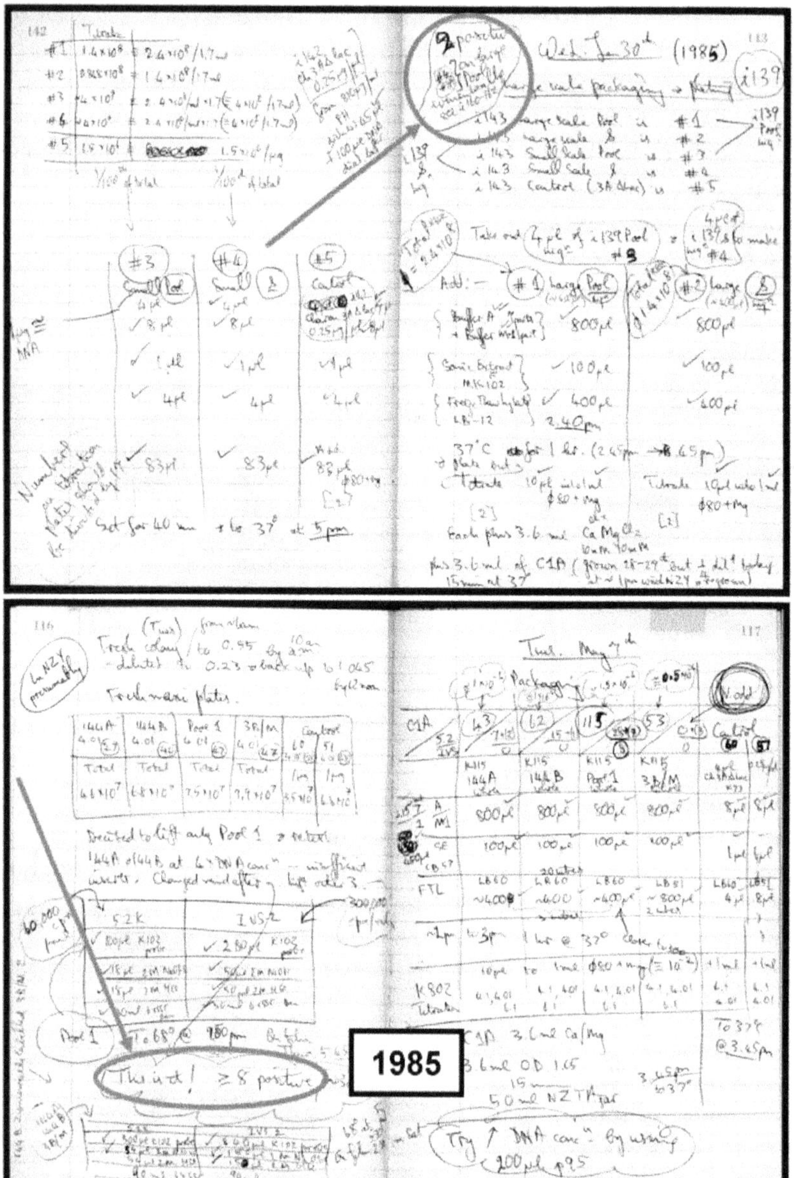

Figure 11.

It took three more months for me to iron out various problems with the gene rescue assay, and for Ron Gregg to generate pools of *individually cloned* Hu11 transformants. But by April we had identified a pool of about 300 cloned Hu11 transformants that gave *three* IVS-2 positive phages. And, in May, DNA from 30 sub-cloned Hu11 transformants from the 300 pool gave us *eight* IVS-2 positive phages (Lower panel Figure 11.). This meant that at least one of the 30 subclones was correctly targeted, and we could now use a *direct* test for recombination (a Southern blot of DNA from each colony) in place of the indirect bacteriophage assay.

On May 18th, 1985 [*"Saturday, yet again!"*], I Southern-blotted Ron's electro-phoresis gel of DNA from 11 of these 30 colonies (Figure 12). On May 20th, I noted on page 134 of my κ notebook that subclone "#20 is it!" – 3 years and 1 month and 7 notebooks after the original idea. In September of 1985, the paper (Smithies *et al.*, 1985) which I imagine the Nobel Committee considered my most important was published – after I was 60!

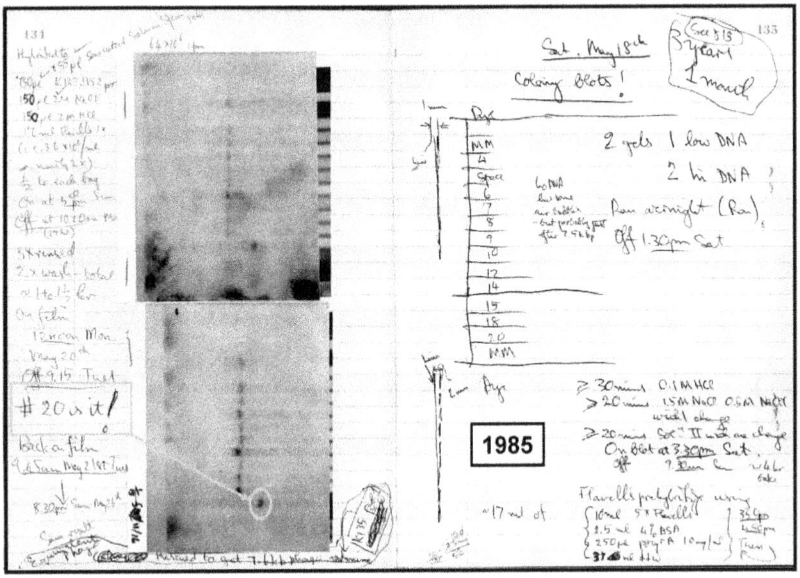

Figure 12.

I have already referred to all who contributed to this paper except one – Sallie Boggs. She was a visiting professor from the University of Pittsburgh. She chose, as her part in the work, to ensure that we had a "back-up" to the bacteriophage assay, in case it did not succeed. To implement this, she carried out Southern blots of DNA from 243 *individual* Hu11 transformants without ever using the phage gene-rescue assay. Although the phage assay, in the end, led to a correctly targeted colony before Sallie found a positive transformant, her work established that the electroporator we had made could introduce single copies of DNA into the cell genome without any other detectable changes in about 80% of transformants (Boggs *et al.*, 1986).

At this point, it was clear that gene targeting was impractical for any near-term use in the gene therapy that I had initially hoped. The frequency of targeting was *too low*. The bacteriophage assay we had used to detect targeting was *desperate* (indeed nobody, including me, ever used the assay again). But these experiments had told us that gene targeting was *possible*. We now knew that we could introduce DNA into a chosen site and alter a target gene in a pre-planned way. So, what to do? Well the first thing was to find a simpler system in which to improve the procedure. And towards this end several investigators in the field independently began experiments with genes that had a directly observable phenotype. Ron Gregg in our group chose the *Hprt* gene, which makes cells resistant to HAT selection when it is normal, and makes them resistant to 6-thioguanine when it is disabled; Mario Capecchi also chose the *Hprt* gene; Raju Kucherlapati chose the *TK* gene. But success was slow in coming.

Then I heard about Martin Evans' work in isolating what we now call embryonic stem (ES) cells and using them to generate mice, and I immediately began to think about using gene targeting in these cells to modify genes in the mouse. Since ES cells grow rapidly and can be cloned from single cells, a low frequency of gene targeting would not be an issue. We could therefore modify a gene in the ES cells, and use the targeted cells to make animal models of human genetic diseases for study and for testing therapeutic procedures. As a step towards this end, in November 1985, Martin personally brought some of his cells to our lab (Figure 13). [*"Students: Don't be shy about asking other scientists for reagents or help!"*]

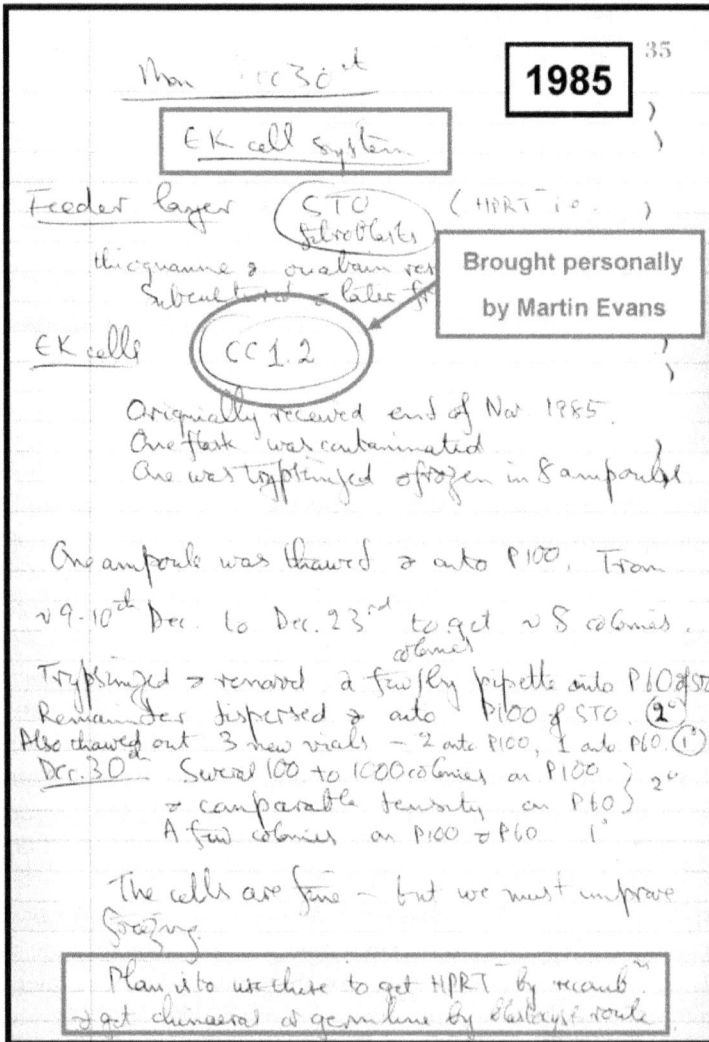

Figure 13.

Martin also put us in touch with Tom Doetschman who had experience with ES cells, which need to be handled correctly if they are to be capable of generating mice. In December of 1987, we published our first use of gene targeting in ES cells – to *correct* a mutation in the *Hprt* gene of E14TG2a ES cells that had been isolated by Hooper *et al.* (1987). The DNA construct, made by Nobuyo Maeda, worked the first time that Tom used it! The big colonies resulting from gene-corrected cells were easy to distinguish from the tiny residues left from cells in which the mutant gene had not been corrected. (Figure 14).

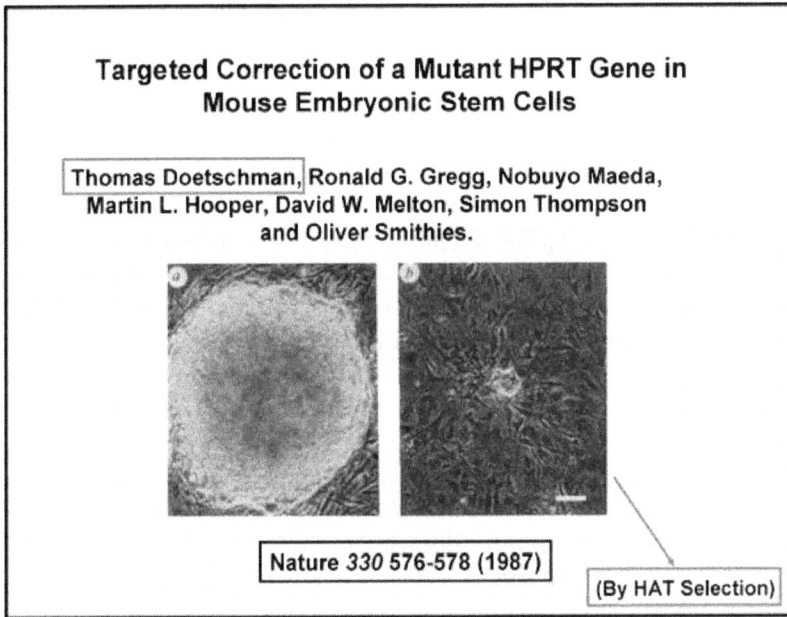

Targeted Correction of a Mutant HPRT Gene in Mouse Embryonic Stem Cells

Thomas Doetschman, Ronald G. Gregg, Nobuyo Maeda, Martin L. Hooper, David W. Melton, Simon Thompson and Oliver Smithies.

Nature *330* 576-578 (1987)

(By HAT Selection)

Figure 14.

Mario Capecchi independently contacted Martin Evans for help with ES cells within weeks of our contacting him. And his group's paper, describing a *knock out* of the normal Hprt gene in ES cells (Thomas and Capecchi, 1987), and ours describing *correction* of a mutant form of the gene (Doetschman *et al.*, 1987), were also within weeks of each other. Both had used drug-selection procedures to isolate the targeted cells, based on the enzymatic activity of HPRT.

However, a procedure was needed for targeting genes that did not have a directly selectable product. A big help would be to have a simplified *recombinant-fragment* assay. The polymerase chain reaction (PCR) described by Kary Mullis at Cold Spring Harbor in 1986 (Mullis *et al.*, 1986) looked to be eminently suitable for this purpose (Left panel, Figure 15), and I began to work on this idea a few months after hearing Kary talk. Again, no suitable apparatus was commercially available. So Hyung-Suk Kim and I made our own PCR machine, which I still use (Right panel Figure 15).

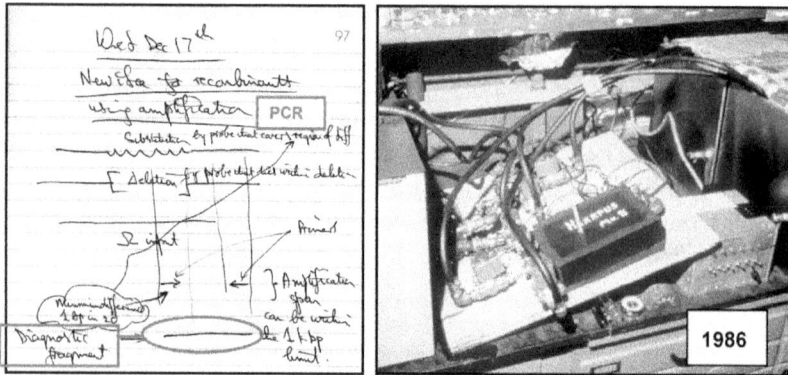

Figure 15.

Time does not permit me to describe many of the animal models that we have since made using gene targeting in ES cells, with the help of our PCR method of detecting the diagnostic recombinant fragment (Kim and Smithies, 1988), together with the powerful positive-negative selection method devised by Mario's group in 1988, as a "general approach for producing mice of any desired genotype" (Mansour *et al.*, 1988). But I can highlight some of them.

Bev Koller, as a post doctoral fellow in my laboratory, was the first to make a mouse model of cystic fibrosis, the most common single gene defect in Caucasians. (Figure 16) (Koller *et al.*, 1991; Snouwaert *et al.*, 1992).

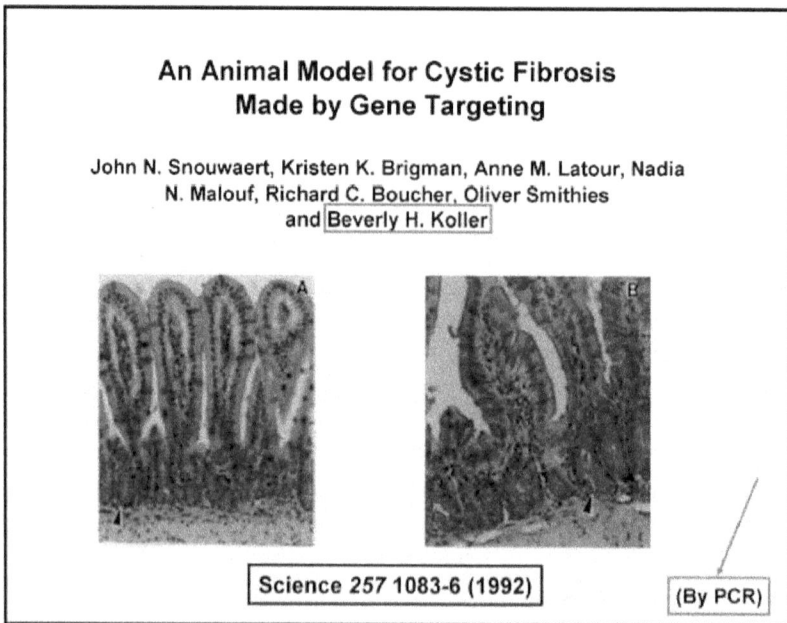

Figure 16.

Nobuyo Maeda and her colleagues made a mouse model of atherosclerosis (Zhang *et al.*, 1992) that became a best-seller at Jackson Laboratories; it is an inspiring model of this genetically complex human disease that causes around 30% of deaths in advanced societies. (Figure 17).

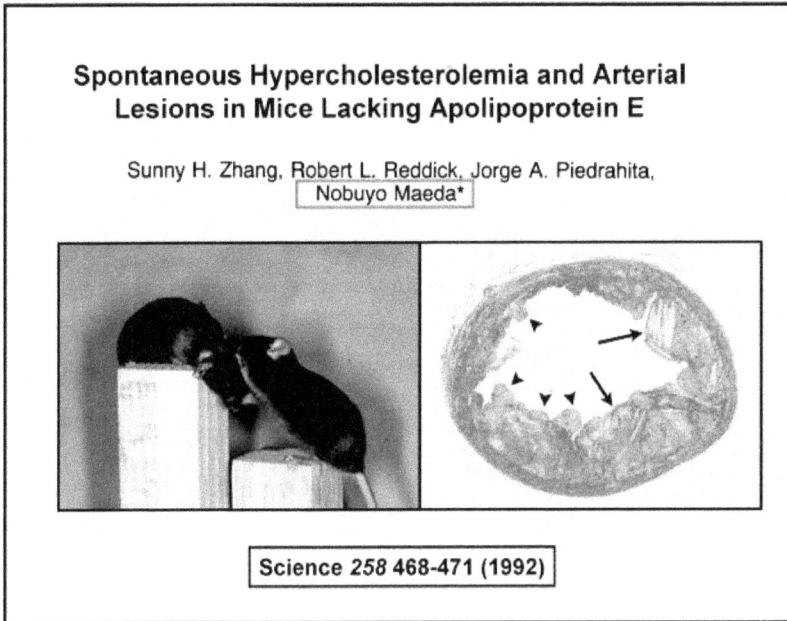

Spontaneous Hypercholesterolemia and Arterial Lesions in Mice Lacking Apolipoprotein E

Sunny H. Zhang, Robert L. Reddick, Jorge A. Piedrahita, Nobuyo Maeda*

Science *258* 468-471 (1992)

Figure 17.

John Krege led me into a very productive investigation of genetic factors important in another very common disease – high blood pressure (Krege *et al.*, 1997; Smithies, 2005). For this work we used a computerized blood pressure measuring apparatus made by John Rogers, who was at that time one of my glider pilot students (Krege *et al.*, 1995). I chose him to make the new machine (Figure 18) because he had told me about a computerized device that he had designed and built to detect the stones left in pitted cherries, which cause lost teeth in the eaters and lawsuits against the suppliers!

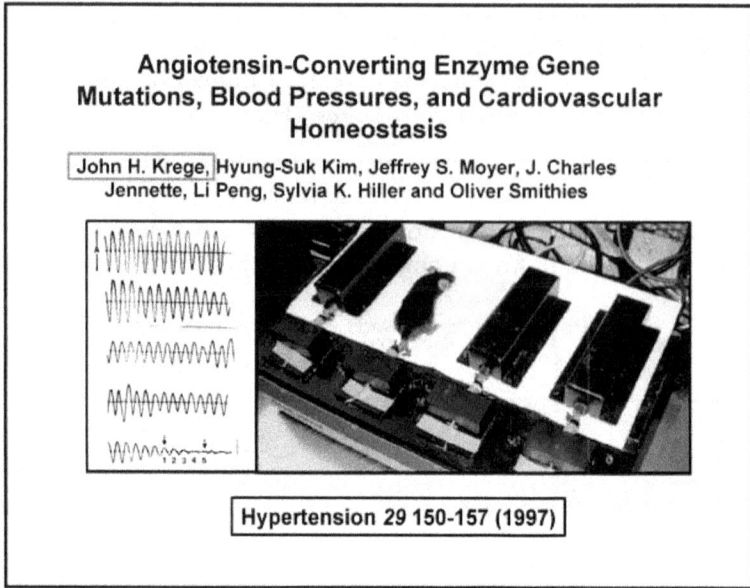

Angiotensin-Converting Enzyme Gene Mutations, Blood Pressures, and Cardiovascular Homeostasis

John H. Krege, Hyung-Suk Kim, Jeffrey S. Moyer, J. Charles Jennette, Li Peng, Sylvia K. Hiller and Oliver Smithies

Hypertension 29 150-157 (1997)

Figure 18.

Marshall Edgell helped me to use a different sort of mouse in computer simulations that explored how genetic factors influence blood pressure (Figure 19) (Smithies *et al.*, 2000).

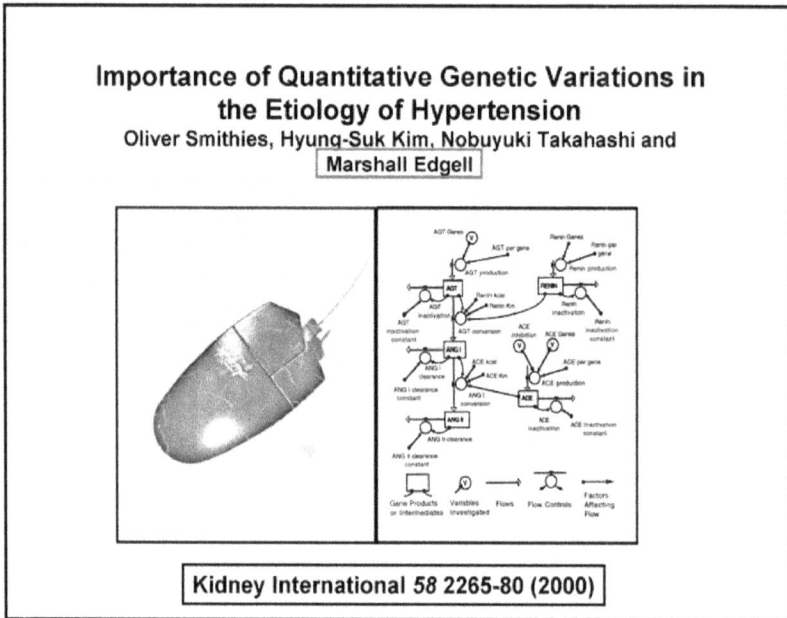

Importance of Quantitative Genetic Variations in the Etiology of Hypertension

Oliver Smithies, Hyung-Suk Kim, Nobuyuki Takahashi and Marshall Edgell

Kidney International 58 2265-80 (2000)

Figure 19.

Devising these and other simulations has helped me to uncover unexpected relationships and has stimulated ideas that I might not have had without this work. In saying this, I stress that the greatest value of these relatively simple computer simulations does not stem from their ability to replicate experimental data, or even make predictions; rather it comes from forcing one to clarify which elements in a complex system are most critical, and how these elements integrate into a logically consistent whole. [*"Students, try a simulation yourself; suitable generic programs for model building are available (for example Stella®) that you can use without being a computer expert".*]

Before closing, I want to mention a previous visit to the Karolinska Institutet on September 6th, 2002. During that visit, I heard Dr. Karl Tryggvason, who is here today, give a fascinating talk on how the kidney separates large molecules from small molecules. But I didn't quite agree with him. And so afterwards, in the corridor, I asked him "Why doesn't it clog?" His response was "That's a good question!" which is the one most of us give when we don't have an answer. Suddenly I thought that I already knew the answer, as a result of having recently written a scientific memoir of my undergraduate tutor, thesis advisor, and lifelong friend, A. G. ("Sandy") Ogston (Smithies, 1999). In one of his papers, Sandy had derived an elegantly simple equation $[f = e^{-\pi(R+r)2n}]$ that very accurately describes the behavior in gels of molecules of different sizes (Ogston, 1958). So, on my return to North Carolina, I wrote a brief communication on the topic and sent it to *Nature* (Figure 20).

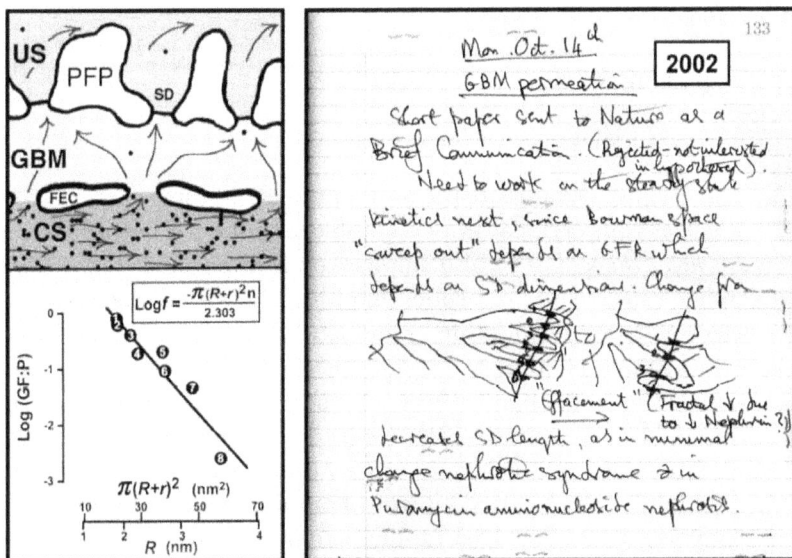

Figure 20.

It was rejected, I'm glad to say, because this caused me to write a better paper that described not only my hypothesis, but also a computer simulation of this aspect of kidney function [*"Another simulation, students!"*], and some testable predictions based on these ideas (Smithies, 2003). My personal scientific efforts are currently directed towards testing the predictions. And the last pages that I turn for you (Figure 21) illustrate the sequencing of a DNA construct made to implement this work.

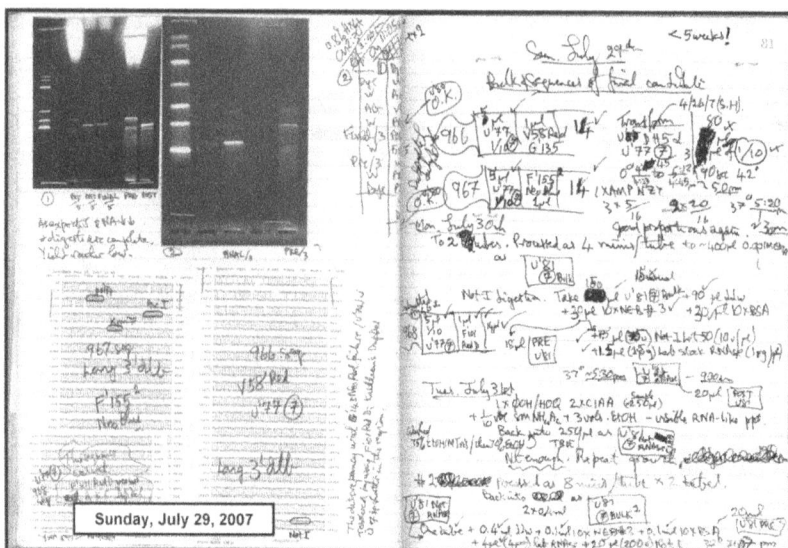

Figure 21.

[*"At 82 it is still possible to work at the weekends!"*]

What's on the *next* page?

I don't know!!

But that's what makes science exciting!!!

Finally, in closing, I emphasize the importance of choosing a branch of science that makes your everyday work enjoyable, as mine has been. [*"Students: when it was not, I changed it!"*] I also emphasize the importance for a scientist to have other interests for diversion (mine is still flying) when science is being fickle. A happy relationship (mine is with my wife Nobuyo Maeda) can also be a source of comfort at such times – and can provide a captive audience with whom to share science's much less frequent times of bliss. Scientific happiness is in sharing ideas and the daily excitement of new results with students, colleagues and other scientists. My adviser, Sandy Ogston, had it right when he summarized his view of our discipline. His words are the theme of my visit to Sweden. They capture better than I can what it means to spend a life doing science.

"For science is more than the search for truth, more than a challenging game, more than a profession. It is a life that a diversity of people lead together, in the closest proximity, a school for social living. We are members one of another."

A. G. Ogston
Australian Biochemical Society Annual Lecture
August 1970, *Search*, Vol. 1, No. 2, 60-63.

REFERENCES

Blattner, F. R., Blechl, A. E., Denniston-Thompson, K., Faber, H. E., Richards, J. E., Slightom, J. L., Tucker, P. W. and Smithies, O. (1978) Cloning human fetal gamma globin and mouse alpha-type globin DNA: preparation and screening of shotgun collections. Science. 202, 1279-1284.

Boggs, S. S., Gregg, R. G., Borenstein, N., and Smithies, O. (1986). Efficient transformation and frequent single-site, single-copy insertion of DNA can be obtained in mouse erythroleukemia cells transformed by electroporation. Experimental Hematology 14, 988-994.

Bridges, C. B. (1936). The bar "gene" a duplication. Science *83*, 210-211.

Connell, G. E., Dixon, G. H., and Smithies, O. (1962). Subdivision of the three common haptoglobin types based on 'hidden' differences. Nature *193*, 505-506.

Doetschman, T., Gregg, R. G., Maeda, N., Hooper, M. L., Melton, D. W., Thompson, S., and Smithies, O. (1987). Targeted correction of a mutant HPRT gene in mouse embryonic stem cells. Nature *330*, 576-578.

Goldfarb, M., Shimizu, K., Perucho, M. and Wigler, M. (1982) Isolation and preliminary characterization of a human transforming gene from T24 bladder carcinoma cells. Nature. *296*, 404-409.

Hinnen, A., Hicks, J. B. and Flink, G. R. (1978). Transformation of yest. Proceedings of the National Academy of Sciences of the United States of America. 75, 1929-1933.

Hooper, M., Hardy, K., Handyside, A., Hunter, S., and Monk, M. (1987). HPRT-deficient (Lesch-Nyhan) mouse embryos derived from germline colonization by cultured cells. Nature *326*, 292-295.

Kim, H. S., and Smithies, O. (1988). Recombinant fragment assay for gene targetting based on the polymerase chain reaction. Nucleic Acids Research *16*, 8887-8903.

Koller, B. H., Kim, H. S., Latour, A. M., Brigman, K., Boucher, R. C., Jr., Scambler, P., Wainwright, B., and Smithies, O. (1991). Toward an animal model of cystic fibrosis: targeted interruption of exon 10 of the cystic fibrosis transmembrane regulator gene in embryonic stem cells. Proceedings of the National Academy of Sciences of the United States of America *88*, 10730-10734.

Krege, J. H., Hodgin, J. B., Hagaman, J. R., and Smithies, O. (1995). A noninvasive computerized tail-cuff system for measuring blood pressure in mice. Hypertension *25*, 1111-1115.

Krege, J. H., Kim, H. S., Moyer, J. S., Jennette, J. C., Peng, L., Hiller, S. K., and Smithies, O. (1997). Angiotensin-converting enzyme gene mutations, blood pressures, and cardiovascular homeostasis. Hypertension *29*, 150-157.

Kucherlapati, R. S., Eves, E. M., Song, K. Y., Morse, B. S. and Smithies, O. (1984) Homologous recombination between plasmids in mammalian cells can be enhanced by treatment of input DNA. Proceedings of the National Academy of Sciences of the United States of America. *81*, 3153-3157.

Kunkel, H. G., and Slater, R. J. (1952). Zone electrophoresis in a starch supporting medium. Proceedings of the Society for Experimental Biology and Medicine Society for Experimental Biology and Medicine, New York, NY *80*, 42-44.

Mansour, S. L., Thomas, K. R., and Capecchi, M. R. (1988). Disruption of the proto-oncogene int-2 in mouse embryo-derived stem cells: a general strategy for targeting mutations to non-selectable genes. Nature *336*, 348-352.

Miller, F. (2002). The importance of being marginal: Norma Ford Walker and a Canadian school of medical genetics. American Journal of Medical Genetics *115*, 102-110.

Mullis, K., Faloona, F., Scharf, S., Saiki, R., Horn, G., and Erlich, H. (1986). Specific enzymatic amplification of DNA *in vitro:* the polymerase chain reaction. Cold Spring Harbor Symposia on Quantitative Biology *51 Pt 1*, 263-273.

Ogston, A. G. (1958). The space in a uniform random suspension of fibres Trans. Faraday Soc. *54*, 1754-1757.

Ogston, A. G. (1970). An open-ended tale. Search, *1*, (2), 60-63.

Potter, H., Weir, L., and Leder, P. (1984). Enhancer-dependent expression of human kappa immunoglobulin genes introduced into mouse pre-B lymphocytes by electroporation. Proceedings of the National Academy of Sciences of the United States of America *81*, 7161-7165.

Slightom, J. L., Blechl, A. E. and Smithies, O. (1980) Human fetal G gamma- and A gamma-globin genes: complete nucleotide sequences suggest that DNA can be exchanged between these duplicated genes. Cell. *21*, 627-638.

Smithies, O. (1955). Zone electrophoresis in starch gels: group variations in the serum proteins of normal human adults. The Biochemical Journal. *61*, 629-641.

Smithies, O. (1962). Molecular size and starch-gel electrophoresis. Archives of biochemistry and biophysics. *Suppl 1*, 125-131.

Smithies, O. (1999). Alexander George Ogston: 30 January 1911-29 June 1996. Biographical memoirs of fellows of the Royal Society. *45*, 351-364.

Smithies, O. (2003). Why the kidney glomerulus does not clog: a gel permeation/diffusion hypothesis of renal function. Proceedings of the National Academy of Sciences of the United States of America *100*, 4108-4113.

Smithies, O. (2005). Many little things: one geneticist's view of complex diseases. Nat. Rev. Genet. *6*, 419-425.

Smithies, O., and Walker, N. F. (1955). Genetic control of some serum proteins in normal humans. Nature *176*, 1265-1266.

Smithies, O., and Walker, N. F. (1956). Notation for serum-protein groups and the genes controlling their inheritance. Nature *178*, 694-695.

Smithies, O., Connell, G. E., and Dixon, G. H. (1962). Chromosomal rearrangements and the evolution of haptoglobin genes. Nature *196*, 232-236.

Smithies, O., Blechl, A. E., Denniston-Thompson, K., Newell, N., Richards, J. E., Slightom, J. L., Tucker, P. W. and Blattner, F. R. (1978). Cloning human fetal gamma globin and mouse alpha-type globin DNA: characterization and partial sequencing. Science. *202*, 1284-1289.

Smithies, O., Koralewski, M. A., Song, K. Y., and Kucherlapati, R. S. (1984). Homologous recombination with DNA introduced into mammalian cells. Cold Spring Harbor symposia on quantitative biology *49*, 161-170.

Smithies, O., Gregg, R. G., Boggs, S. S., Koralewski, M. A., and Kucherlapati, R. S. (1985). Insertion of DNA sequences into the human chromosomal beta-globin locus by homologous recombination. Nature *317*, 230-234.

Smithies, O., Kim, H. S., Takahashi, N., and Edgell, M. H. (2000). Importance of quantitative genetic variations in the etiology of hypertension. Kidney international *58*, 2265-2280.

Snouwaert, J. N., Brigman, K. K., Latour, A. M., Malouf, N. N., Boucher, R. C., Smithies, O., and Koller, B. H. (1992). An animal model for cystic fibrosis made by gene targeting. Science *257*, 1083-1088.

Sturtevant, A. H. (1925). The effects of unequal crossing over at the bar locus in *Drosophila*. Genetics *10*, 117-147.

Szostak, J. W. and Wu, R. (1979). Insertion of a genetic marker into the ribosomal DNA of yeast. Plasmid. 2, 536-554.

Thomas, K. R., and Capecchi, M. R. (1987). Site-directed mutagenesis by gene targeting in mouse embryo-derived stem cells. Cell *51*, 503-512.

Tice, S. C. (1914). A new sex-limited character in *Drosophila*. Biol. Bull. *26*, 221-230.

Zavodny, P. J., Roginski, R. S., and Skoultchi, A. I. (1983). Regulated expression of human globin genes and flanking DNA in mouse erythroleukemia–human cell hybrids. Progress in clinical and biological research *134*, 53-62.

Zeleny, C. (1919). A change in the bar gene of *Drosophila* involving further decrease in facet number and increase in dominance. Jour. Gen. Physiol. *2*, 69-71.

Zhang, S. H., Reddick, R. L., Piedrahita, J. A., and Maeda, N. (1992). Spontaneous hypercholesterolemia and arterial lesions in mice lacking apolipoprotein E. Science *258*, 468-471.

Portrait photo of Oliver Smithies by photographer Ulla Montan.

Physiology or Medicine 2008

Harald zur Hausen

"for his discovery of human papilloma viruses causing cervical cancer"

and the other half jointly to

Françoise Barré-Sinoussi and Luc Montagnier

"for their discovery of human immunodeficiency virus"

THE NOBEL PRIZE IN PHYSIOLOGY OR MEDICINE

Speech by Professor Jan Andersson of the Nobel Assembly at Karolinska Institutet.
Translation of the Swedish text.

Your Majesties, Your Royal Highnesses, Ladies and Gentlemen,

This year's Nobel Prize in Physiology or Medicine rewards two discoveries of viruses that cause severe global diseases.

Human history has been subjected to and shaped by major epidemics, such as bubonic plague, cholera, tuberculosis, smallpox, measles and influenza. These communicable diseases have contributed to the downfall of peoples and cultures. They recognise no national boundaries, and when they first appear among a group of people they are often fatal. All new epidemics have generated anxiety and panic reactions due to uncertainty about how the infectious agent spreads. Our foremost antidote is and has been – knowledge.

One part of this year's Prize goes to Harald zur Hausen, who sought to understand what causes cervical cancer, the world's second most common cancer among women. Half a million are affected every year. When zur Hausen asserted that human papilloma virus – which was known to induce skin warts – caused this cancer, not many people believed him. He predicted that the virus existed in different types and changed shape in tumour cells. He realised that no virus was formed there, but that certain virus genes instead integrated into cervical cell genomes, thereby giving rise to the cancer over time.

It was not possible to culture the virus, but zur Hausen instead had to prove his hypothesis by using small bits of single-stranded DNA from the actual wart viruses. He used these bits of DNA as bait to be able to attract virus DNA copies, first from genital warts and later also from cervical cancer cells. After more than a decade of persistent work, he isolated different types of human papilloma virus that caused more than 70 per cent of all cervical cancer. Virus genes were present in the DNA of every cancer cell and re-programmed the cells to grow uncontrollably.

This discovery has led to vaccines against these viruses which provide protection against these infections. It has also lead to improved methods for predicting which women are in the risk zone.

The other part of this year's Prize is related to the discovery of the human immunodeficiency virus (HIV). In 1981 there were a number of alarming reports that young, previously healthy people were dying of pneumonia or unusual tumours. This was the beginning of an epidemic whose enigmatic nature and impact on different cultures defied all previously described epidemics. It started during an era when we thought we had conquered all major infectious diseases. There were now antibiotics, vaccines, good hygiene and improved living conditions that would keep us immune. Institutes for infectious disease control would be shut down. But with everyone talking about AIDS, new concerns spread in many countries.

Françoise Barré-Sinoussi and Luc Montagnier speculated that AIDS was caused by an unknown retrovirus, a kind of tumour virus from the animal kingdom that had more recently also been found in humans. But how could a virus with a size of one ten thousandth of a millimetre trigger all these different manifestations of disease?

Barré-Sinoussi and Montagnier discovered HIV in the lymph nodes of recently infected patients. It was being produced in large quantities in white blood cells and it induced the death of cells of a kind that was missing in patients suffering from AIDS. HIV was like a chameleon. Each patient carried a unique virus variant, which could hide in the host cell genome. Barré-Sinoussi and Montagnier's discovery explained the progression of the disease and its communicability, enabling them to establish the connection between HIV and AIDS. Methods for discovering infected people and blood products were developed, and HIV/AIDS patients quickly gained access to effective anti-retroviral drugs that helped suppress their disease. So far 60 million people in the world have been infected with HIV, and a total of 25 million have died of AIDS. A growing number of people are receiving anti-retroviral drugs, even in developing countries. Another cause for celebration is that the epidemic is now declining among young people in the countries where it began. However, the best prophylactic against HIV/AIDS is and remains true love, generating longstanding relationships.

Sehr geehrter Herr Professor zur Hausen,

Ihre Entdeckung hat zu der Erkenntnis geführt, daß Viren bei Menschen Krebs erregen können, und daß wir neue Möglichkeiten zur Vorbeugung von Krebs gefunden haben, der durch humane Papillomaviren hervorgerufen wird.

Professeurs Barré-Sinoussi et Montagnier,

Votre découverte nous a révélé l'existence d'un nouveau virus, le VIH, et cette connaissance a ouvert la voie à de nouveaux traitements contre la maladie du VIH/SIDA.

On behalf of the Nobel Assembly at Karolinska Institutet it is my privilege and pleasure to express our warmest congratulations and our deepest admiration as I now ask you to step forward to receive the Nobel Prize from the hands of His Majesty the King.

HARALD ZUR HAUSEN

Born in 1936, I experienced the Second World War as a child in the city of Gelsenkirchen-Buer. This area was heavily bombed, but fortunately all members of my family survived the war and post-war period. As a child I remember my own intensive interest in biology, birds, other animals and flowers and was determined at an early age to become a scientist. Since schools were closed due to the bombing raids in 1943, my elementary school training was full of gaps. When I entered "Gymnasium" at the age of 10 in 1946, during the first year these gaps were evident and created some difficulties for me. After the first year there, however, although not being the top pupil, I went to school without any major problems. In 1950 my parents moved to Northern Germany where I finished high school in 1955 with the "Abitur".

After briefly considering whether to study biology or medicine, I opted for medicine and initiated my studies at the University of Bonn. The first two years were particularly hard, since I simultaneously decided to attend lectures and courses in biology as well. The first examination after 5 semesters ("Physikum") was passed without any problems with remarkably good grades. This created some self-confidence for the forthcoming semesters, which I spent at the University of Hamburg for one year and the (at that time) Medical Academy in Düsseldorf. At the end of 1960 I graduated there in medicine and also finished my MD thesis.

Although I remained firmly determined to continue in science, I wanted to receive a licence to practice medicine. This required at that time two years of medical internship. It brought me for short periods of time into surgery, internal medicine and for the remaining time into gynaecology and obstetrics. The last part fascinated me tremendously, although it turned out to be physically highly demanding. When I left the hospital and started to work in Medical Microbiology and Immunology at the University of Düsseldorf, for the first and only time I had some doubts whether this was the correct decision. For a short while I considered returning to the life of a practising physician; after a couple of months, however, I became more fascinated by early experimental studies. Initially I started to work on virus-induced chromosomal modifications and at the same time received relatively solid training in diagnostic bacteriology and virology, both of them at that time in an early stage of development.

During my 3½ years in Düsseldorf, I became increasingly aware of the limitations in my scientific education and decided to search for a post-doctoral position elsewhere, preferably in the United States. I received an

interesting offer from Werner and Gertrude Henle at the Children's Hospital
of Philadelphia, where Werner headed the Division of Virology. In 1964 I
got married and our first son Jan Dirk arrived one year later. Within the
same year we decided to accept the offer from Philadelphia; in the end of
December 1965 I arrived there and started work at the beginning of 1966.

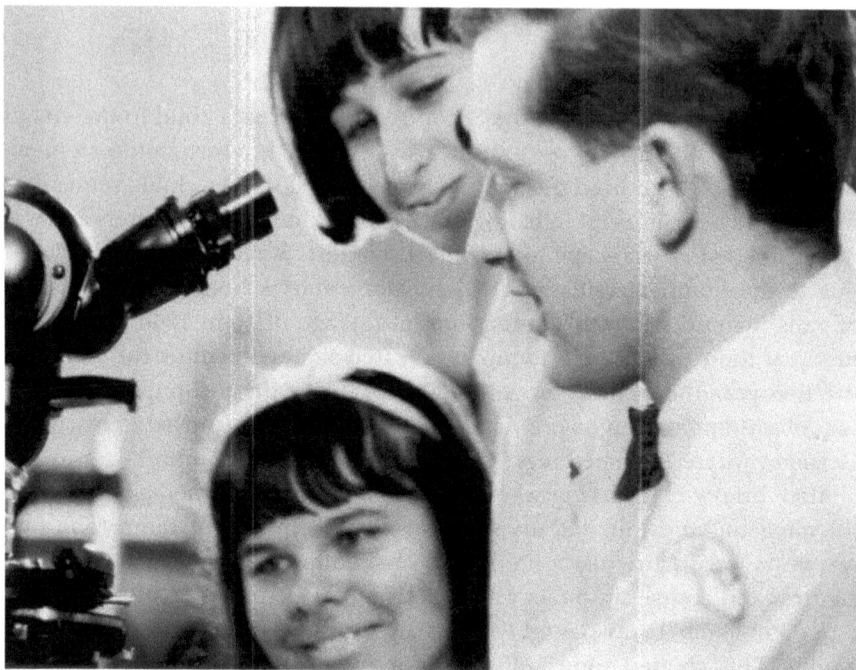

Figure 1. Harald zur Hausen in 1967 in the laboratory of the Children's Hospital of
Philadelphia with two of the technicians.

The Henle's laboratory was deeply interested in the newly discovered
Epstein-Barr virus (EBV), and the whole team was actively engaged in developing serological tests for this virus and in studying its epidemiology. They
had noted early that Burkitt's lymphoma patients developed high antibody
titres against viral antigens. I felt very much compelled to work with this
agent, but noted at the same time my lack of familiarity with the rapidly
developing molecular biological methods. I urged Werner Henle to permit
me to work with a different agent, namely adenovirus type 12, hoping that
this relatively well established system would permit me to become acquainted
with molecular methods. He reluctantly agreed. I started to work eagerly on
the induction of specific chromosomal aberrations in adenovirus type 12-in-
fected human cells, simultaneously studying a DNA-replication disturbance
of individual chromosomes in human lymphoblastoid and lymphoma cell
lines, and, to please my mentor, I demonstrated electron microscopically the
presence of EBV particles directly in individual serologically antigen-positive
Burkitt's lymphoma cells. During my years in Philadelphia the immortalising
function of EBV was demonstrated for human B-lymphocytes, and the role

of this virus as a causative agent of infectious mononucleosis was conclusively established.

In 1968 I received an attractive offer from Eberhard Wecker, who headed the newly opened Institute for Virology at the University of Würzburg, Germany. He offered me the establishment of my own independent group and granted me his support for a quick start in the German academic system. I accepted this offer and moved with my family in March 1969 back to Germany. Here I decided to change my topics completely to EBV research. The intention was to prove that EBV DNA persists in every tumour cell of Burkitt's lymphoma and does not establish a persistent infection there, as assumed at that time by a number of my former colleagues. With the aid of Werner Henle in Philadelphia and George Klein in Stockholm I received a large number of Burkitt's lymphoma cell lines and tumour biopsies. The biopsies also included material from nasopharyngeal carcinomas, where serological assays also suggested an involvement of EBV infections.

The major problem, the purification of sufficient quantities of EBV DNA from a low number of spontaneously virus-producing cells, was quickly solved. By the end of 1969 I had the first data available that the non-EBV-producing Burkitt's lymphoma cell line Raji contained multiple copies per cell of EBV DNA. Shortly thereafter it was also possible to demonstrate EBV DNA in Burkitt's lymphoma and nasopharyngeal cancer biopsies. It seems that this was the first demonstration of persistent tumour virus DNA in human malignancies.

In nasopharyngeal carcinomas, composed of a mixture of epithelial tumour cells and lymphocytic infiltrates, it was intensively discussed whether the EBV DNA might rest in the lymphocytic infiltrates. By using in-situ hybridisations, in 1973 we were able to document the presence of EBV DNA in the epithelial tumour cells.

In 1972 I was appointed chairman of the newly established Institute of Clinical Virology in Erlangen-Nürnberg. With the move to this city I planned to change my scientific direction. Cervical cancer had long been suspected of being caused by an infectious agent. In the late 1960s Herpes simplex type 2 (HSV-2) emerged as the prime suspect based on some seroepidemiological observations. Since our previous EBV work led to the identification of EBV DNA in specific human cancers, I had asked my colleague Heinrich Schulte-Holthausen to use the same technique to search for HSV-2 sequences in cervical cancer biopsies. All attempts, however, failed.

During the previous years I had studied a large number of anecdotal reports describing malignant conversion of genital warts into squamous cell carcinomas. Since genital warts had been shown to contain typical papillomavirus particles, this triggered the suspicion that the genital wart virus might represent the causative agent for cervical cancer. Based on this hypothesis we initiated our papillomavirus programme in Erlangen. With the aid of the local Dermatology Hospital we received a large number of wart biopsies. Viral particles could be extracted from plantar warts and in 1974 we published our first report, demonstrating a cross-hybridisation of the plantar wart virus DNA with some warts, but by far not with all of them. Genital warts and cervical cancer biopsies were negative. This was our first hint that there exist different types of papillomaviruses. In the following years our group, as well as the group around Gérard Orth in Paris, were able to identify the plurality of the human papillomavirus family by isolating a steadily increasing number of novel types.

In 1977 I was appointed as chairman of the Institute of Virology of the University of Freiburg, Germany. Most members of my group in Erlangen joined me in moving to Freiburg. Here we continued intensively our studies on human papillomaviruses.

Late in 1979 my co-workers Lutz Gissmann and Ethel-Michele de Villiers successfully isolated and cloned the first DNA from genital warts, HPV-6. It was initially disappointing not to detect this DNA in cervical cancer biopsies. HPV-6 DNA, however, turned out to be helpful in isolating another closely related genital wart papillomavirus, HPV-11, initially from a laryngeal papilloma. By using HPV-11 as a probe, one out of 24 cervical cancer biopsies turned out to be positive. In addition, in other biopsies some faint bands became visible, permitting the speculation that they might represent hints of the presence of related, but different HPV types in these cancers. Two of my former students; Mathias Dürst and Michael Boshart, were asked to clone these bands. Both of them were successful. In 1983 we were able to document the isolation of HPV-16, in 1984 the isolation of HPV-18 DNA. We noted from the beginning that HPV-16 DNA was present in about 50% of cervical cancer biopsies, HPV-18 in our early experiments in slightly more than 20%, including several cervical cancer cell lines, among them the HeLa line.

Within the first two years after isolating HPVs 16 and 18 it became clear that these viruses must play an important role in cervical cancer development: viral DNA was commonly found in an integrated state, indicating the clonality of the tumour. In addition, part of the viral genome frequently became deleted in the process of integration. Two viral genes, E6 and E7, were consistently transcribed in the cancer cells. Precursor lesions of cervical cancer also contained these viruses and expressed the respective genes. Early contacts with pharmaceutical companies for the development of HPV vaccines failed, in view of a market analysis conducted by one of them which indicated that there would be no market available. Fortunately, this changed in later years.

My period in Freiburg permitted me to also work on other aspects of tumour virology: I discovered the potent activity of some phorbol esters in inducing latent Epstein-Barr virus DNA. This procedure also proved to be successful for other persistent Herpes-type viruses. In addition, I isolated a novel lymphotropic polyomavirus from African Green Monkey lymphoblasts. Up to 20% of sera from human adults also revealed neutralising antibodies to this virus. Our attempts to isolate a human correlate, however, failed. I also identified a novel adeno-associated virus, now labelled AAV-5, from my own skin scrapings. In collaboration with my colleague Jörg Schlehofer, we were also able to demonstrate that herpes simplex virus, but also other herpes-, adeno-, and vaccinia virus infections of polyoma- or papillomavirus DNA harbouring cells, resulted in amplification of the DNA of the latter.

The early hypothesis that cervical cancer was caused by papillomaviruses, the successful isolation and characterisation of the two most frequent HPV types in this cancer and the subsequent steps leading to a better understanding of the mechanism of HPV-mediated carcinogenesis and eventually to the development of a preventive vaccine were cited as the prime reasons for awarding one half of the Nobel Prize for Medicine or Physiology to me in 2008.

In 1983 I was appointed as the Scientific Director of the German Cancer Research Centre (Deutsches Krebsforschungszentrum) in Heidelberg, a national research centre. Besides the major task of reorganising this research centre, I tried to maintain some time for laboratory research and continued jointly with Frank Rösl to analyse intracellular and extracellular control mechanisms preventing the activity of viral oncogenes in proliferating epithelial cells.

In 2003, after 20 years, I retired from the scientific directorship of the German Cancer Research Centre. Subsequently, I kept a laboratory in the virus building of the Cancer Centre and continue up to now to act as Editor-in-Chief of the International Journal of Cancer. I started this commitment at the beginning of 2000.

Figure 2. Harald zur Hausen in 2008.

In retrospect, I have devoted my scientific life mainly to the question to what extent infectious agents contribute to human cancer, trusting that this will contribute to novel modes of cancer prevention, diagnosis and hopefully later on also to cancer therapy. I am of course pleased to see that at least part of this programme has been successful. I am grateful to a large number of my former co-workers, who skilfully contributed to the programme. In addition, I most gratefully acknowledge the contributions of my wife, Ethel-Michele de Villiers, who is also a scientist and tumour virologist, for her never-ending support.

THE SEARCH FOR INFECTIOUS CAUSES OF HUMAN CANCERS: WHERE AND WHY

Nobel Lecture, December 7, 2008

by

HARALD ZUR HAUSEN

Deutsches Krebsforschungszentrum, Im Neuenheimer Feld 280, D-69120 Heidelberg, Germany.

Slightly more than 20% of the global cancer burden can currently be linked to infectious agents, including viruses, bacteria and parasites. This manuscript analyses reasons for their relatively late discovery and highlights epidemiological observations that may point to the involvement of additional infectious agents in specific human cancers. Emphasis is placed on haematopoietic malignancies, breast and colorectal cancers, but also basal cell carcinomas of the skin and lung cancers in non-smokers.

INTRODUCTION

Present state of the global cancer burden
Currently a large number of infectious agents have been identified which either cause or contribute to specific human cancers (reviewed in zur Hausen, 2006). They include two members of the herpes virus family, Epstein-Barr virus and human herpes virus type 8, high risk and low risk human papillomaviruses (HPV), Hepatitis B and C viruses, a recently identified human polyomavirus, Merkel cell polyomavirus (Feng *et al.*, 2008), the human T-lymphotropic retrovirus type 1 (HTLV-1), and human immunodeficiency viruses (HIV) types 1 and 2. In addition, human endogenous retroviruses have been suspected to play a role in human cancers. Besides viruses, other pathogens have also been identified. They include the bacterium *Helicobacter pylori*, a major contributor to gastric cancer, and parasitic infections, here in particular *Schistosoma haematobium*, a major cause of bladder cancer in Egypt, and liver flukes. The latter, *Opisthorchis viverinni* and *Clonorchis sinensis*, are important factors for cholangiocarcinomas and hepatocellular carcinomas in South-eastern Thailand and Southern China. Figure 1 shows an estimate of the present contribution of infectious agents to global cancer incidence.

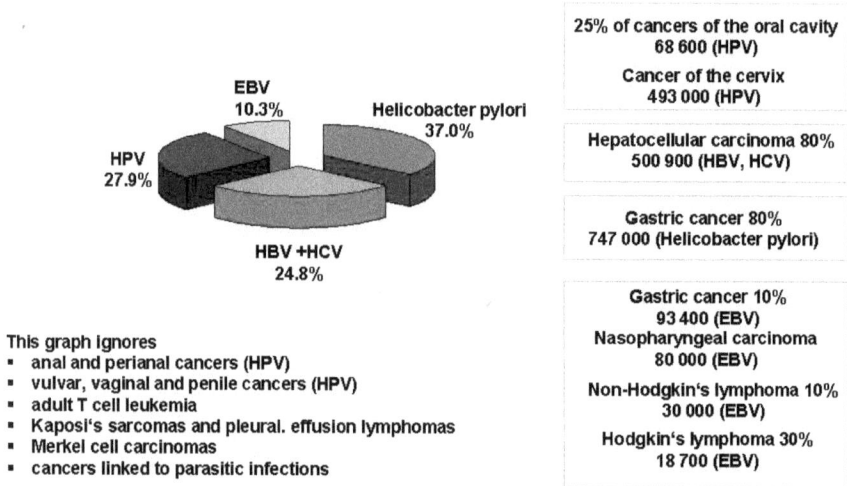

Cancers due to 5 infections correspond to 18.6% of total cancer incidence

EBV
10.3%

Helicobacter pylori
37.0%

HPV
27.9%

HBV +HCV
24.8%

25% of cancers of the oral cavity
68 600 (HPV)

Cancer of the cervix
493 000 (HPV)

Hepatocellular carcinoma 80%
500 900 (HBV, HCV)

Gastric cancer 80%
747 000 (Helicobacter pylori)

Gastric cancer 10%
93 400 (EBV)
Nasopharyngeal carcinoma
80 000 (EBV)

Non-Hodgkin's lymphoma 10%
30 000 (EBV)

Hodgkin's lymphoma 30%
18 700 (EBV)

This graph ignores
- anal and perianal cancers (HPV)
- vulvar, vaginal and penile cancers (HPV)
- adult T cell leukemia
- Kaposi's sarcomas and pleural. effusion lymphomas
- Merkel cell carcinomas
- cancers linked to parasitic infections

Figure 1. Estimated annual global cancer incidence due to infections (with inclusion of data from Parkin *et al.*, 2002; modified from zur Hausen, 2006).

It is important to note that there exist vast gender differences in the global role of papillomaviruses in human cancers. This is mainly due to the role of this virus family in the induction of cancer of the cervix. More than 50% of cancers linked to infections in females are caused by HPV infections. In males only approximately 4.3% of cancers have been linked to this virus family.

PROBLEMS IN IDENTIFYING INFECTIOUS AGENTS INVOLVED IN HUMAN CANCER INDUCTION

1. Why has it been so difficult to identify infectious agents as causative factors for human tumours?

The search for an infectious cause of at least some human cancers dates back to the second half of the nineteenth century (reviewed in zur Hausen, 2006). Yet the first hints for a role of infectious agents in human cancers date back to the beginning of the 20th century, when Schistosoma infections in Egypt and liver flukes in Eastern Europe and Asia were suspected to play a role in the development of bladder and liver cancers. In spite of intensive search, it took approximately 65 additional years before further evidence was obtained, namely by linking a specific virus, Epstein-Barr virus, to two human cancers, Burkitt's lymphoma and nasopharyngeal carcinoma. During the past three or four decades progress has been more rapid, currently linking about 20% of the global cancer incidence to infectious events.

Why has it been so difficult to identify infectious agents as causative factors for human cancers? Several reasons seem to provide an explanation:

1. Because no human cancer arises as the acute consequence of infection. The latency periods between primary infection and cancer development are frequently in the range of 15 to 40 years. The X-chromosome-linked lymphoproliferation (XLLP) represents a rare exception. Based on a specific host cell mutation, Epstein-Barr virus here causes an acute lymphoproliferative disease.

2. Besides some rare exceptions, no synthesis of the infectious agents occurs in cancer cells.

3. Most of the infections linked to human cancers are common in human populations; they are ubiquitous. They were present during the whole human evolution process. Yet only a small proportion of infected individuals develops the respective cancer type.

4. Mutations in host cell genes or within the viral genome are mandatory for malignant conversion.

5. Chemical (e.g. aflatoxin) and physical carcinogens (e.g. ultraviolet light in Epidermodysplasia verruciformis) usually act as mutagens. They facilitate the selection of specific mutations and frequently act synergistically with carcinogenic infectious agents.

6. Some infectious agents act as indirect carcinogens, without persistence of their genes within the respective cancer cells (HIV, *Helicobacter pylori*, *Schistosoma haematobium*, Hepatitis C and B).

Among all these factors, the ubiquity of most of these infections and the long time periods required for malignant transformation were the main reasons for the remarkable difficulties in identifying their carcinogenic functions.

2. Epidemiology provided hints for a successful search
a. Geographic coincidence
Geographic coincidence of a specific infection (Hepatitis B) and of liver cancer led to the original suspicion that this infection may predispose the patient to the subsequent development of hepatocellular carcinomas (reviewed in zur Hausen, 2006). The additional contribution of a chemical carcinogen was also suspected, based on similar observations. Figure 2 reveals the geographic distribution of Hepatitis B virus infections and hepatocellular carcinomas. Geographic clustering of specific cancers may, however, also result from other causes: Countries with a high rate of heavy smokers also experience a high incidence of lung cancer. The intensive solar exposure of Caucasian populations in Australia, South Africa and South America is responsible for a high percentage of skin cancer patients.

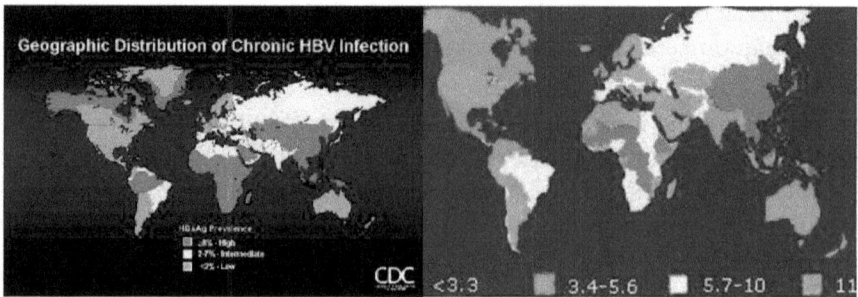

Figure 2. Geographic distribution of Hepatitis B infections (left) and of hepatocellular carcinomas (right). Modified from figures provided by CDC and Globocan 2002.

b. Regional clustering of cases

Regional clustering of specific cancer types triggered some investigations on a potential role of infectious agents in these malignant proliferations. Burkitt's lymphoma in equatorial Africa represents one of the most illustrative examples. Burkitt noted the apparent dependence of tumour incidence on climatic conditions and altitude and described the regional correlation with holoendemic *Plasmodium falciparum* infections (Burkitt, 1962). As a consequence he speculated that the tumour might be due to a viral infection, transmitted by an arthropod vector, possibly the same one carrying malaria parasites.

Nasopharyngeal carcinoma, occurring at high frequency in specific regions of South East Asia, represents another example. Adult T-cell leukaemia in the coastal regions of Southern Japan, cholangiocarcinomas in South East Thailand and bladder cancer in the Nile Delta or along the Nile River also raised early suspicions of an infectious origin. These observations resulted in speculations; they could not prove the underlying hypothesis by themselves.

c. Dependence on sexual contacts

If one disregards the occurrence of scrotum cancer in chimney sweepers, the early studies of Rigoni-Stern in Verona, Italy, pointing in 1842 to the role of sexual contacts in the causation of cervical cancer, represent a particularly interesting example of suspected contact transmission of a human cancer. It took another 140 years before viral infections were identified as causing this frequent cancer in women. These observations led to the identification of additional anogenital and oral cancers linked to the same virus infections.

d. Cancers arising under immunosuppression

Epidemiological surveys identified immunosuppression as a condition resulting in the appearance of remarkably specific forms of cancer. Many of those malignancies have by now been shown to be caused by reactivated viruses, whose oncogenic potential is usually suppressed by immunological reactions. The most prominent tumours arising here are Epstein-Barr virus caused B-cell lymphomas, Kaposi's sarcomas linked to human herpesvirus type 8 reactivation, and Merkel cell carcinomas of the skin associated with a novel polyomavirus. The initial discovery of the viral origin of cervical cancer and its precursor lesions was not based on the moderately enhanced incidence under immunosuppression. Specific types of common warts as a non-malignant proliferative condition also occur at high frequency in immunosuppressed patients, mainly containing of genus-Beta papillomaviruses. The viral origin of basal and squamous cell carcinomas of the skin, frequently found in these patients, remains controversial.

MECHANISTIC ASPECTS OF CANCER INDUCTION BY INFECTIONS

Figure 3 lists identified mechanisms by which infections may contribute to cancer development. The expression of specific viral oncogenes as a mandatory precondition for the maintenance of the malignant phenotype has been identified as directly contributing to human carcinogenesis (reviewed in zur Hausen, 2006). A novel mode of direct viral carcinogenesis has probably been identified in Merkel cell carcinomas, where functional inactivation of the helicase part of the large T-antigen of the Merkel cell polyomavirus renders the viral DNA replication-incompetent (Shuda *et al.*, 2008). Viral DNA persisting in normal tissues seems to retain replication competence.

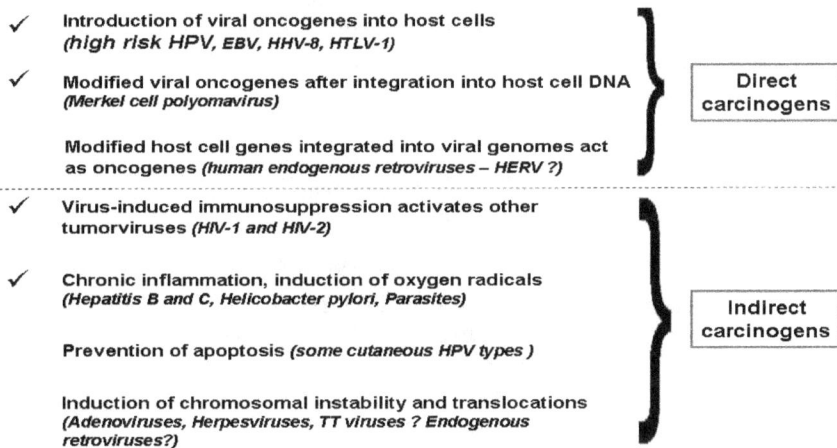

✓ Introduction of viral oncogenes into host cells
   (high risk HPV, EBV, HHV-8, HTLV-1)

✓ Modified viral oncogenes after integration into host cell DNA   **Direct carcinogens**
   (Merkel cell polyomavirus)

   Modified host cell genes integrated into viral genomes act
   as oncogenes *(human endogenous retroviruses – HERV ?)*

✓ Virus-induced immunosuppression activates other
   tumorviruses *(HIV-1 and HIV-2)*

✓ Chronic inflammation, induction of oxygen radicals   **Indirect carcinogens**
   (Hepatitis B and C, Helicobacter pylori, Parasites)

   Prevention of apoptosis *(some cutaneous HPV types)*

   Induction of chromosomal instability and translocations
   *(Adenoviruses, Herpesviruses, TT viruses ? Endogenous
   retroviruses?)*

Figure 3. Summary of identified mechanisms by which infections either directly or indirectly contribute to carcinogenesis. Mechanistic contributions of infections to human cancers have been marked.

The most prominent indirect infectious carcinogens are agents causing immunosuppression or inducing reactive oxygen species via inflammatory reactions. Whereas the mechanism of immunosuppression induced by human immunodeficiency viruses (HIV) or after organ transplantation is reasonably well understood, the accurate mechanism by which Hepatitis B and C viruses, *Helicobacter pylori* and carcinogenic parasitic infections contribute to cancer still remains somewhat obscure.

WHERE IS IT WORTHWHILE TO SEARCH FOR AN INFECTIOUS ETIOLOGY OF HUMAN CANCERS NOT YET LINKED TO INFECTIONS?

When we summarise infectious agents that have been discovered during the past 15 years, it is interesting to note that several novel viruses belonging to potentially carcinogenic virus families have been identified even during the past two years (Figure 4). This raises the suspicion that additional links to novel or already identified infectious agents to cancers, hitherto not linked to infections, will become apparent. Thus it appears worthwhile to search for cancer-related epidemiologic observations that may point to the involvement of infectious agents in cancers hitherto not linked to infections. The following will summarise some hypotheses and considerations based on these reports.

Year	Virus	Symptoms	Natural Host
1994	Sabia virus	Hemorrhagic fever	Rodents
1994	Hum. Herpesvirus 8	Kaposi's sarcoma	Humans
1994	Hendravirus	Encephalitis	Bats, horses
1997	Influenza H5N1	Avian flue	Birds
1997	TT viruses	?	Humans
1998	Nipah virus	Encephalitis	Bats, pigs
2003	SARS Coronavirus	SARS	Chinese bushcat
2005	Bocavirus (parvovirus)	Acute wheezing	Humans
2005	New coronavirus	Respiratory symptoms	Humans
2007	KI-polyomavirus	?	Humans
2007	WU-polyomavirus	?	Humans
2008	MC-polyomavirus	Merkel-tumor	Humans
2008	Lymphotrop. polyomavirus	Periph. blood PML patients	Humans

Within the same time period at least 30 novel types of human papillomaviruses have been identified

Figure 4. "New" human pathogenic viruses, 1994–2008. The light arrows identify important human pathogens or a whole novel virus family (TT viruses). The dark arrows point to established or potentially oncogenic virus isolates.

1. Cancers occurring under immunosuppression

A review published in 2006 by Vajdic and colleagues demonstrates a large number of cancers occurring at increased frequency under immunosuppression after kidney transplantation. Kaposi's sarcoma, mainly found in HIV-infected patients, stands out and is found about 200 times more frequently in these patients compared to non-infected controls (Figure 5). The most interesting part of Figure 5 appears to be the 7–8 fold higher rate of vulvar and penile cancer in comparison to cancer of the cervix. The vast majority of cervical cancers are caused by high risk human papillomavirus (HPV) infections. In vulvar and penile cancers only 30–50% seems to be linked to the same HPV infections. The etiology of 50–70% of these cancers is unknown. Interestingly, the age distribution of HPV-positive and HPV-negative vulvar and penile cancers differs in that the negative tumours regularly occur in older age groups. Thus, the negative group should require attention as possible candidates for an unknown viral etiology. Unidentified types of HPVs or novel polyomaviruses may represent interesting candidates. Salivary gland, eye, thyroid and tongue cancers should also deserve attention.

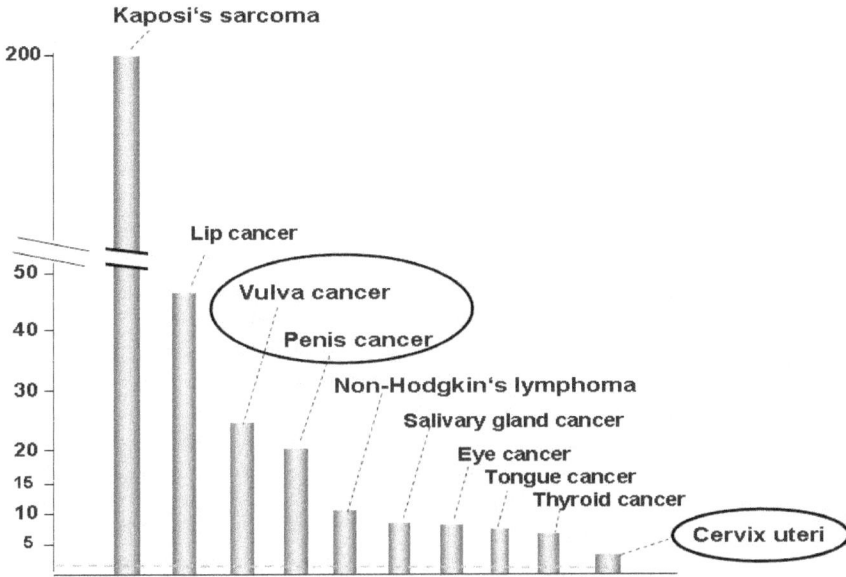

Figure 5. Some of the most frequently occurring cancers occurring after kidney transplantation (modified from Vajdic *et al.*, 2006). The dotted line indicates the incidence in immunocompetent patients.

2. Cancers not elevated or even reduced after immunosuppression

2a. Breast cancer as an example

Some cancers do not show an increased incidence during immunosuppression. Indeed, immunosuppression may even possess a protective effect against some of these tumours. Those cancers are shown in Figure 6. Besides prostate, rectal and brain tumours, human breast cancer represents a particularly intriguing malignancy, because murine mammary cancer is also not increased under immunosuppression. This latter tumour is caused by a retrovirus infection, murine mammary tumour virus (MMTV).

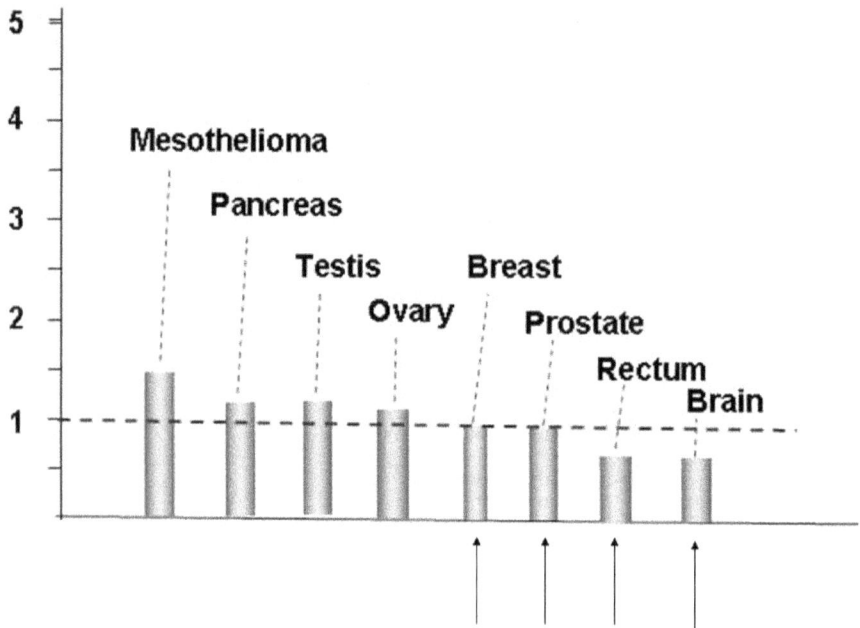

Figure 6. Cancer incidence marginally or not affected during immunosuppression after kidney transplantation (modified from Vajdic *et al.*, 2006).

In murine mammary tumours, the mechanism of a slightly protective effect exerted by immunosuppression is partially understood (see review zur Hausen, 2006). It is schematically outlined in Figure 7. The primary infection occurs via the milk of the infected mother. The virus reaches the Peyer's patches where it infects B- and T-lymphocytes. Superantigen induction in the infected cells leads to reactive T-cell depletion and immunotolerance. The superantigen expressing cells produce high quantities of infectious MMTV; this substantially increases the risk for the infection of mammary tissue. Specific integration of the MMTV proviral DNA in the mammary cells emerges as the prime risk factor for the resulting mammary carcinomas. Immunosuppression of such infected animals apparently interferes with the emergence of superantigen-producing T- and B-lymphocytes and, as a consequence, suppresses virus production which in turn decreases the risk for cancer development.

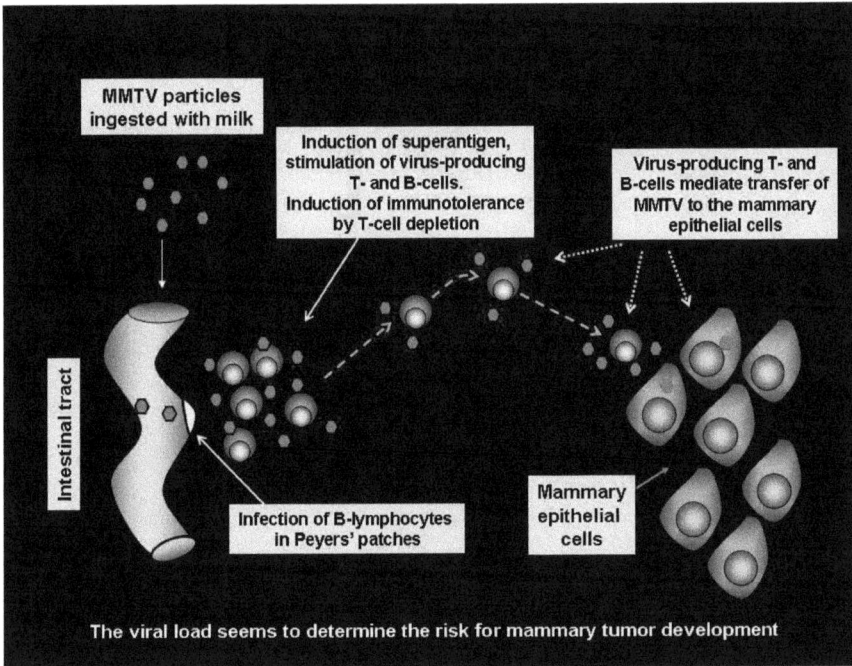

Figure 7. Schematic outline of events following infection of newborn mice with murine mammary tumour virus (modified from zur Hausen, 2006).

Is it possible that a similar mechanism contributes to human mammary cancer? A few data seem to support this notion. They may point to a possible involvement of a specific subgroup of human endogenous retroviruses (HERV) in this malignancy. At least 8% of our genome consists of retroviral sequences acquired in the course of human evolution. Although the vast majority of these sequences no longer reveal functional open reading frames, members of one subgroup HERV-K, which entered our germ line approximately 800,000 years ago, are still able to code for complete, though non-infectious virus particles. Retroviral gag and env transcripts of the 22q11.21 region are found in these particles (Ruprecht *et al.*, 2008). Correction of stop codons in HERV-K sequences resulted, moreover, in the re-constitution of infectious HERV-K viruses (Dewannieux *et al.*, 2006, Lee and Bieniasz, 2007). HERV-K expression also becomes activated by other virus infections: HIV infections activate HERV-K sequences (Laderoute *et al.*, 2007). Similarly Epstein-Barr virus infections result in the induction of HERV-K superantigen (Sutkowski *et al.*, 2004, Meylan *et al.*, 2005, Hsiao *et al.*, 2006). Epstein-Barr virus containing Burkitt's lymphoma cells occasionally reveal particles strongly resembling retroviral type A structures upon induction by the tumour-promoting phorbol-ester TPA (zur Hausen, unpublished). Typical structures are shown in Fig. 8.

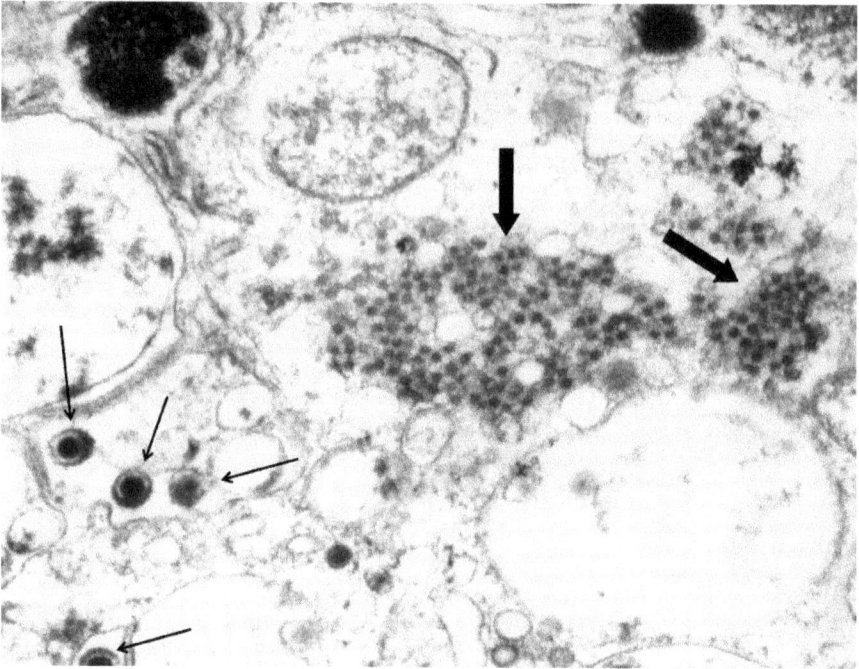

Figure 8. Epstein-Barr virus particles (small arrows) and two clusters of A-type particle-like structures (big arrows) in a TPA-treated Burkitt's lymphoma cell.

Some recent reports may further stress a potential role of reactivated HERV-K viruses in the pathogenesis of human breast cancer: an antigen-specific immune response was demonstrated in breast cancer patients (Wang-Johanning *et al.*, 2008). In addition, breast cancer patients, HIV-associated lymphomas, non-HIV-associated lymphomas and HIV-associated Hodgkin's lymphomas reveal about seven-fold elevated concentrations of HERV-K (HML-2) RNA in their plasma when compared to healthy controls. (Contreras-Galindo *et al.*, 2008). The RNA titres in lymphoma patients in remission returned to control values.

Although the available data seem to support a potential role of endogenous retroviruses in human breast cancer, they certainly do not prove it. Other agents may also contribute to at least a proportion of these cancers. A possible link between red meat consumption and breast cancer and a potential involvement of other viral factors will be discussed in connection with a subsequent paragraph. Nevertheless, human breast cancer remains an interesting candidate for a viral etiology.

3. Cancer incidence influenced by infections
The risk of some cancers seems to be influenced by other infections which neither directly contribute to carcinogenesis nor induce long-lasting immunosuppression.

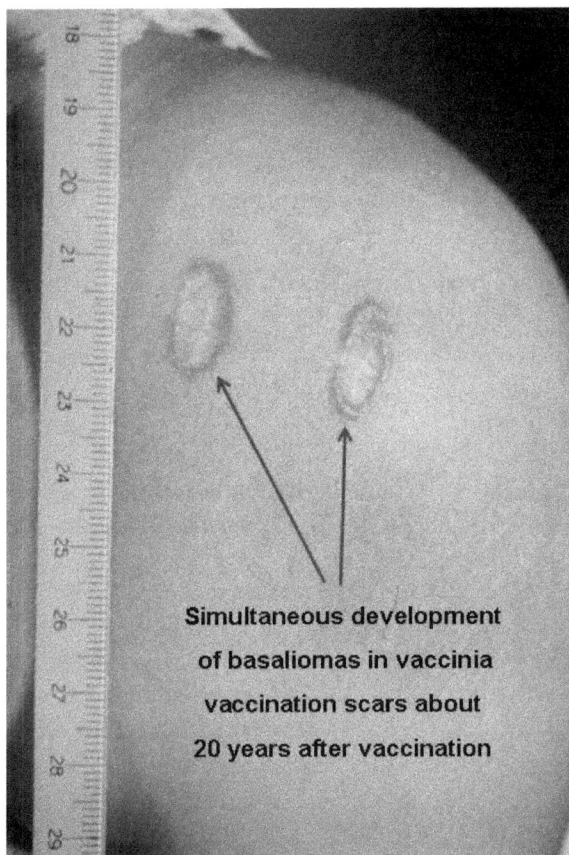

Figure 9. Multifocal basal cell carcinomas in small pox vaccination scars.

3a. Basal cell carcinomas in pox scars

Figure 9 reveals an example of multiple basal cell carcinomas arising after 20 years in a smallpox vaccination scar. This does not represent a solitary observation, since a large number of basal cell carcinomas, but also melanomas, squamous cell carcinomas, and a few more rarely occurring malignancies (dermatofibrosarcoma protuberans, fibrosarcoma and malignant fibrous histiocytomas) have also been reported to occur in smallpox vaccination scars (Waibel and Walsh, 2006) They are summarised in Figure 10.

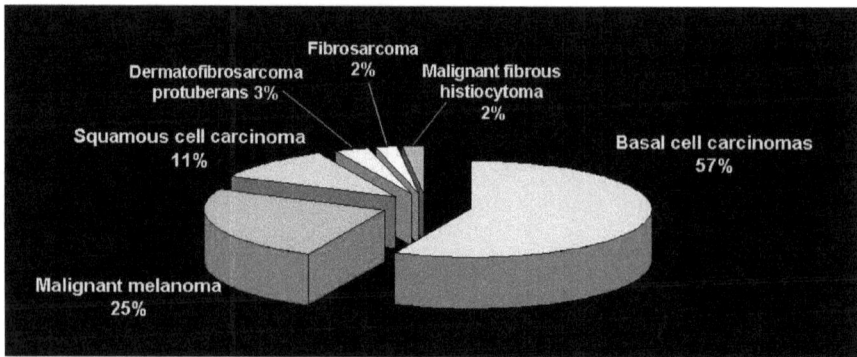

Figure 10. Malignant tumours arising in vaccinia virus vaccination scars (modified from Waibel and Walsh, 2006).

Prior to the eradication of smallpox infections, vaccines against these infections were prepared by inoculating vaccinia virus into the scarified skin of calves and harvesting the skin crusts containing the vaccinia virus particles. It is possible that these preparations contained contaminating bovine viruses. Previously it has been demonstrated that vaccinia virus infections cause amplification of persisting polyoma type virus genomes (Schlehofer *et al.,* 1986). This may increase the likelihood for contaminations with bovine members of this virus family. Persistent papillomavirus DNA would be also affected in cells replicating vaccinia virus (Schmitt *et al.,* 1989).

The published data permit several interpretations:

• Vaccinia virus infection of calf skin resulted in the activation of specific cattle viruses whose subsequent inoculation into humans as contamination represented a risk factor for subsequent local cancer development;

• Vaccinia virus infection of the human skin resulted in local activation of human potentially oncogenic viruses, increasing the risk of cancer development 20–60 years later;

• Early inflammatory reactions induced by this vaccination resulted in mutational events resulting in some cases in the simultaneous appearance of multifocal cancers.

Although still other interpretations remain possible, and basal cell carcinomas have also occasionally been observed in other non-vaccination scars, the observations described here should promote studies on a possible viral role in the initiation of these malignant proliferations.

3b. Haematopoietic malignancies

As shown in Figure 11, a number of human viruses turn out to be oncogenic when inoculated into newborn rodents. Intracerebral infections by JC virus are able to induce astrocytomas in adult owl monkeys (London *et al.,* 1978). For obvious reasons the reverse question, whether specific animal viruses are also able to induce tumours in humans, has not yet been carefully investigated (zur Hausen, 2001). Yet we are living in close contact with domestic animals and regularly handle their products. This is particularly interesting because contact with cattle and consumption of red meat have been identified as risk factors for specific human malignancies. Contact with cattle has also frequently been considered as risk factor for haematopoietic malignancies, in particular childhood acute lymphocytic leukaemias (reviewed in zur Hausen, 2009).

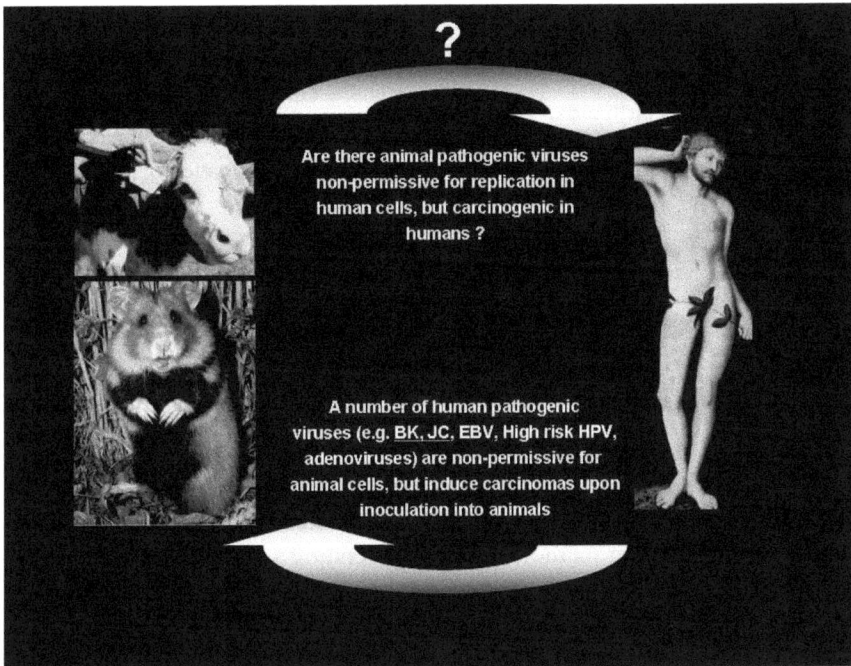

Figure 11. Some human viruses are carcinogenic for several animal species. Do animal viruses exist that are potentially carcinogenic in humans?

3b 1. Risk and protective factors

In the following, the reasons for considering childhood leukaemias as potential candidates for an infectious etiology will be briefly summarised. A more detailed account has been published recently (zur Hausen, 2009). Some protective factors as well as several risk factors for this malignancy are presented in Figure 12.

Repeatedly reported <u>protective</u> <u>factors</u> for childhood leukemias:	<u>Risk factors</u> for childhood leukemias
Multiple infections in early childhood	Rare infections during the first year of life
Underprivileged social status	High socioeconomic status
Crowded household, many siblings	
Inverse risk with birth order	Prenatal chromosomal translocations
More than 6 months of breastfeeding	Agricultural occupation of parents

Figure 12. Protective and risk factors for childhood acute lymphoblastic leukaemia.

Rare infections during the first year of life are frequently reported as risk factor for childhood leukaemias (reviewed in zur Hausen, 2009). Conversely, multiple infections during this period emerge as a protective factor. These observations are underlined by correlative data: a high socioeconomic state represents a risk factor, whereas crowded household conditions and many siblings emerge as protective factors. Cattle farming has been reported as an additional risk factor, whereas more than six months of breast feeding seem to reduce the risk.

Two additional sets of data deserve discussion: the frequent occurrence of specific chromosomal translocations in leukaemic cells, often observed already prenatally (Greaves, 2003). The same types of chromosomal alterations have also been found in healthy individuals, though here their frequency appears to be very low. Another striking observation originates from the description of occasional small clusters of leukaemic cases, specifically in regions where an influx of urban populations occurred in previously rural areas (reviewed in Kinlen, 2004).

3b. 2. Possible explanations
Three main hypotheses have been published to explain the epidemiological findings: Greaves (summarised in 2006) speculated that there exists an insufficient maturation state of the immune system in case of low exposure to infections. Preceding chromosomal translocations as the first event, followed by delayed infection *"with an unspecified agent"* should increase the risk for subsequent leukaemic conversion. Alternatively, Kinlen (1995) proposed that sudden mixing of a population of low exposure to a putative leukaemogenic agent (particularly in rural areas) with another population originating from urban areas previously highly exposed to the incriminated agent, could promote an epidemic of the relevant infection. These hypotheses were

supplemented by a further speculation, assuming that the protective effect of multiple infections during the first year of childhood was due to the reduction of the load with a putative leukaemogenic agent by interferon production as outlined in figure 13 (zur Hausen and de Villiers, 2005; zur Hausen, 2009).

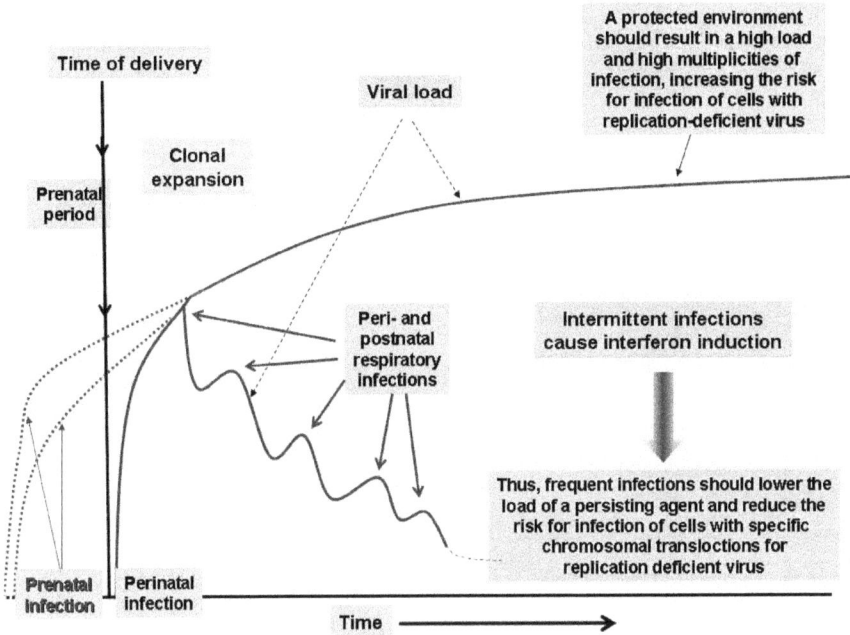

Figure 13. Schematic outline of the target cell conditioning hypothesis. Interferon synthesis resulting from multiple infections in early childhood reduces the load of a persisting potentially leukaemogenic agent and thus reduces the risk for malignant proliferation (adapted from zur Hausen and de Villiers, 2005).

Reports on supertransforming properties of specifically replication-incompetent SV40 and murine polyomaviruses (Small *et al.*, 1982; Roberge, C. and Bastin, 1988), in addition to the recent demonstration of replication-incompetent Merkel cell polyomavirus in Merkel cell carcinomas (Shuda *et al.*, 2008) resulted in an attempt to combine the three hypotheses, assuming that replication-incompetent polyoma-type viruses and high multiplicities at the time of initial infection represent an important precondition for an increased leukaemogenic risk. The generation of replication-incompetent viral progeny seems to depend on high multiplicities of infection and the co-infection of cells with both, replication-competent and incompetent genomes. The sole subsequent infection of a susceptible cell with a replication-incompetent genome may lead to the outgrowth of a leukaemic clone. Susceptibility of a cell to this malignant conversion would require the previous or subsequent acquisition of a specific chromosomal translocation. These translocations also occur in healthy individuals, though at low frequency (Liu *et al.*, 1994; Bell *et al.*, 1995; Fuscoe *et al.*, 1996; Roulland *et al.*, 2006). They represent risk factors but are clearly not sufficient for cell transformation. They should activate the oncogene of the replication-incompetent virus. A synopsis of this hypothesis is presented in Figure 14.

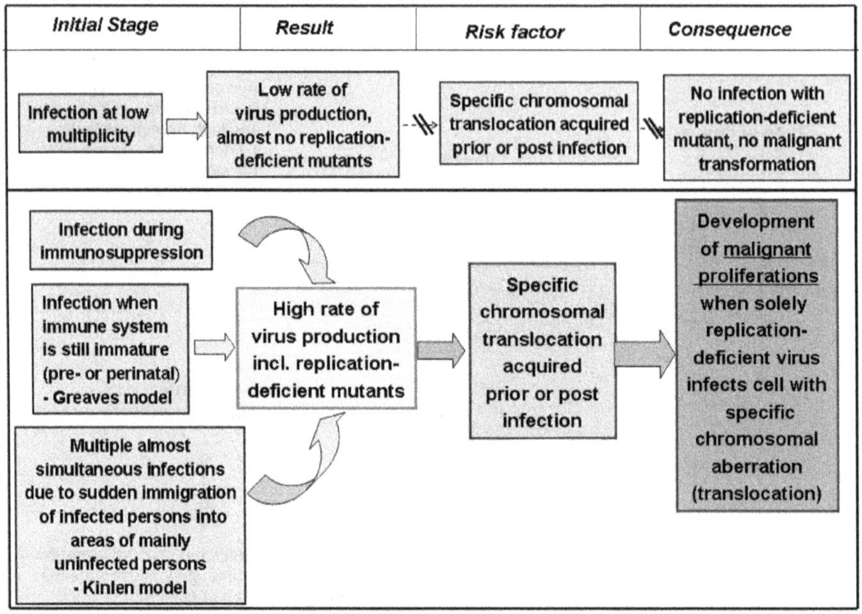

Initial Stage	Result	Risk factor	Consequence
Infection at low multiplicity	Low rate of virus production, almost no replication-deficient mutants	Specific chromosomal translocation acquired prior or post infection	No infection with replication-deficient mutant, no malignant transformation
Infection during immunosuppression / Infection when immune system is still immature (pre- or perinatal) - Greaves model / Multiple almost simultaneous infections due to sudden immigration of infected persons into areas of mainly uninfected persons - Kinlen model	High rate of virus production incl. replication-deficient mutants	Specific chromosomal translocation acquired prior or post infection	Development of malignant proliferations when solely replication-deficient virus infects cell with specific chromosomal aberration (translocation)

Figure 14. Synopsis of the target cell conditioning model for childhood leukaemia.

A polyoma-type virus infection would fit best for this model, although members of structurally related virus families might be also considered. Since a number of reports document elevated risks in families of cattle farmers and for individual in close contact with cattle (reviewed in zur Hausen, 2009), at least part of childhood leukaemias could be due to a native cattle virus. This virus should be replication-incompetent for human cells, but its oncogene may become activated in cells with specific chromsosomal modifications. Since a number of reports also suggest human occupational risks of persons with communicative contacts (e.g. teachers, hairdressers), other types of similar infections may be spread by human-human contacts (reviewed in zur Hausen, 2009).

It remains an interesting question to which extent other haematopoietic malignancies, like acute and chronic myelogenous leukaemias, chronic lymphatic leukaemias, B- and T-cell non-Hodgkin lymphomas, Epstein-Barr virus-negative Hodgkin lymphomas, and multiple myelomas could be included in these considerations. Yet undefined polyomavirus-like particles have been electronmicroscopically demonstrated in trichodysplasia of a patient with non-Hodgkin's lymphoma (e.g. Osswald *et al.*, 2007).

4. Cancers potentially linked to animal – human transmission
  Colorectal, breast and lung cancers

A large number of reports consistently describe an increased risk for colorectal cancers related to a high consumption of red meat (reviewed in Kuhnle *et al.*, 2007; Santarelli *et al.*, 2008). Recently this has also been noted for lung cancer in non-smokers (Cross *et al.*, 2007; Hu *et al.*, 2008; Lam *et al.*, 2009) and, to a more limited degree and less consistently, also for breast cancer (Taylor *et al.*, 2007; Egeberg *et al.*, 2008; Hu *et al.*, 2008; Linos *et al.*, 2008; Mignone *et al.*, 2009). A correlation seems to exist in countries with a high rate of red meat consumption and a high risk for colorectal and breast cancer. Common and frequently cited interpretations of these observations are dietary factors. Carcinogenic N-nitroso compounds, heterocyclic amines and heterocyclic aromatic hydrocarbons arise during cooking, broiling or meat curing. Some of these compounds require metabolic activation prior to conversion into a carcinogenic form, as initially described by Sugimura and colleagues (Ohgaki *et al.*, 1986). In addition, potentially carcinogenic nitrosyl haem and nitroso thiols have been reported to be significantly increased in faeces following a diet rich in red meat (Cross *et al.*, 2002).

In contrast to red meat, consumption of white meat, and here specifically chicken and other poultry meat, has not been found to be associated with an elevated risk of colorectal or other cancers. It has been reported, however, that fried, grilled or smoked chicken meat contains equally high concentrations of heterocyclic aromatic hydrocarbons and other carcinogens arising in the preparatory steps prior to consumption (Yano *et al.*, 1988; Kazerouni *et al.*, 2001; Reinik *et al.*, 2007). If this holds up and if no other hitherto unknown carcinogens are found specifically in red meat, these observations may require a fresh look at previous interpretations. In meat prepared medium or raw (Figure 15), temperatures in the central portions do not exceed 55 to 65° Celsius. At least members of the polyoma- and papillomavirus families readily survive these temperatures without significant loss of their infectivity (Lelie *et al.*, 1987; Sauerbrei and Wutzler, 2008). The only known bovine polyomavirus was initially identified as a contamination of foetal bovine sera, thus, it must have been present in the peripheral blood of yet unborn or newborn calves. Existing members of the polyomavirus family have been poorly studied in our domestic animals. These viruses are commonly non-oncogenic in their natural hosts, but reveal carcinogenicity only in heterologous tissues. At present, six different genotypes of polyomaviruses have been identified in humans, but only one in cattle.

Figure 15. Red meat cooked "raw".

It is tempting to speculate that a hitherto unidentified bovine infectious agent with pronounced thermostability, replication-incompetent for human cells and possibly structurally related to the polyomavirus family, may play a role in colorectal cancer, potentially also in lung cancers of non-smokers and in breast cancer. This could be interpreted to mean that the described chemical carcinogens arising during cooking or curing processes are not sufficient for the induction of the respective cancers. In the case of red meat consumption they may, however, interact with viral agents, present in red, but not in white meat.

CONCLUSIONS

Although we know that at present, slightly more than 20% of global cancer incidence is linked to infectious events, some epidemiological observations suggest that this percentage will increase in the future. The recognition that no cancer linked to infections develops without additional modifications within the host cell genome permits the speculation that even cancers with well established chromosomal modifications deserve a careful analysis for additional involvement of infectious agents. Prime malignancies suggested here as candidates for potential links with infections are haematopoietic malignancies, particularly childhood lymphoblastic leukaemias, Epstein-Barr virus-negative Hodgkin's lymphomas, basal cell carcinomas of the skin, and breast, colorectal and a subgroup of lung cancers. Although still hypothetical, this proposal is accessible to experimental verification. Even if only one of these speculations turns out to be correct, this would have profound implications for the prevention, diagnosis and hopefully also therapy of the respective malignancy.

REFERENCES

Bell, D. A., Liu, Y., and Cortopassi, G. A., "Occurrence of bcl-2 oncogene translocation with increased frequency in the peripheral blood of heavy smokers", *J. Natl. Cancer Inst.* 1995; **87:** 223–224.

Cross, A. J., Pollock, J. R., Bingham, S. A., "Red meat and colorectal cancer risk: the effect of dietary iron and haem on endogenous N-nitrosation", *IARC Sci Publ.* 2002; **156:** 205–206.

Burkitt, D. A., "Children's cancer dependent on climatic factors", *Nature* 1962; **194:** 232 –234.

Contreras-Galindo, R., Kaplan, M. H., Leissner, P., Verjat, T., Ferlenghi, I., Bagnoli, F., Giusti, F., Dosik, M. H., Hayes, D. F., Gitlin, S. D., Markovitz, D. M., "Human endogenous retrovirus K (HML-2) elements in the plasma of people with lymphoma and breast cancer", *J. Virol.* 2008; **82:** 9329–9336.

Cross, A. J., Leitzmann, M. F., Gail, M. H., Hollenbeck, A. R., Schatzkin, A., Sinha, R., "A prospective study of red and processed meat intake in relation to cancer risk", *PLoS Med.* 2007; **4:** e325.

Dewannieux, M., Harper, F., Richaud, A., Letzelter, C., Ribet, D., Pierron, G., Heidmann, T., "Identification of an infectious progenitor for the multiple-copy HERV-K human endogenous retroelements", *Genome Res.* 2006; **16:** 1548–1556.

Egeberg, R., Olsen, A., Autrup, H., Christensen, J., Stripp, C., Tetens, I., Overvad, K., Tjønneland, A., "Meat consumption, N-acetyl transferase 1 and 2 polymorphism and risk of breast cancer in Danish postmenopausal women", *Eur. J. Cancer Prev.* 2008; **17:** 39–47.

Feng, H., Shuda, M., Chang, Y., Moore, P. S., "Clonal integration of a polyomavirus in human Merkel cell carcinoma", *Science* 2008; **319:** 1096–1100.

Fuscoe, J. C., Setzer, R. W., Collared, D. D., Moore, M. M., "Quantification of t(14;18) in the lymphocytes of healthy adult humans as a possible biomarker for environmental exposures to carcinogens", *Carcinogenesis* 1996; **17:** 1013–1020.

Greaves, M., "Pre-natal origins of childhood leukemia", *Rev. Clin. Exp. Hematol.* 2003; **7:** 233–245.

Greaves, M., "The causation of childhood leukemia: a paradox of progress?", *Discov. Med.* 2006; **6:** 24–28.

Hsiao, F. C., Lin, M., Tai, A., Chen, G., Huber, B. T., "Cutting edge: Epstein-Barr virus transactivates the HERV-K18 superantigen by docking to the human complement receptor 2 (CD21) on primary B cells", *J. Immunol.* 2006; **177:** 2056–2060.

Hu, J., La Vecchia, C., DesMeules, M., Negri, E., Mery, L., Canadian Cancer Registries Epidemiology Research Group, "Meat and fish consumption and cancer in Canada", *Nutr. Cancer* 2008; **60:** 313–324.

Kazerouni, N., Sinha, R., Hsu, C. H., Greenberg, A., Rothman, N., "Analysis of 200 food items for benzo[a]pyrene and estimation of its intake in an epidemiologic study", *Food Chem. Toxicol.* 2001 May; **39**(5): 423–436.

Kinlen, L. J., "Epidemiological evidence for an infective basis in childhood leukaemia", *Br. J. Cancer* 1995; **71:** 1–5.

Kinlen, L. J., "Childhood leukemia and population mixing", *Pediatrics* 2004 Jul; **114** (1): 330–331.

Kuhnle, G. G., Bingham, S. A., "Dietary meat, endogenous nitrosation and colorectal cancer", *Biochem. Soc. Trans.* 2007; **35:** 1355–1357.

Lam, T. K., Cross, A. J., Consonni, D., Randi, G., Bagnardi, V., Bertazzi, P. A., Caporaso, N. E., Sinha, R., Subar, A. F., Landi, M. T., "Intakes of red meat, processed meat, and meat mutagens increase lung cancer risk", *Cancer Res.* 2009; **69:** 932–939.

Lelie, P. N., Reesink, H. W., Lucas, C. J., "Inactivation of 12 viruses by heating steps applied during manufacture of a hepatitis B vaccine", *J. Med. Virol.* 1987; **23:** 297–301.

Linos, E., Willett, W. C., Cho, E., Colditz, G., Frazier, L. A., "Red meat consumption during adolescence among premenopausal women and risk of breast cancer", *Cancer Epidemiol. Biomarkers Prev.* 2008; **17:** 2146–2151.

Lee Y.N., Bieniasz P.D., "Reconstitution of an infectious human endogenous retrovirus", *PLoS Pathog.* 2007; **3**(1): e10.

Liu, Y., Hernandez, A. M., Shibata, D., Cortopassi, D. A., "BCL2 translocation frequency rises with age in humans", *Proc. Natl. Acad. Sci. USA.* 1994; **91:** 8910–8914.

London, W. T., Houff, S. A., Madden, D. L., Fuccillo, D. A., Gravell, M., Wallen, W. C., Palmer, A. E., Sever, J. L., Padgett, B. L., Walker, D. L., ZuRhein, G. M., Ohashi, T., "Brain tumors in owl monkeys inoculated with a human polyomavirus (JC virus)", *Science* 1978; **201:** 1246–1249.

Meylan, F., De Smedt, M., Leclercq, G., Plum, J., Leupin, O., Marguerat, S., Conrad, B., "Negative thymocyte selection to HERV-K18 superantigens in humans", *Blood* 2005; **105:** 4377–4382.

Mignone, L. I., Giovannucci, E., Newcomb, P. A., Titus-Ernstoff, L., Trentham-Dietz, A., Hampton, J. M., Orav, E. J., Willett, W. C., Egan, K. M., "Meat consumption, heterocyclic amines, NAT2, and the risk of breast cancer", *Nutr. Cancer* 2009; **61:** 36–46.

Ohgaki, H., Hasegawa, H., Kato, T., Suenaga, M., Ubukata, M., Sato, S., Takayama, S., Sugimura, T., "Carcinogenicity in mice and rats of heterocyclic amines in cooked foods", *Environ. Health Perspect.* 1986; **67:** 129–134.

Osswald, S. S., Kulick, K. B., Tomaszewski, M. M., Sperling, L. C., "Viral-associated trichodysplasia in a patient with lymphoma: a case report and review", *J. Cutan. Pathol.* 2007; **34:** 721–725.

Parkin, D. M., Bray, F., Ferlay, J., Pisani, P., "Global cancer statistics, 2002", *CA Cancer J. Clin.* 2002; **55:** 74–108.

Reinik, M., Tamme, T., Roasto, M., Juhkam, K., Tenno, T., Kiis, A., "Polycyclic aromatic hydrocarbons (PAHs) in meat products and estimated PAH intake by children and the general population in Estonia", *Food Addit. Contam.* 2007; **24:** 429–437.

Roberge, C., Bastin, M., "Site-directed mutagenesis of the polyomavirus genome: replication-defective large T mutants with increased immortalization potential", *Virology* 1988; **162:** 144–150.

Roulland, S., Lebailly, P., Lecluse, Y., Heutte, N., Nadel, B., Gauduchon, P., "Long-term clonal persistence and evolution of t(14;18)-bearing B cells in healthy individuals", *Leukemia* 2006; **20:** 158–162.

Ruprecht, K., Ferreira, H., Flockerzi, A., Wahl, S., Sauter, M., Mayer, J., Mueller-Lantzsch, N., "Human endogenous retrovirus family HERV-K(HML-2) RNA transcripts are selectively packaged into retroviral particles produced by the human germ cell tumor line Tera-1 and originate mainly from a provirus on chromosome 22q11.21", *J. Virol.* 2008; **82:** 10008–10016.

Santarelli, R. L., Pierre, F., Corpet, D. E., "Processed meat and colorectal cancer: a review of epidemiologic and experimental evidence", *Nutr. Cancer* 2008; 60: 131–144.

Sauerbrei, A., Wutzler, P., "Testing thermal resistance of viruses", *Arch. Virol.* 2009; **154** (1): 115–119.

Schlehofer, J. R., Ehrbar, M., zur Hausen, H., "Vaccinia virus, herpes simplex virus, and carcinogens induce DNA amplification in a human cell line and support replication of a helpervirus dependent parvovirus", *Virology* 1986; **152:** 110–117.

Schmitt, J., Schlehofer, J. R., Mergener, K., Gissmann, L., zur Hausen, H., "Amplification of bovine papillomavirus DNA by N-methyl-N'-nitro-N-nitrosoguanidine, ultraviolet irradiation, or infection with herpes simplex virus", *Virology* 1989; **172:** 73–81.

Shuda, M., Feng, H., Kwun, H. J., Rosen, S. T., Gjoerup, O., Moore, P. S., Chang, Y., "T antigen mutations are a human tumor-specific signature for Merkel cell polyomavirus", *Proc. Natl. Acad. Sci. USA* 2008; **105:** 16272–16277.

Small, M. B., Gluzman, Y., Ozer, H. L., "Enhanced transformation of human fibroblasts by origin-defective simian virus 40", *Nature* 1982; **296:** 671–672.

Sutkowski, N., Chen, G., Calderon, G., Huber, B. T., "Epstein-Barr virus latent membrane protein LMP-2A is sufficient for transactivation of the human endogenous retrovirus HERV-K18 superantigen", *J. Virol.* 2004; **78:** 7852–7860.

Taylor, E. F., Burley, V. J., Greenwood, D. C., Cade, J. E., "Meat consumption and risk of breast cancer in the UK Women's Cohort Study", *Br. J. Cancer* 2007; **96:** 1139–1146.

Vajdic, C. M., McDonald, S. P., McCredie, M. R., van Leeuwen, M. T., Stewart, J. H., Law, M., Chapman, J. R., Webster, A. C., Kaldor, J. M., Grulich, A.E., "Cancer incidence before and after kidney transplantation", *JAMA.* 2006; **296:** 2823–2831.

Wang-Johanning, F., Radvanyi, L., Rycaj, K., Plummer, J. B., Yan, P., Sastry, K. J., Piyathilake, C. J., Hunt, K. K., Johanning, G. L., "Human endogenous retrovirus K triggers an antigen-specific immune response in breast cancer patients", *Cancer Res.* 2008; **68:** 5869–5877.

Waibel, K. H., Walsh, D. S., "Smallpox vaccination site complications", *Int. J. Dermatol.* 2006; **45:** 684–688.

Yano, M., Wakabayashi, K., Tahira, T., Arakawa, N., Nagao, M., Sugimura, T., "Presence of nitrosable mutagen precursors in cooked meat and fish", *Mutat. Res.* 1988; **202:** 119–123.

zur Hausen, H., "Proliferation-inducing viruses in non-permissive systems as possible causes of human cancers", *Lancet* 2001; **357:** 381–384.

zur Hausen, H., "Infections causing human cancers", Wiley-VCH, Weinheim – New York, Publ. 2006.

zur Hausen, H., "Childhood leukemias and other hematopoietic malignancies: interdependence between an infectious event and chromosomal modifications", *Int. J. Cancer* 2009; DOI 10.1002/ijc.24365.

zur Hausen, H., de Villiers, E. M., "Virus target cell conditioning model to explain some epidemiologic characteristics of childhood leukemias and lymphomas", *Int. J. Cancer* 2005; **115** (1): 1–5.

Portrait photo of Harald zur Hausen by photographer Ulla Montan.

FRANÇOISE BARRÉ-SINOUSSI

I was born in July 1947 in the 19th arrondissement of Paris, the city which remains my home today. My childhood holidays were, however, spent in the Auvergne countryside in central France, where I was content to spend my days outdoors, observing the wonders of the natural living world. Even the smallest of insects could capture my attention for hours. This fascination for the natural world was perhaps the earliest indication of the future direction my life would take.

During my school years, my passion for science was reflected in my grades, which were by far better in scientific subjects than in languages and philosophy. Having completed my baccalauréat in 1966, I was initially undecided between medicine and biomedical sciences as the subject for my university studies. I finally decided to opt for an undergraduate degree at the Faculty of Sciences at the University of Paris. My choice was ultimately dictated by the pragmatic reasoning that a degree in Natural Sciences was shorter and less expensive than a degree in Medicine, and I was keen to not have to burden my family with unnecessary further expenses to support me during my studies. Towards the end of my degree, I seriously questioned the possibility of research as a career option. It was therefore important for me to gain laboratory experience to clarify these doubts about my future. I contacted a large number of both private and public laboratories offering to volunteer part-time at the bench. My search for a host laboratory proved fruitless for many months. It was only when a friend of mine from university suggested contacting a group with whom she had been collaborating that I finally found a laboratory willing to host me as a volunteer. The group was led by Jean-Claude Chermann at the Institut Pasteur site at Marne-la-Coquette. Chermann was studying the relationship between retroviruses and cancers in mice. Very early on he transmitted so much passion and enthusiasm for the research I was doing, that, although I was supposed to continue attending classes for my degree, I spent all my time in the lab and only made an appearance at the university site to pass the necessary exams. Very quickly after my arrival in the Chermann group, Jean-Claude proposed a PhD project. My project analysed the use of a synthetic molecule which inhibited the reverse transcriptase to control leukaemia induced by Friend virus. This synthetic molecule, named HPA23, proved capable of inhibiting reverse transcriptase activity of Friend virus in culture. Pre-clinical tests showed that the molecule was capable of delaying the progression of the disease in mice. I completed my PhD relatively rapidly, as Jacques Monod, director of the Institut Pasteur at the time, had decided to move all external sites of the institute (including the Marne-la-Coquette site) back to the main campus in the 15th arrondissement of Paris. The move of the laboratory to the main campus would have proved a confusing time, and I was eager to complete my PhD before the move. I was awarded my PhD in 1974 by the Faculty of Sciences at the University of Paris.

During my time as a PhD student, the group was visited by Dr Dan Haapala and Dr Robert Bassin, two researchers from the National Cancer Institute (NCI) of the National Institutes of Health (NIH) in the United States, for a sabbatical research period. Furthermore, a member of this lab had been in our group teaching us the technique for the detection of reverse transcriptase, soon after the discovery of this enzyme by David Baltimore and Howard Temin. Following these contacts, I decided to join Bob Bassin for a post-doctoral research fellowship at the NIH in Bethesda in the mid-70s. My research project was a challenging one, aiming to identify the viral target of the Fv1 gene product implicated in the genetic restriction of murine leukaemia virus replication. Although the project was difficult, my experience at the NIH was a truly enriching one, albeit relatively short. I only remained one year in the United States, as during my PhD I had met my future husband, whom I later married in 1978. In addition, while in the US I discovered that I had been awarded an INSERM (National Institute for Health and Medical Research in France) position to return to Jean-Claude Chermann's laboratory (which had in the meanwhile moved to the central Pasteur campus) in the unit of Professor Luc Montagnier.

The group, which was slowly expanding in size in the late 70s and early 80s, was one of the few groups which continued to study the link between retroviruses and cancers. Indeed, many others had turned their attention to oncogenes, whose crucial role had been illustrated in the mid-70s by J. Michael Bishop and Harold Varmus. My research project at the time was to study the natural control of retroviral infections in the host, in particular the role of interferon in controlling endogenous retroviruses, and the functional implication of retroviral sequences on the metastatic potential of tumour cells in mouse models.

In late 1982, Luc Montagnier was contacted by Françoise Brun-Vézinet, a virologist working at the Bichat Hospital in Paris. Françoise Brun-Vézinet was working closely with Willy Rozenbaum, one of the first clinicians in France to observe the alarming new epidemic, which seemed to be affecting certain homosexuals. Willy had observed a number of cases in his ward, and had made the link with the Centres for Disease Control (CDC) report which had been published in 1981. After this first contact, Luc Montagnier asked me whether I was interested in working on this new project to determine whether a retrovirus could be responsible for the disease. After discussion with Jean-Claude Chermann, we accepted the proposal. We had previously detected mouse mammary tumour virus (MMTV) sequences in the lymphocytes of breast cancer patients, and we were familiar with the technique of reverse transcriptase activity detection. It would have been a relatively routine procedure to detect the presence of a retrovirus, and we were obviously keen to determine whether a retrovirus was present in patients affected by this newly described disease (later to be known as AIDS).

In late December 1982, meetings were held between our group at the Institut Pasteur and Willy Rozenbaum and Françoise Brun-Vézinet. The clinical observations suggested that the disease attacked the immune cells, but the strong depletion of CD4 lymphocytes greatly hindered the isolation of the virus from these rare cells in patients with AIDS. We therefore decided to use a lymph node biopsy from a patient with generalised lymphadenopathy. We waited until the new year to obtain the first patient biopsy from which lymphocytes were isolated and cultured. The cell culture supernatant was regularly tested for reverse transcriptase activity. The first week of sampling did not show any reverse transcriptase activity, but in the second week I detected weak enzymatic activity, which increased significantly a few days later. The reverse transcriptase activity level dropped dramatically however, as the T lymphocytes in the culture were dying. To save the culture, with the hope of preserving the virus, we decided to add lymphocytes from a blood donor to the cell culture. This idea proved successful, and as we had hoped, the virus – which was still present in the cell culture – started to infect the newly added lymphocytes and we were soon again able to detect significant reverse transcriptase activity. We named this newly isolated virus lymphadenopathy associated virus (LAV). At this point it was important to visualise the retroviral particles, and Charles Dauguet, in charge of the microscopy platform at Pasteur, provided the first images of the virus in February 1983.

The isolation, amplification and characterisation of the virus rapidly ensued, and the first report was published in *Science* in May 1983. In the same month, I presented our findings at the annual Cold Spring Harbor Meeting, after which I was invited by researchers at the CDC and by others at the NIH in Bethesda to discuss the results in further detail. During the following months, we continued to characterise this newly isolated virus, and a collaboration with molecular biologists at the Institut Pasteur determined the genome sequence. The collective efforts by researchers in our group and others, and by clinicians, brought together sufficient data to convince the scientific community and the relevant authorities that LAV (later to be named human immunodeficiency virus, HIV) was the etiological agent of AIDS.

Figure 1. Detection of HIV infection course at Institut Pasteur in Bangui, Central African Republic, 1987.

The year 1983 marked the beginning of my career in HIV research at the Institut Pasteur, which still continues to this day. I remained at the Institut Pasteur, even after the departure of Jean-Claude Chermann in 1987, and I was finally appointed as head of the Biology of Retroviruses Unit in 1992. My professional life has been intrinsically linked with collaboration with resource-limited countries. My first visit to an African country was in 1985, on the occasion of a World Health Organisation (WHO) workshop in Bangui (Central African Republic). This visit was an eye-opening experience. The culture shock and dire conditions impressed me greatly and instilled in me a desire and necessity to collaborate with resource-limited countries. My first visit to Vietnam in 1988 was the first of many visits, and the first collaborative steps with Asian countries. This long-lasting collaboration with Africa and Asia has resulted in continual exchanges between young scientists from the respective countries and researchers in Paris.

Figure 2. A visit to the National Institute of Hygiene and Epidemiology, Hanoi, Vietnam, 1988.

My unit at the Institut Pasteur was re-confirmed in 2005 and re-named the Regulation of Retroviral Infections Unit. The unit hosts approximately 20 people at any one time, consisting of students, post-docs and permanent research staff. Currently the unit is interested in defining the immune correlates of protection against HIV infection for vaccine research and the correlates of protection against AIDS for immunotherapy. Along these lines, the unit is focusing its research on the mechanisms of host control of HIV infection, both at the cell level and at the level of the immune response. We are studying examples of natural protection against infection, such as HIV-exposed but uninfected individuals (EU) and the placental barrier against HIV in-utero transmission; or of natural protection against disease, such as HIV controllers (HIC) and animal models of non-pathogenic infection (African Green Monkey, AGM /SIVagm).

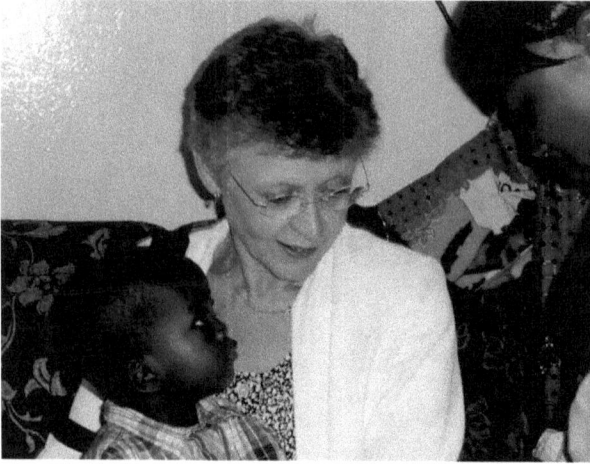

Figure 3. A recent visit to Yaoundé, Cameroon, 2008.

Although I anticipate continuing my professional endeavours largely un-
changed by the Nobel Prize, I hope that this recognition will provide the
necessary spark to spur international efforts in the fight against HIV/AIDS.

HIV: A DISCOVERY OPENING THE ROAD TO NOVEL SCIENTIFIC KNOWLEDGE AND GLOBAL HEALTH IMPROVEMENT

Nobel Lecture, December 7, 2008

by

Françoise Barré-Sinoussi

Institut Pasteur, Unit of Regulation of Retroviral Infections, Department of Virology, 25 rue du Docteur Roux, 75724 Paris, Cedex 15, France.

THE EARLY DAYS

The story begins more than twenty-five years ago, when the initial clinical observations of a alarming new epidemic were made. In June 1981 clinicians in the United States first reported a number of cases of *Pneumocystis carinii* in homosexual men[1]. Subsequently the first cases of what would later be known as AIDS were observed in France. At the time, I was working at the Institut Pasteur with Luc Montagnier and Jean-Claude Chermann. In December 1982, we were contacted by clinicians in France who provided us with a lymph node biopsy from an AIDS patient, with the aim of isolating the etiological agent causing the disease.

The hypothesis at the time was that a retrovirus might be the etiological agent responsible for AIDS. The only human retrovirus known at that time was the Human T-lymphotropic virus (HTLV), known to cause transformation of T cells, and arguably it would have been possible to culture the cells from the lymph node biopsy, and simply observe for T cell transformation. Luckily we did not assume that HTLV was necessarily the cause of the disease, and we decided to sample the culture supernatant every three to four days to screen for reverse transcriptase activity. Indeed we started to observe reverse transcriptase activity, which decreased shortly after that, in correlation with cell death. Initially we were concerned about possible toxicity related to tissue culture components, but following the addition of fresh lymphocytes and fresh components to the culture, the same cell death phenomenon was observed in correlation with the detection of reverse transcriptase activity. We thus realised that the virus itself was responsible for this phenomenon.

The isolation of this new human retrovirus (at the time known as LAV, lymphoadenopathy associated virus) was first reported and published in May 1983[2]. In this first report we described that LAV could be propagated in peripheral blood mononuclear cells and in cord blood lymphocytes. We also described the viral protein p25, and importantly we determined that there was no, or weak, cross reactivity with HTLV-1 proteins, indicating that we were dealing with a new human virus. In the same report we demonstrated the presence of antibodies against LAV in a second patient affected by AIDS. The report of the virus was, however, just the beginning.

The isolation of the virus was not sufficient to convince the scientific community of the implication of the virus in AIDS. It was therefore essential to further characterise the virus and establish a clear link between the virus and the disease to persuade the scientific community and the relevant authorities that the newly isolated virus was the etiological agent responsible for the emerging epidemic. In 1983 we decided to immediately halt all other research projects which were ongoing in the laboratory (including determining whether MMTV sequences could be associated with breast cancer – a hypothesis still valid today) and to mobilise a network of efficient collaborations with clinicians, immunologists and molecular biologists. In order to determine whether this newly isolated virus was truly responsible for the disease affecting AIDS patients, we quickly developed a serological test to perform sero-epidemiological studies[3]. Crucially, this same test was subsequently used as a diagnostic tool for blood testing. The development of the diagnostic test was made possible by a strong and efficient partnership with the private sector, namely Sanofi Pasteur.

In 1984 R. Gallo and colleagues and J. Levy and co-workers published reports confirming the identification of the causative agent of AIDS[4, 5].

FROM BED-SIDE TO BENCH TO BED-SIDE

These early years of HIV research were the reflection of clinical observations, which prompted basic research in the laboratory, which in turn resulted in the development of clinical tools. The identification and the initial characterisation of the virus led to the first diagnostic tests that permitted blood testing for the virus, and consequently the prevention of transmission by blood and blood derivatives. The increasing knowledge of the virus, its proven link to AIDS and its modes of transmission spurred the first programmes for voluntary counselling and testing and subsequently prevention of sexual transmission. The knowledge that the virus infected cells bearing the CD4 receptor[6–8] and the fact that HIV could be cytopathogenic to CD4 lymphocytes was the basis for CD4 cell monitoring in HIV infected patients. The characterisation of viral reverse transcriptase activity provided the rationale for first using azidothymine (AZT) as therapy for HIV patients[9] and, importantly, as the first therapeutic approach to prevent mother-to-child transmission of HIV. The efficiency of AZT alone was rapidly reported to be limited in regards to the first observation of resistance to monotherapy. These observations led to the development in the early 1990s of combined therapy, also known as highly active antiretroviral therapy (HAART). The cloning and sequencing of HIV[10, 11], performed by molecular biologists at the Institut Pasteur, provided the necessary knowledge of the basis of the test to determine the viral load and to monitor resistance to therapy (Figure 1).

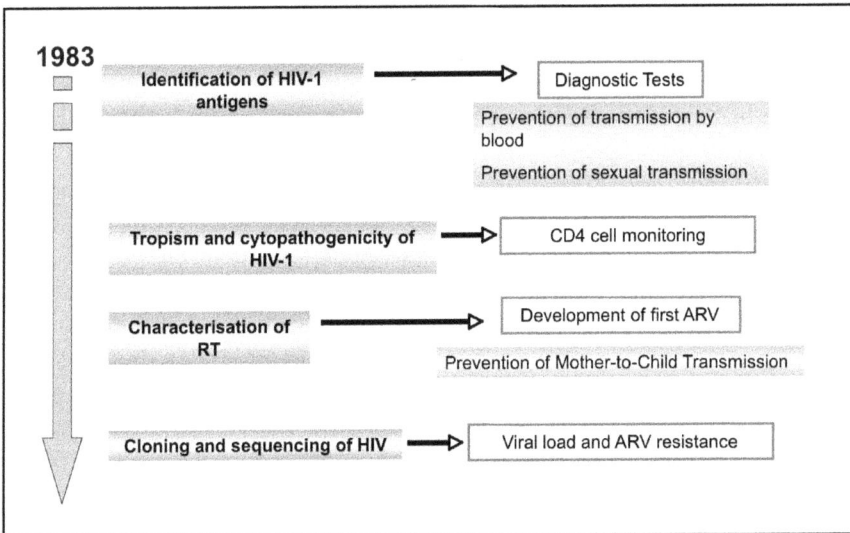

Figure 1. HIV research: from bed-side to bench and back to bed-side.

The cloning and sequencing of HIV also permitted researchers to gain insight into the tremendous diversity and the origin of HIV. Early studies indicated that HIV-1 may have resulted from the introduction of a virus from chimpanzee to humans. A collaboration with the Centre Pasteur Cameroun identified a new group of HIV-1, isolated from a woman with AIDS: the HIV-1 group N, distinct from groups O and M. Sequencing of simian viruses revealed a chimpanzee virus very closely related to the HIV-1 N group, in particular in specific genes[12, 13].

DETERMINANTS OF HIV PATHOGENESIS

Since the discovery of the virus, we have learnt that HIV infection is much more complex than we initially thought, and the mechanisms leading to AIDS pathogenesis are still today not entirely understood. We have, however, gained significant insight into the virus and the evolution of the disease it causes; for example we now know that soon after infection by HIV, the virus integrates into the host cells, establishing permanent reservoirs. Studies have shown that the evolution of the infection and the progression of the disease are linked to the viral load: indeed the peak of viral load correlates with a sharp decrease in the number of CD4 cells, which are partially restored in relation to the decrease of the viral load. In recent years, we have made important progress in understanding that HIV infection causes chronic immune activation. Very early after exposure to the virus, HIV infected patients are characterised by markers of generalised and persistent T cell activation. Given the close correlation between immune activation and the disease outcome, markers of T cell activation may emerge as better prognosis markers for disease progression than viral load and CD4 cell counts. It is also known today that HIV infection results in early massive depletion of CCR5+ CD4+ T memory cells in the gastrointestinal tract, associated with microbial translocation in the gut. Further studies will be necessary to determine whether the microbial translocation is a cause or consequence of massive T cell depletion[14, 15].

The evolution and progression of the disease caused by HIV is closely linked to a number of determinants of both the virus itself and the host. Indeed each particular path of disease progression is determined by a delicate interplay between viral and host factors. The virus itself greatly varies: in its tropism and replicative capacity, as well as the intrinsic immunosuppressive properties of some viral proteins, all influencing HIV pathogenesis. A number of host factors are also important in determining the distinct paths of disease progression in different HIV infected individuals. The adaptive immune responses (including CD8 cell and CD4 cell responses) are finely tuned in each separate individual, resulting in differential control of the infection. Equally different genetic polymorphisms of receptors, ligands and key immune proteins all result in specific modulations of the host response to HIV infection. Recently it has become apparent that humans and non-human primates possess a number of proteins capable of restricting HIV/SIV infection, and therefore play crucial roles in intracellular innate immunity. In brief, therefore, the specific evolution of the disease caused by HIV is the result of an intricate interplay between the virus and the host.

NATURAL MODELS OF HIV/AIDS PROTECTION

Natural models provide key insight into the mechanisms which underlie protection against HIV infection and disease progression.

A very small number of individuals (known as exposed uninfected, EU) appear to be resistant to infection, despite repeated exposure to HIV. A study, performed in collaboration with Vietnam, analysed two groups of intravenous drug users (IDU) who routinely exchanged needles for drug injection. Despite both groups being infected with hepatitis viruses and HTLV, one group was consistently negative for HIV in both serological and PCR tests. To address the mechanisms involved in this natural protection against infection by HIV, the innate immune responses of the two groups were compared. The two groups were characterised by a significant difference in the activity of the natural killer (NK) cells; in fact the EU group of Vietnamese IDU typically showed an increased level of NK cell activity[16]. We further analysed the NK cell repertoire associated with this increased activity. The analysis revealed an increased ratio of specific NK receptors in EU compared to control or HIV infected individuals. Furthermore, the functional activated NK cells in EU express significantly higher levels of the CD161 receptor than control individuals, suggesting that a specific subset of NK cells in EU might be involved in their protection against HIV infection.

Another insightful model, which permits us to shed light on some of the mechanisms providing protection against HIV, is provided by the few individuals who, despite being infected by the virus, are capable of controlling its replication. These rare individuals (known as HIV controllers, or elite controllers) are defined by having been infected for more than 10 years while presenting undetectable plasma viral RNA, despite being naïve of antiretroviral therapy. In 2007 we reported that the CD8 T lymphocytes of HIV controllers possess HIV suppressive capacities, being able to suppress HIV replication in the CD4 T cells[17, 18]. The majority of HIV controllers included in this first study possess CD8 T cells capable of eliminating infected CD4 T cells. Further analyses showed that the CD8 T cells of HIV controllers were activated, but to a lesser extent than in HIV progressors. Although the majority of HIV controllers appear to possess suppressive CD8 T cells, some individuals are not characterised by this trait, while still efficiently controlling the virus. This observation indicates that there is likely more than one immune mechanism contributing to the tight control of HIV replication in these lucky few individuals.

Animal models represent a key component of many domains of biomedical research. HIV and its closely related counterpart, the Simian Immunodeficiency Virus (SIV), are strictly primate-specific viruses, so simian models which are susceptible to SIV infection play a key role in understanding infection and disease progression. Most naturally infected African primates, like the African Green Monkey (AGM) do not develop AIDS, in contrary to the Asian Rhesus Macaque; they therefore provide a unique model to investigate protective mechanisms against AIDS[19, 20]. Interestingly, SIV is capable of replicating in both pathogenic and non-pathogenic simian models, suggesting that viral replication is not the only key determinant of an AIDS outcome. Both models show a massive depletion of $CCR5^+$ $CD4^+$ cells in the gastrointestinal tract. Partial restoration of $CD4^+$ cells and the absence of apoptosis, however, ensure that the intestinal mucosa remains intact in the non-pathogenic model, therefore impeding microbial translocation. Comparative analyses between the two primate models show that T cell activation during the chronic phase of viral infection is a key difference between the non-pathogenic and the pathogenic infections. Indeed chronic T cell activation is the hallmark of HIV-1 infection in humans and pathogenic SIV infection in Rhesus Macaques. The profiles of T cell activation are determined by the innate immune responses, mediated in particular by dendritic cells. The Rhesus Macaque model shows a more persistent recruitment of plasmacytoid dendritric cells (PDC, the principal producers of interferon-α) to the lymph nodes[21, 22]. More recently we have used microarrays to compare the gene profiles of the two primate models. This technique has allowed us to observe that interferon type 1 pathways are activated in both models, but only transiently in the non-pathogenic model in the very early phases of infection. This observation contributes to the hypothesis that key events which determine the disease progression occur in the very early stages of infection. Important differences are observed in the early acute phase of SIV infection in the non-pathogenic and pathogenic models, including a higher induction of pro-inflammatory cytokines in the pathogenic model (Figure 2).

marker		SIVagm Non-pathogenic	SIVmac Pathogenic
T cell activation		+	+
Intestinal Mucosa	CCR5+CD4+ depletion	+++	+++
	Microbial translocation	-	+
Cytokines	TNF-α, IL-6, IL-12,	-	++
	TGF-β1	+	+
	Smad7	-	+
Recruitment of PDC to LN		+	++

Figure 2. Non-pathogenic versus pathogenic SIV infection – eary acute phase markers.

The early acute phase of HIV infection appears, therefore, to be crucial in determining disease progression. Given the importance of this very early phase following infection, the role of the innate immune system, our body's first line of defence against infections, should strongly be considered.

More detailed insight into the role of the innate immune system in HIV has been gained by an *in vitro* model of a co-culture of natural killer (NK) cells and dendritic cells (DC), which are natural target cells of HIV infection. This *in vitro* model has highlighted that the expression of a number of NK cell surface receptors, in particular the CD85j receptor, is reduced when the cells are in contact with infected DCs. We also observed that the subpopulation of NK cells which express the CD85j receptor is capable of strongly suppressing HIV replication in DCs[23]. This suppression is only partially relieved by incubation with monoclonal antibodies against HLA class I (the natural ligands of CD85j) but strongly abolished by incubation with a recombinant CD85j protein, suggesting that the interaction between the CD85j receptor and a peculiar (as of yet unknown) ligand might result in signalling pathways necessary for suppression.

Future research will need to focus on understanding in greater detail the complex cross-talk between DC, NK and T cells during HIV infection, as well as the modulation of signalling pathways and soluble factors which can influence disease progression (Figure 3).

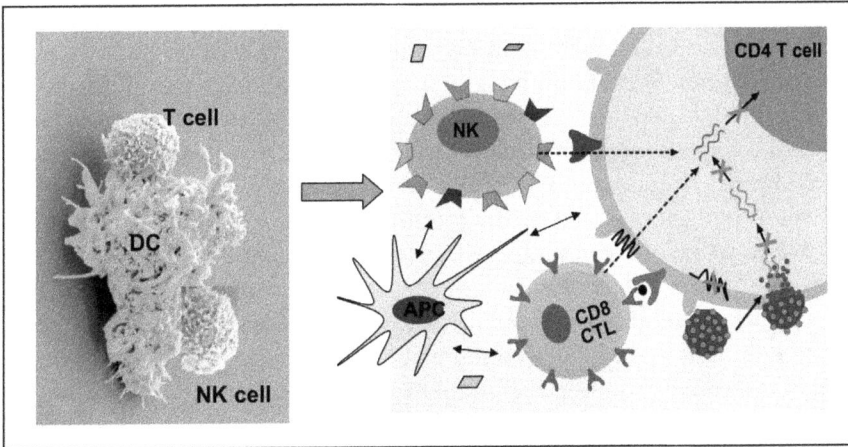

Figure 3. Potential mechanisms of HIV control. The control of HIV may be determined by a complex interplay between innate and adaptive immune systems. Natural killer (NK) cells might control HIV by directly eliminating infected cells or by providing the optimal cytokine environment. HIV-specific CD8 cells with cytotoxic properties could also play a key role in controlling HIV infection. Other control mechanisms include the production of neutralising antibodies and intracellular restriction factors which counteract the HIV replicative cycle.

HIV/AIDS: THE STATE OF THE EPIDEMIC IN 2008 AND FUTURE DIRECTIONS

At the end of 2007, 33 million people were living with HIV, 2.7 million were newly infected and a further 2 million died of AIDS[24]. These numbers demonstrate that HIV/AIDS still remains a crucial global health issue. Since the discovery of HIV, enormous progress has been made in the development of antiretroviral (ARV) drugs, and in expanding their availability to those who require treatment. In 2002, only approximately 2% of people needing treatment received it; this figure has since increased to 30% in 2008. Despite this massive increase in access, for every new person starting ARV treatment, 2–3 new cases of HIV infection are reported. The benefits of ARV are irrefutable; their importance is clearly illustrated in Botswana where the introduction of ARV in 2002 has been followed by a decrease in the number of AIDS-related deaths in the country.

Despite the immense benefits of ARV treatment, many issues still need to be tackled. ARV treatment remains a life-long commitment with consequent economic limits. Moreover, life-long treatment is associated with the emergence of drug resistance, metabolic disorders and cancers. In the absence of a cure for HIV, it is essential to continue investigating and promoting all prevention measures. The accomplishments of prevention programmes for HIV infection have been proven by the success in limiting sexual transmission (mainly through the promotion of condom use) and in limiting mother-to-child transmission of HIV by improving diagnosis and introducing ARV treatment to pregnant HIV+ women. Research on prevention methods continues, and recently we have also learnt that male circumcision can diminish the risk of infection, and therefore could be considered as part of a comprehensive prevention plan. Prevention programmes which are currently being investigated include pre- and post-exposure prophylaxis, and further research directed on specific anti-viral microbicides, which so far have only provided disappointing results, could lead to beneficial outcomes.

Despite innumerable advances in the field of HIV research over the last twenty-five years, many new aspects of HIV infection and immunopathogenesis are still emerging. It is clear that there are currently many areas of HIV research which require investigation, and many domains will benefit from a strong emphasis on basic research. Further insight into the very early stages of HIV infection, including the establishment of viral reservoirs and the immune responses induced, is crucial for the development of novel therapeutic and vaccinal strategies. Future directions for new intervention strategies which should be investigated in further detail include, for example, the identification of new targets and the use of siRNA to restrict HIV infection. Recent years have highlighted the central role of chronic immune activation in disease progression; if microbial translocation is confirmed to be responsible, at least in part, for this immune activation, drugs limiting microbial translocation need to be considered as alternative therapeutic strategies.

Co-infection between HIV and other diseases should not be overlooked. Co-infections in HIV+ patients are currently an acute public health issue, and further clinical attention and basic research are required to address this delicate problem. A salient example is provided by the co-infection of HIV and tuberculosis: in sub-Saharan Africa and other resource-limited regions, tuberculosis is a major cause of death among people living with HIV.

An effective vaccine against HIV represents the ultimate prevention approach. HIV, however, exhibits several scientific challenges. HIV is a virus characterised by a high degree of genetic variability, enabling the virus to evade the human immune system. Furthermore, the very early establishment of viral reservoirs, characteristic of HIV, hinders the development of an effective vaccine. To add to the complexity, HIV is transmitted not only by cell-free virus, but also by cell-to-cell contact, and it is still unclear what mechanisms are necessary to impede cell-to-cell transmission. To achieve progress in the development of an effective vaccine, it is essential to understand the complex dysfunctions of the immune system which are more rapidly induced by HIV than the specific immune response following infection. A key step in the development of an effective vaccine will be the identification and definition of the viral determinants responsible for early pathogenic signals and mechanisms to prevent such harmful pathogenic pathways[25–27] (Figure 4). In addition to the indispensable advances in basic science, the development of an effective HIV vaccine requires innovative and creative strategies, within the context of a clearly defined international agenda to promote collaboration.

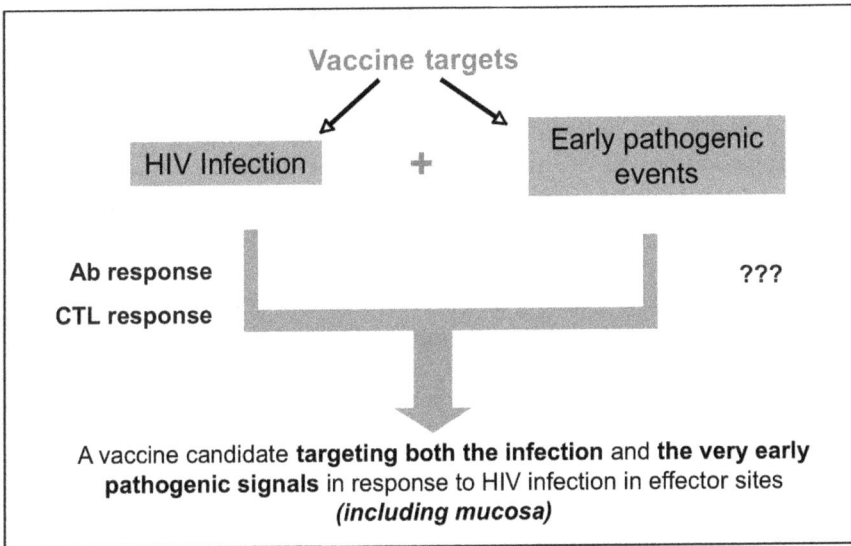

Figure 4. HIV vaccine research: re-thinking future strategies.

BENEFITS BEYOND HIV/AIDS

HIV can be a powerful tool for unravelling future scientific knowledge. Research on HIV has significantly contributed to the understanding of the delicate relationship between viruses and hosts. Continued effort to understand this complex virus has also revealed novel cellular partners which may be involved in controlling infection. HIV can also help identify new receptors and ligands and novel signalling pathways. Future research on understanding the immune responses induced by HIV will provide more information on the complex cross-talk between innate and adaptive immunity. This cross-talk between innate and adaptive immunity is crucial for the induction of effective immune responses to infections, and the elucidation of these mechanisms will prove important for other diseases as well.

Given the complexity of developing an effective HIV vaccine, we need to think of new innovative and creative concepts and strategies for vaccines. These new strategies may likely prove useful for the development of vaccines against other infectious diseases or even cancer.

The use of lentiviral vectors in gene therapy for a number of diseases, including cystic fibrosis and muscular dystrophy, has already been widely investigated. Despite some setbacks, new promising results are encouraging. The field is still at an early stage, but further research may provide new therapeutic strategies.

IMPROVING GLOBAL HEALTH: A KEY ROLE OF THE FIGHT AGAINST HIV/AIDS

Importantly, the fight against HIV is also a key element in the improvement of global health. It is now widely recognised that investment in health is one of the key components of sustainable development. International efforts in close collaboration with national programmes in HIV/AIDS have promoted an overall amelioration of global health, which is not limited to people living with HIV. Global interventions in HIV/AIDS reinforce local infrastructure, contribute to capacity building and increase human resources, contributing greatly to better overall health in a country (Figure 5).

Figure 5. Benefits beyond HIV: global health systems improvement.

The effect of HIV/AIDS interventions on general public health improvement is exemplified by the case of Cambodia. In 1994–95, we started working in collaboration with Cambodia. At the time, human resources were extremely limited, if at all existent. We and others quickly started to train people, who were very keen and determined to learn about progress in science and medicine. In 1995, the first Voluntary and Confidential Counselling and Testing (VCCT) site was created at the Institut Pasteur in Cambodia. Over the years, we and others have been working to improve the quality of human resources in the country, the quality of health infrastructures and the number of VCCT sites has greatly increased to more than 200 in the entire country in 2008. A similar trend was noted in clinical sites for treatment of opportunistic infections and antiretroviral therapy. Theses sites are present in the entire country, covering a large part of its territory, and the services provided are not only HIV monitoring and treatment but also care for other infections. Today, approximately thirty thousand patients are under antiretroviral therapy and Cambodia may well be one of the first countries to reach the objective of universal access by the year 2010. Such progress would not have been possible without the strong political determination of the government.

Although the road ahead is still long, we are on the right path to achieve a world without AIDS. This goal will be reached by following a model of research, echoing the tradition of Louis Pasteur: continued basic and clinical research, investment of both public and private sectors, public health interventions and the participation, which should be strongly acknowledged, of people living with HIV/AIDS.

REFERENCES

1. Gottlieb, M. S. *et al.*, "Pneumocystis carinii pneumonia and mucosal candidiasis in previously healthy homosexual men: evidence of a new acquired cellular immunodeficiency", *N Engl J Med* 305, 1425–31 (1981).
2. Barré-Sinoussi, F. *et al.*, "Isolation of a T-lymphotropic retrovirus from a patient at risk for acquired immune deficiency syndrome (AIDS)", *Science* 220, 868–71 (1983).
3. Brun-Vezinet, F. *et al.*, "Detection of IgG antibodies to lymphadenopathy-associated virus in patients with AIDS or lymphadenopathy syndrome", *Lancet* 1, 1253–6 (1984).
4. Gallo, R. C. *et al.*, "Frequent detection and isolation of cytopathic retroviruses (HTLV-III) from patients with AIDS and at risk for AIDS", *Science* 224, 500–3 (1984).
5. Levy, J. A. *et al.*, "Isolation of lymphocytopathic retroviruses from San Francisco patients with AIDS", *Science* 225, 840–2 (1984).
6. Dalgleish, A. G. *et al.*, "The CD4 (T4) antigen is an essential component of the receptor for the AIDS retrovirus", *Nature* 312, 763–7 (1984).
7. Klatzmann, D. *et al.*, "T-lymphocyte T4 molecule behaves as the receptor for human retrovirus LAV", *Nature* 312, 767–8 (1984).
8. Klatzmann, D. *et al.*, "Selective tropism of lymphadenopathy associated virus (LAV) for helper-inducer T lymphocytes", *Science* 225, 59–63 (1984).
9. Broder, S. *et al.*, "Effects of suramin on HTLV-III/LAV infection presenting as Kaposi's sarcoma or AIDS-related complex: clinical pharmacology and suppression of virus replication *in vivo*", *Lancet* 2, 627–30 (1985).
10. Alizon, M. *et al.*, "Molecular cloning of lymphadenopathy-associated virus", *Nature* 312, 757–60 (1984).
11. Wain-Hobson, S., Sonigo, P., Danos, O., Cole, S. & Alizon, M., "Nucleotide sequence of the AIDS virus, LAV", *Cell* 40, 9–17 (1985).
12. Simon, F. *et al.*, "Identification of a new human immunodeficiency virus type 1 distinct from group M and group O", *Nat Med* 4, 1032–7 (1998).
13. Apetrei, C., Robertson, D. L. & Marx, P. A., "The history of SIVS and AIDS: epidemiology, phylogeny and biology of isolates from naturally SIV infected non-human primates (NHP) in Africa", *Front Biosci* 9, 225–54 (2004).
14. Cadogan, M. & Dalgleish, A. G., "HIV immunopathogenesis and strategies for intervention", *Lancet Infect Dis* 8, 675–84 (2008).
15. Douek, D. C., Roederer, M. & Koup, R. A., "Emerging Concepts in the Immunopathogenesis of AIDS", *Annu Rev Med* (2008).
16. Ravet, S. *et al.*, "Distinctive NK-cell receptor repertoires sustain high-level constitutive NK-cell activation in HIV-exposed uninfected individuals", *Blood* 109, 4296–305 (2007).
17. Saez-Cirion, A. *et al.*, "HIV controllers exhibit potent CD8 T cell capacity to suppress HIV infection ex vivo and peculiar cytotoxic T lymphocyte activation phenotype", *Proc Natl Acad Sci U S A* 104, 6776–81 (2007).
18. Saez-Cirion, A., Pancino, G., Sinet, M., Venet, A. & Lambotte, O., "HIV controllers: how do they tame the virus?", *Trends Immunol* 28, 532–40 (2007).
19. Silvestri, G., "AIDS pathogenesis: a tale of two monkeys", *J Med Primatol* 37 Suppl 2, 6–12 (2008).
20. Liovat, A. S., Jacquelin, B., Ploquin, M. J., Barré-Sinoussi, F. & Muller-Trutwin, M. C., "African non human primates infected by SIV – why don't they get sick? Lessons from studies on the early phase of non-pathogenic SIV infection", *Curr HIV Res* 7, 39–50 (2009).
21. Diop, O. M. *et al.*, "Plasmacytoid dendritic cell dynamics and alpha interferon production during Simian immunodeficiency virus infection with a nonpathogenic outcome", *J Virol* 82, 5145–52 (2008).
22. Kornfeld, C. *et al.*, "Antiinflammatory profiles during primary SIV infection in African green monkeys are associated with protection against AIDS", *J Clin Invest* 115, 1082–91 (2005).

23. Scott-Algara, D. *et al.*, "The CD85j+ NK cell subset potently controls HIV-1 replication in autologous dendritic cells", *PLoS ONE* 3, e1975 (2008).
24. UNAIDS, *Global Report on AIDS Epidemic.* (2008).
25. Barouch, D. H., "Challenges in the development of an HIV-1 vaccine", *Nature* 455, 613–9 (2008).
26. Haynes, B. F. & Shattock, R. J, "Critical issues in mucosal immunity for HIV-1 vaccine development", *J Allergy Clin Immunol* 122, 3–9; quiz 10–1 (2008).
27. Walker, B. D. & Burton, D. R., "Toward an AIDS vaccine", *Science* 320, 760–4 (2008).

Portrait photo of Françoise Barré-Sinoussi by photographer Ulla Montan.

LUC MONTAGNIER

I was born on August 18, 1932 in Chabris, a "bourg", larger than a village but smaller than a town, located in Berry south of the Loire Valley. This was – and still is – a region of agriculture with some renowned products such as welsh rabbit, goat cheeses and white asparagus. It was the place where my mother had grown up but, in fact, I never lived there.

On my father's side, his parents came from Auvergne, a province in the centre of France, made of rich plains and old volcanoes, the latter probably being at the origin of my family name: Montagnier, the man living in mountains.

In his youth, my father had caught a terrible disease: streptococcal arthritis, ending in irreversible lesions in the aortic valves. He was therefore declared unfit for military service and had to find a sedentary job: he became an accountant and excelled in this profession, which implied, at that time, mainly hand-written work. He started working in the Poitiers area and then moved a little farther north to Châtellerault, a small city between Tours and Poitiers.

As an only child, I was cherished by my mother, a housewife, but two events dominated this pre-war period, of which I keep a vivid memory:

I was badly injured by a high speed car while crossing a main road: multiple wounds of which I keep some visible scars. After two days in a coma, I emerged as if I was born again, at the age of 5 (Figure 1).

Figure 1. Luc Montagnier at the age of 5.

...and two years later came the declaration of war in 1939, while the whole family was harvesting grapes in the vineyards of my mother's brother. I still remember the images in a newspaper of Warsaw ruins after a bombing by German planes.

And then, in 1940, came the "real" war: the German invasion, my parents and I leaving their house (close to a risky railway station), fleeing on the roads in a little car, and finally more exposed to German bombing during this "exodus" than if we had stayed home.

The first year of German occupation was terrible, in that we had no food reserves and most of the time we were starving. I was a rather puny boy and during the four years of the war did not gain a gram! The "ersatz" did not stimulate my appetite, when I was dreaming of chocolate and oranges! My father had chronic enterocolitis and, worse, my grandfather (his father) was diagnosed with rectal cancer. He died in 1947 after terrible suffering and each time I visited him, I could see the inexorable progression of the disease. This affected me so much that it is probably one reason why I decided later to study medicine and to start research on cancer.

In June 1944, our house (so close to the railway) was partly destroyed – this time by an Allied bombing. I keep a mixed feeling of this year of the liberation of France. It was a great relief but I could not forget also the vision of so many dead people, civilians and soldiers, and the images of skinny deportees released from concentration camps. I will hate wars and their atrocities for the rest of my life.

At high school I did well, being usually ahead of my classmates. This is when I became curious about scientific knowledge, having left behind my religious Catholic belief.

Following the example of my father, who was tinkering in his leisure days with electric batteries, I set up a chemistry laboratory in the cellar of the new house which was requisitioned to accommodate us. There, I enthusiastically produced hydrogen gas, sweet-smelling aldehydes and nitro compounds (not nitro-glycerine!) that had the unfortunate habit of blowing up in my face.

I was delighted to read – in popularised books – the impressive progress of physics, especially atomic physics. Being good in physics and chemistry – but not as good in maths – I decided not to prepare to compete for the "Grandes Ecoles" but instead to register both at the School of Medicine and the Faculty of Sciences in Poitiers. My goal was in fact to start a research carrier in human biology, but there was no such specialty in Poitiers, either in Medicine or in Sciences. Since both the Faculty and School were within walking distance, I could spend the morning at the hospital and the afternoon attending courses in botany, zoology and geology, which were the main disciplines of the degree course in Sciences.

Fortunately the new Professor of Botany, Pierre Gavaudan, was a very atypical professor in that his scientific interests went far beyond the classification of plants. In fact, I owe him for having opened me a large window on what was the beginning of a new Biology, the DNA double helix, the *in vitro* synthesis of proteins by ribosomes and the structure of viruses.

At the same time, I was installing at home a device combining a time-lapse movie camera and a microscope, thanks to a gift by my father. This allowed me to do my first research work. I was studying a phenomenon known since 1908 as the phototaxy of chloroplasts: the property of some algae living at the surface of ponds to orient their large unique chloroplast according to the intensity of light; if the light was too intense, the chloroplast turned inside the tubular cell to present its edge. In dark or weaker light, the chloroplast, a flat plate, exposed its larger surface. The phenomenon took a few minutes, which could be analysed by time-lapse cinematography. Using different glass filters, I could show that it was not the wavelength absorbed by the chlorophyll (red light) which regulated the orientation of the chloroplasts but indirectly some yellowish pigments absorbing the blue light. I was very proud, at the age of 21, to defend this work as a small thesis at the Faculty of Sciences of Poitiers. I was asked by my mentor, Pierre Gavaudan, to do research also on a literature-based subject: the L-forms of bacteria. This allowed me to make my first incursion – not the last – into the world of filtering bacteria. I could only find the references on this controversial subject at the library of the Institut Pasteur in Paris. This was indeed the time when I left Poitiers for Paris, where I was able to complete my medical studies as well as explore some aspects of biology closer to human beings, particularly neurophysiology, virology and oncology.

Having been hired as an assistant at the Sorbonne at the age of 23, I started learning old-fashioned technologies derived from Alexis Carrel's work on chick embryo heart cultures, as well as that of human cell lines in monolayers. Although my research was not productive at all, I keep from this period a solid expertise of Pasteurian technologies for working in perfectly sterile conditions without the use of antibiotics.

In 1957, the first description of infectious viral RNA from the tobacco mosaic virus by Fraenkel-Conrat and Gierer and Schramm determined my vocation: to become a virologist using the modern approach of molecular biology.

I started with the foot and mouth virus and then, in Kingsley Sanders' laboratory at Carshalton near London, I was proud to identify for the first time an infectious double-stranded RNA from cells infected with the murine encephalomyocarditis virus, a small single-stranded RNA virus. This demonstrated for the first time that RNA could replicate like DNA by making a base-paired complementary strand.

In order to perfect my knowledge of oncogenic viruses, I moved from Carshalton to Glasgow where a new Institute of Virology had been recently inaugurated, headed by a remarkable virologist, Michael Stocker, and where many high-ranking visitors, among them Renato Dulbecco, were spending sabbatical years.

Working on a small oncogenic DNA virus, polyoma, I could show there, with I. Macpherson, a new property of transformed cells, that of growing in soft agar. Using this technique, it was easy to detect the transforming capacity of polyoma virus and its DNA. We showed that naked DNA alone carried all the oncogenic potential of the virus. This now looks pretty obvious, but it was not so at that time.

Back to France at the Institut Curie, I extended this finding to a number of cancer cells, transformed or not by oncogenic RNA or DNA viruses. However, this property allowed me to distinguish some *in vitro* steps in the process of transformation, which were correlated with some modifications of the plasma membrane and of the carbohydrate layer surrounding it.

A great mystery remained at that time: that of the replication of the oncogenic RNA viruses, now known as retroviruses. Howard Temin (Figure 2) had proposed the hypothesis of a DNA intermediate, but other possibilities could be considered. I myself tried to find a double-stranded RNA specific of the Rous sarcoma virus, a virus able to infect and transform chick embryo cells. I indeed isolated double-stranded RNA sequences, but they were of cellular origin and existed at the same level in non-infected cells! With Louise Harel, I later showed that this RNA was partly coming from repetitious sequences of DNA. In retrospect, it could at least in part represent the recently identified interfering RNAs involved in the negative control of messenger RNA translation.

Figure 2. Receiving an award plate of the American Society of Pathology from Howard Temin's hands in 1985.

In 1969–70, the isolation of an RNA-polymerase associated with the viral particles of the vesicular stomatitis virus led to the idea that perhaps a key enzyme was also associated with the oncogenic RNA viruses. Indeed, Howard Temin and Mizutani, and independently David Baltimore, discovered in 1970 a specific enzyme associated with Rous sarcoma virus (RSV), the reverse transcriptase (RT), capable of reversely transcribing the viral RNA into DNA.

At about the same time, Hill and Hillova in Villejuif, France, demonstrated that the DNA extracted from RSV transformed cells was infectious and carry the genetic information of the viral RNA, confirming that the enzyme was working faithfully in infected cells.

I myself, with P. Vigier, confirmed and extended this discovery by showing that the infectious DNA was associated with the chromosomal DNA of the cells, showing integration of the proviral DNA, as earlier postulated by Temin.

Work on the chicken RSV was extended to similar viruses in mammals, so that many researchers at that time believed that RT activity was a new, highly sensitive tool for detecting similar viruses in human leukaemia and cancer. This was stimulated by the generously funded virus-cancer program launched by America's National Institutes of Health. Unfortunately, the hunt for human retroviruses was basically unsuccessful but led to important basic work on the molecular biology of animal retroviruses.

In 1972, I was asked by Jacques Monod, then head of the Institut Pasteur, to create a research unit in the newly created Department of Virology of the Institute. I accepted, and this new laboratory allowed me to develop new avenues of research within the general theme of Viral Oncology, the ultimate goal remaining the detection of viruses involved in human cancers.

Thus, I became interested in the mechanism of action of interferon and its role in its expression of retroviruses. I came into this field after having demonstrated the biological activity of interferon messenger RNA in collaboration with two world-renowned experts in the field, Edward and Jacqueline De Maeyer.

From 1973 on, Ara Hovanessian and his co-workers joined my unit and brought a new dimension: the complex biochemical mechanism sustaining the antiviral activity of this remarkable group of cellular proteins.

In 1975, two other researchers joined my unit and brought their expertise on murine retroviruses: J. C. Chermann and his collaborator, Françoise Barré-Sinoussi (Figure 3). The latter mastered particularly the detection of retroviruses by their RT activity. I convinced them to participate in a joint study inside the unit to look again for retroviruses in human cancers. We started in 1977 with blood samples coming from different Paris hospitals and biopsy specimens.

Figure 3. HIV discoverers in the park of the Institut Pasteur Annex in Garches, near Paris, during a break of a "100 guards meeting" in 1987. From left to right: Jonas Salk, Jean-Claude Gluckman, Jean-Claude Chermann, Luc Montagnier, Robert Gallo, Françoise Barré-Sinoussi, Willy Rozenbaum, Charles Mérieux.

Two advances made in other laboratories boosted this search:

In Villejuif, France, Ion Gresser had prepared a potent antiserum neutral-ising any molecule of alpha endogenous interferon produced by individual cells. This interferon, we realised, was produced by mouse cells induced to express some of their endogenous retroviruses. Its blockade by the antiserum increased by up to 50 times the production of endogenous retroviruses in the culture medium. We could conclude that, despite the fact that endogenous retroviruses have been integrated in the genome of vertebrates for millions of years, their expression is still controlled by the interferon system, like that of exogenous viruses.

At about the same period, the discovery by Denis Morgan and Frank Ruscetti in Dr. Gallo's laboratory of a growth factor allowing the *in vitro* mul-tiplication of human T lymphocytes (TCGF, then named interleukin 2, Il2) made it possible to propagate T lymphocytes in sustained cultures.

We knew at that time that some retroviruses involved in mouse mammary tumour formation (MMTV) could not only be expressed in the tumour cells but also in the circulating lymphocytes.

Taking advantage of these two advances, we started a search for retrovi-ruses in human cancers. Using anti-interferon serum and Il_2, we focused particularly on the T lymphocyte cultures from breast cancer patients.

Indeed, in 1980, we were able to detect a DNA sequence close to that MMTV, not only in the cells of an inflammatory breast cancer (from a North African woman), but also in her cultured T lymphocytes. A second patient showed similar results.

Unfortunately, the molecular tools we had at that time could not tell us whether we were dealing with endogenous retroviral sequences or with an exogenous virus. Nowadays, having access to more powerful technologies, I am planning to reinitiate these studies.

But in 1983, the same approach, the use of anti-interferon serum, and the use of long term cultures of T lymphocytes greatly facilitated the isolation of HIV.

My involvement in AIDS began in 1982, when the information circulated that a transmissible agent – possibly a virus – could be at the origin of this new mysterious disease. At that time there were only a few cases in France, but they attracted the interest of a group of young clinicians and immunolo-gists. They were looking for virologists, especially retro-virologists, as a likely hypothesis was that HTLV – the only human retrovirus known so far, recently described by R. C. Gallo – could be involved. Retrovirus causing leukaemia in rodents often also causes a wasting syndrome, which could be the result of secondary immune depression. This was also the case of patients suffering from leukaemia induced by HTLV.

A member of the working group, Françoise Brun-Vézinet, was a former student of the virology course that I was then directing. She called me up to organise the search for the putative retrovirus from a patient presenting with an early sign of the disease, lymphodenopathy. The patient was a young gay man who had been travelling to the USA and who was consulting Dr. Willy

Rozenbaum – one of the leaders of the working group – for a swollen lymph node in the neck.

The reasoning was that if we were to find a virus at this early stage of the disease, it could be more a cause than a consequence of the immune depression.

Another incentive to start this research was a request from the producers of hepatitis B virus vaccine in the industrial subsidiary of the Institut Pasteur. They were using plasmas from American blood donors and were concerned by the risk of transmission of the AIDS agent through their procedure of viral antigen purification.

The lymph node biopsy arrived on January 3, 1983, a date which I remember well because it was also the first day of the virology course at the Institut Pasteur, which I had to introduce. I could only dissect the small hard piece at the end of the day. I dissociated the lymphocytes with a Dounce glass homogeniser and started their stimulation in culture with a bacterial mitogen, Protein A, known as an activator of B and T lymphocytes, since I did not know which fraction of lymphocytes could produce the putative virus. Three days later, I added the T cell growth factor I had obtained from a colleague working in the laboratory of Jean Dausset.

The T cells grew well. As previously established in a protocol for the search of retrovirus in human cancers, it was decided with my associates, Françoise Barré-Sinoussi and Jean-Claude Chermann, to measure the RT activity in the culture medium every 3 days. On day 15, Françoise showed me a hint of positivity (incorporation of radioactive thymidine in polymeric DNA), which was confirmed the following week.

We had evidence of a retrovirus, but this was just the beginning of a series of questions:
• Was it close to HTLV or not?
• Was it a passenger virus or, on the contrary, the real cause of the disease?

In order to answer these basic questions, we had to characterise the virus biochemically and immunologically, and to do that, we needed to propagate it in sufficient amounts. Fortunately, the virus could be easily propagated on activated T lymphocytes from adult blood donors. No cytopathic effect was observed with this first isolate, but unlike HTLV infected cultures, no transformed immortalised cell lines could emerge from the cultures, which always died after 3–4 weeks as do normal lymphocytes.

By contrast, subsequent isolates I made from culture of lymphocytes of sick patients with AIDS were cytopathic for T lymphocytes culture and – we discovered later – could be cultivated in larger amounts in tumour cell lines derived from leukaemia.

Shortly after the virus isolation, my co-workers and I were able to show that it was not immunologically related to HTLV, and in electron microscopy, it was very different from HTLV viral particles. In fact, as soon as June 1983, I noticed the quasi-identity of our virus with the published electron microscopy pictures of the visna virus in sheep, the infectious anaemia virus in horses and the bovine lymphocytic virus: it was a retrolentivirus, a sub-family of viruses causing long-lasting disease in animals without immunodeficiency.

This indicated clearly that we were dealing with a virus very different from HTLV, and my task was now to organise a team of researchers to accumulate evidence that this new virus was indeed the cause of AIDS.

It was an exciting period, since every Saturday morning when we had a meeting in my office, new data were brought by my associates favouring the causative role of the virus. The viral isolates were called LAV, for Lymphadenopathy Associated Virus, when it was isolated from patients displaying swollen lymph nodes, a frequent sign of the early phase of infection. The isolates made from patients with full-blown AIDS were called Immuno Deficiency Associated Viruses (IDAV). The latter generally grew better in T lymphocyte culture and induced the formation of large syncitia, resulting from the fusion between several infected cells. Some of them – we found out later – could also multiply in continuous cell lines of B or T cell origin. The latter property greatly facilitated the mass production of the virus for commercial use.

By September 1983, I was able to make a synthesised presentation of all our data favouring a causal link between the virus and the disease at a meeting on the HTLV organised by L. Gross and R. Gallo at Cold Spring Harbor.

This presentation was received with scepticism by a small audience (it was a late night session) and the HTLV theory still prevailed. Mentally, most attendants were not prepared to accept the idea of a second family of retroviruses (lentiretroviruses) existing in humans and causing immune deficiency, and having no counterpart in animals!

This situation is not infrequent in science, since new discoveries often raise controversy. The only problem is that it was a matter of life and death for blood transfused people and haemophiliacs, since a serologic blood test using our virus antigen was already working at laboratory scale but awaited industrial and commercial development.

This came in 1985, after two other teams of researchers, first that of Dr. Gallo at the NIH in early 1984 and that of Jay Levy in San Francisco, confirmed and extended our findings. In particular, Dr. Gallo and his associates gave more strength to the correlation between the virus and the disease, improved the detection of the antibody response and were able to grow several viral strains, including ours, in T cell lines of cancer origin. Meanwhile, my co-workers showed the tropism of the virus for CD4T cells and identified the CD4 surface molecule as the main receptor to the virus.

The rest of the story is described in the next chapter. I would just like to illustrate how I discovered what I believe are two important phenomena for explaining the destruction of the immune system induced by HIV.

During the latent phase of the infection, no virus is found in the blood. It is mostly localised in lymphocytes of lymphatic tissues and yet, we found that most of the lymphocytes present in the blood are sick! In 1987, a young visitor from Sweden, Jan Alberts, came to my lab. He wanted to cultivate human lymphocytes in a serum-free synthetic medium and to learn some technologies about HIV culture. The surprise came when we compared the viability in his medium of lymphocytes from healthy donors and those from HIV infected patients, even in their early asymptomatic stage of infection. While the former could survive several days without dying, the majority (more than 50%) of the latter died very quickly. Addition of interleukin 2 partially prevented their death.

When we used normal culture medium supplemented with foetal calf serum, the same difference was observed, although the survival time of the lymphocytes from infected patients was longer.

It did not take very long before three of my collaborators found the reason for such deaths: apoptosis. This is an active process by which the cell "decides" to die in a clean way, without releasing too many toxic compounds into the medium.

It is a physiological way of preventing abnormal proliferation of activated lymphocyte clones, but here the phenomenon was enormous and bore not only on the main cellular target of HIV infection, CD4+ T-lymphocytes, but also on cells which were not infectable by the virus, such as CD8+ T-lymphocytes, B-lymphocytes, monocytes, natural killer cells… Clearly, it was a general phenomenon, the culture simply revealing a predisposition to apoptosis of the majority of circulating blood cells, although most of them were not infected. Indeed, my collaborator Marie-Lise Gougeon found a very good relation between *in vitro* apoptosis and the *in vivo* observed drop of CD4 T cells in patients.

We have spent a lot of time trying to find the origin of this massive apoptosis, without finding a completely satisfactory explanation: the most likely is the intensive oxidative stress existing in patients since the beginning of their infection. This is also a finding I am very proud of: although oxidative stress has been – and still is – completely overlooked by AIDS researchers, it is likely to aggravate the wrong activation of the immune system at the origin of its decline and also it triggers inflammation through the production of cytokines.

Of course, the next question arises: what are the factors causing oxidative stress: viral proteins, fragments of viral DNA, co-infection with mycoplasmas? Even after 25 years, we still do not know the complete answer. But the phenomenon does exist and needs to be treated, while most AIDS clinicians do not care about it at all!

The treatment by combined antiretroviral therapy has, without doubt, changed the prognosis of this lethal disease, from a death sentence to an almost "normal" life. However, the virus is still there, ready to multiply when the treatment is interrupted, and not all HIV infected patients in the developing world have access to it. And the epidemics still kill 2–3 million people each year. It is thus absolutely necessary to resolve these problems. Basic research, as well as clinical research, has to be continued.

In addition, I realised in the 1990s that research should not only be localised in the wealthy laboratories of the developed countries, but also in southern countries where a lot of patients were suffering from AIDS and many other diseases like tuberculosis and malaria.

Too many examples showed that collaboration between northern and southern research laboratories is unequal, the south providing serum samples to be analysed in the north. This "safari" concept is wrong. There are now many young researchers trained in northern laboratories who would like to return to their own countries, but are prevented from doing so because laboratories and adequate structures are missing. Moreover, one has to be in the regions where disease proliferates to realise how complex the reality is.

This is why I joined with the former Director General of UNESCO, Federico Mayor, in initiating a foundation aimed at creating centres for research and prevention in African countries. Although the task was difficult, this concept was met with enthusiasm from colleagues and medical doctors and also found the support of governments, particularly in Côte d'Ivoire and Cameroon.

I wish that based on these pilot experiments, a whole network of similar centres could cover all the countries of the developing world where the populations are badly hit by epidemics.

Another lesson I drew from my AIDS experience was the weakening effect of oxidative stress on the immune system and its pro-inflammatory role in many chronic diseases, such as Parkinson's, Alzheimer's and rheumatoid arthritis: a likely consequence of chronic infections? Or both consequence and cause? There are many questions, which can be resolved only by hard work and innovative thinking. I hope to be able to continue both.

25 YEARS AFTER HIV DISCOVERY: PROSPECTS FOR CURE AND VACCINE

Nobel Lecture, December 8, 2008

by

LUC MONTAGNIER

World Foundation for AIDS Research and Prevention, UNESCO, 1, rue Miollis, 75732 Paris Cedex 15, France.

The impressive advances in our scientific knowledge during the last century allow us to have a much better vision of our origin on earth and our situation in the universe than our ancestors. Life probably started on earth around three and a half billion of years ago, and a genetic memory emerged early, based on an extraordinarily stable molecule, the DNA double helix, bearing a genetic code identical for all living organisms, from bacteria to men. We are thus the heirs of myriads of molecular inventions, which have accumulated over millions – sometime billions – of years. Environmental pressure has of course both maintained these inventions and also modulated them over the generations, through the deaths of individuals and sexual reproduction. For the last 30,000 years, our biological constitution has not changed: a hypertrophic cortical brain, a larynx to speak and a hand to manipulate. But for the last 10,000 years, another memory has emerged, which make our species quite different from the others: this is the cultural memory which transmits knowledge and societal organisation from generation to generation, through the use of language, writing and more recently virtual means of communication.

This revolution occurred in various sites on the earth almost simultaneously through sedentarisation of human populations by agriculture, leading to several civilisations. Each human being thus receives two pieces of luggage: genetic memory at birth and cultural memory during all his life, and he will become a real human only if he is benefiting from both. For the last three centuries, particularly in the 20th century, our scientific knowledge has increased exponentially and has diffused all over the world.

We have a tendency to consider ourselves as pure spirits, but the hard reality still reminds us of our biological nature: each of us is programmed to die and, during his life, is exposed to diseases. At the dawn of this new century, we are still facing two major health problems:

- New epidemics related to infectious agents (mostly bacteria and viruses)
- Chronic diseases (mostly cancers, cardiovascular, neurodegenerative, arthritic, auto-immune diseases, diabetes) linked to the increase in life expectancy and environmental changes related to human activities.

This presentation will obviously focus on one new epidemic, AIDS, but we should not forget that there are other persistent and life-endangering epidemics, especially in tropical countries, such as malaria and tuberculosis.

Moreover, other new epidemics should not be ruled out as human activities generate more favourable factors:

- Lack or loss of hygiene habits
- lack of water
- globalisation and acceleration of exchange and travel
- atmospheric and chemical pollution leading to oxidative stress and immune depression
- malnutrition, drug abuse and ageing, also leading to immune depression
- global warming leading to new ecological niches for insect vectors
- changes in sexual behaviours.

This last factor and immune depression caused by malnutrition, drug abuse and increased co-infections, are probably the causes of the emergence of AIDS as a global epidemic, affecting most if not all continents, recently including the Polynesian islands.

The causative agent existed in Africa before the emergence of the epidemics in Central Africa and North America in the 1970s. As there exist related viruses apparently well tolerated in non-human primates, it is tempting to consider AIDS as a zoonosis, resulting from the transmission to humans of related viruses infecting primate species without causing disease.

But let us first recall the circumstances of the discovery of HIV in my laboratory at the Institut Pasteur (Figure 1).

Figure 1. The Pasteur team involved in the discovery of HIV in 1983–1984.

AIDS as a pathologically distinct entity was first identified in June 1981 by members of the CDC (particularly Harold Jaffe and James Curran) after reports received from two medical doctors, Michael Gottlieb in Los Angeles and Alvin Friedman-Kien in New York, of clusters of opportunistic infections and Kaposi sarcoma occurring in young gay men and related to sexual intercourse.

Following the publication of this report in the CDC Bulletin, similar cases were described in Western European countries and particularly in France by a group of young clinicians and immunologists led by Jacques Leibovitch and Willy Rozenbaum.

It was soon recognised that a similar disease, characterised at the biological level by a profound depression of cellular immunity and clinically by infections previously described in chemically or genetically immunodepressed patients, also existed in haemophiliacs and blood transfused patients.

The case of haemophiliacs gave a clue as to the nature of the transmissible agent: these AIDS patients had received purified concentrates of factor 8 or 9, made from pools of donated blood which had been filtrated by bacteriological filters.

This purification process should have eliminated any soluble toxic compound and the filtration should have retained bacterial or fungal agents: only viruses could be present in the preparations given to patients. This is why I became interested in a search for viruses; but what kind of viruses? Many viruses have immunodepressive activity, in order to persist in their hosts. This is particularly the case of herpes viruses (cytomegalovirus) and retroviruses. A putative candidate was the Human T Leukemia virus (HTLV) described by R. C. Gallo and Japanese researchers.

Having more expertise on retroviruses (see biography Chapter I), we embarked on the search for an HTLV-like virus, at the suggestion of the French working group and also encouraged by Institut Pasteur Production, an industrial subsidiary of the Institute producing a hepatitis B vaccine from pool of plasmas from blood donors.

Knowing that retroviruses are usually expressed in activated cells, I set up classical conditions to culture activated lymphocytes, using first a bacterial activator of both T and B lymphocytes, Protein A, since I did not know in what subset of cells the virus was hiding out.

The reasoning at that stage was that we should look first at lymphocytes from swollen lymph nodes, supposedly the site where viruses accumulate in the early phase of infection.

On January 3, 1983, I received a biopsy of a patient with cervical adenopathy, a symptom already recognised as an early sign of AIDS. After dissection of the sample into small pieces and their dissociation into single cells, the lymphocytes were cultured in nutrient medium in the presence of Protein A and anti-interferon serum.

In fact, after addition of Interleukin 2, only T lymphocytes were multiplying well and produced a small amount of virus detected by its reverse transcriptase activity, measured by my associate Françoise Barré-Sinoussi. Only some 9 months later could I also show growth of the virus in B-lymphocytes transformed by Epstein-Barr virus (4).

The viral growth ceased as the cellular growth started declining, but we could propagate the virus in cultures of lymphocytes from adult blood donors as well as in lymphocytes from cord blood. This allowed characterisation of the virus, and showed for the first time that it was different from HTLVs. A p24-25 protein could be immuno-precipitated by the serum of the patient and not by antibodies specific to the p24 gag protein of HTLV1, kindly provided by Dr. R. C. Gallo.

Electron microscopy of sections of the original lymph node biopsy, as well as those from infected cultured lymphocytes, showed rare viral particles with a dense conical core, similar to the retrolentiviruses of animals (infectious anaemia virus in horses, visna virus in sheep, etc.), but different from HTLV. Unlike the case of HTLV, we never saw the emergence of permanent transformed lines from the infected lymphocyte cultures (Figure 2).

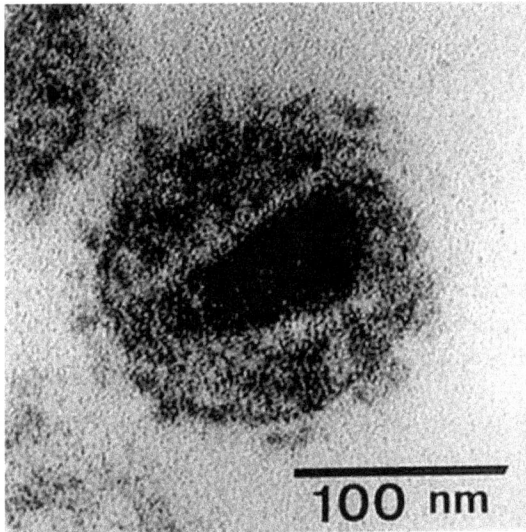

Figure 2. HIV through the electron microscope.

These results were published in a *Science* paper in May 1983 (1), together with two papers by the Gallo and Essex groups in favour of HTLV being the cause of AIDS. During the following months, more data accumulated in my laboratory showing that this new virus was not a passenger virus, but was really the best candidate to be the cause of AIDS.

1) The same type of virus was isolated from patients of different origins: gay men with multiple partners, haemophiliacs, drug abusers, Africans.

First Viral Isolates of the Viral Oncology Unit

Patient initials		Origin	Clinical conditions	Cytopathic effect
Bru	♂	Gay man, caucasian	Pre-AIDS	−
Loi	♂	Haemophiliac, caucasian	AIDS	+
Lai	♂	Gay man, caucasian	AIDS (Ks)	++
Eli	♀	Zaïre, african	AIDS	+

Table 1.

2) Besides immune-precipitation of viral proteins (p25, P18), serums from patients with lymphadenopathy syndrome and a fraction of the serums from patients with advanced AIDS, were positive in an ELISA test using proteins from partially purified virus (2).

3) *In vitro*, the virus was shown to infect only CD4+ T lymphocytes and not the CD8+ subset (3).

4) A cytopathic effect was observed with isolates made from patients with late symptoms of AIDS. Particularly the third isolate made from a young gay man with Kaposi Sarcoma (Lai) caused the formation of large syncitia, presumably due to the fusion of several infected cells (Figure 3). Attempts to grow the first isolate Bru in T cell lines isolated from patients with leukaemia or lymphoma were unsuccessful. However, we discovered later (5) that the Bru isolate was contaminated with the Lai isolate, which by contrast could be grown in T cell lines (CEM, HUT78) in laboratories which received our Bru isolate at their request.

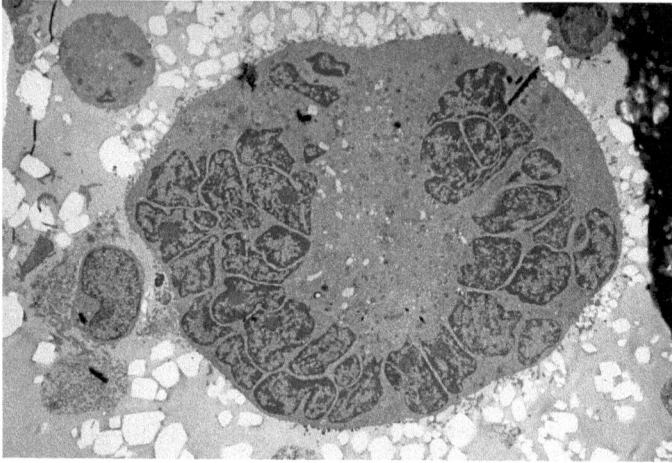

Figure 3. Electron micrography picture of a giant cell (syncitium) resulting of the fusion of many lymphocytes expressing the HIV fusion protein (6).

In fact, a few laboratory isolates were shown to grow in mass quantities in T cell lines, facilitating analysis of the virus and its use for detection of antibodies by commercial blood tests.

Our data, which I presented in September 1983 at a meeting on HTLV in Cold Spring Harbor (6), were met with scepticism. Only in the spring of 1984 did the description of a quasi identical virus under the name of HTLV III by the R. C. Gallo group (7) convince the scientific community that this new retrolentivirus was the cause of AIDS. The Jay Levy group in San Francisco also isolated the same kind of virus (8), followed by many other laboratories.

However, a few opponents led by P. Duesberg argued and are still arguing that there is no real demonstration that the virus does exist and is the cause of AIDS according to Koch's postulates.

In fact, the proviral DNA of the virus, renamed HIV (Human Immunodeficiency Virus) by an international nomenclature committee, was cloned and sequenced (9–11), showing the classical gene structure of animal retroviruses which Dr. Duesberg helped himself to uncover earlier.

But in addition, new genes (Tat, Nef), important in regulation of the expression of the viral genetic information, were recognised from the DNA sequencing, making the viral genome probably the most complex known in the retrovirus family (Figure 4). HIV and its primate cousins are therefore a well-characterised entity only composed of DNA sequences, none existing in the human genome.

HIV-1

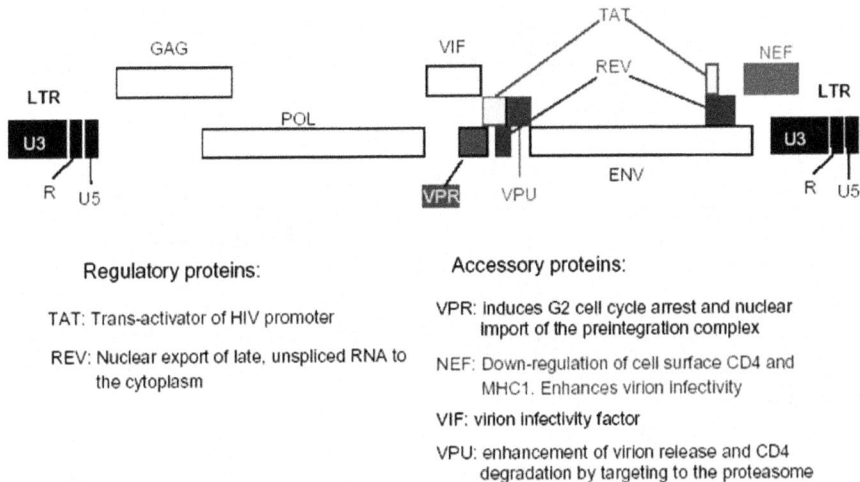

Regulatory proteins:

TAT: Trans-activator of HIV promoter

REV: Nuclear export of late, unspliced RNA to the cytoplasm

Accessory proteins:

VPR: induces G2 cell cycle arrest and nuclear import of the preintegration complex

NEF: Down-regulation of cell surface CD4 and MHC1. Enhances virion infectivity

VIF: virion infectivity factor

VPU: enhancement of virion release and CD4 degradation by targeting to the proteasome

Figure 4. Genome structure of HIV1: gag, pol, env are the gene codes for the structural proteins.

A posteriori, two facts should have provided to the few remaining sceptics final proof that HIV is the culprit in AIDS:

1) Transmission of AIDS by blood transfusion has practically disappeared in countries where the detection of HIV antibodies in blood donors has been implemented;

2) The inhibition of virus multiplication by a combination of specific inhibitors of the viral enzymes (reverse transcriptase, protease), has greatly improved the clinical conditions of patients. Mutations in the genome of HIV inducing resistance to these inhibitors has led to relapses and aggravation of the patients' condition.

In 1986, thanks to our collaboration with Portuguese colleagues, we isolated a second virus (which I named HIV2), from West African patients hospitalised in a Lisbon hospital (12). They all had the signs of AIDS but had no antibodies against our first virus. In fact, they had only antibodies to the most variable protein of HIV, the surface glycoprotein. The patients had lost antibodies against the well-conserved internal proteins of HIV2 which show common epitopes with their counterparts of HIV1, unlike the glycoprotein (Figure 5).

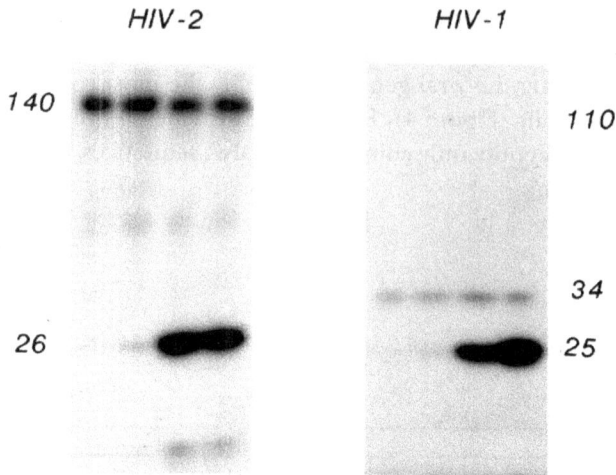

Figure 5. Immunoprecipitation of radioactively labelled proteins of HIV1 and HIV2 by a serum of an AIDS patient infected by HIV2; note in the HIV2 panel the precipitation of the envelope protein.

The isolation of HIV1 (6) and HIV2 (12) viruses from AIDS patients in Africa made us realise that we were dealing with a large epidemic of heterosexually transmitted viruses.

Evidence that HIV was not transmitted by casual contacts came from our study in a French boarding school where HIV infected haemophilic children were in close contact, day and night, with HIV negative non-haemophilic children: none of the latter was found HIV positive (13).

The isolation of the virus causing AIDS allowed the implementation of rational prevention measures and also the beginning of a search for efficient viral inhibitors.

The first candidate, azidothymidine, was an efficient inhibitor of HIV reverse transcriptase in *in vitro* experiments (Broder, Mitsuya *et al.*); and however, its use in AIDS patients, first looking promising, was later recognised as disappointing (14).

In fact, the treatment readily induced mutants of the virus resistant to AZT and did not extend the life span of the patients. The main obstacle to treatment with a single or two inhibitors was the capacity of the virus to mutate, which also impedes the design of an efficient vaccine and explains the complexity of the pathophysiology of AIDS.

Only a combination of three inhibitors proved to be efficient in the clinical outcome. Since 1996, clinicians are using HAART (Highly Active Antiretroviral Therapy) to treat patients with high virus load and low CD4+ T cell number, preventing them most of the time from contracting lethal opportunistic infections (15).

Some milestones in the Research of AIDS

1981	Identification of the disease in the USA
1983	First isolation of HIV
1984	Confirmation of HIV as the causal agent of AIDS – Biological and molecular characterization
1985	First blood test to eliminate transmission of HIV by blood transfusion
1986	Isolation of HIV-2
1986	First use of AZT as an antiretroviral drug
1991	Apoptosis as a mechanism of cell death in AIDS
1995	Decrease of HIV perinatal transmission with AZT
1995	Demonstration of high rate of HIV replication during the silent period of infection
1996	Identification of HIV main co-receptors
1996–97	Generalization of HAART in developed countries

Table 2.

HIV VARIABILITY

In fact, in order to escape the immune reactions of their hosts, most viruses have a strategy of changing their immunogenic epitopes. In the case of HIV, a conjunction of several factors put this at an unprecedented level.

I have listed below the factors which seem to be most responsible for this variability.

1. Errors of reverse transcription
2. Genetic recombination
3. Incomplete neutralisation by Vif of the activity of the APOBEC3G cellular gene
4. Oxidative stress

The first is that the replicative enzyme, reverse transcriptase (RT), has no editing compensation, so that the transcription errors may reach $1/10^5$ nucleotides, far from $1/10^9$ of the cellular DNA polymerases.

However some other retroviruses, such as HTLV, do not show this variation rate, since once integrated, the proviral DNA remains replicated by the cellular DNA replicative machinery. The difference could be explained by the fact that the HIV infected cells die, so that the virus can maintain itself only by many cycles of new infections involving each time reverse transcription of its RNA into DNA. However, in *in vitro* infection of cell lines, also involving a cytopathic effect and many cycles of re-infection, the virus seems to be stable, in the absence of immunoselective pressure.

Another variation factor is genetic recombination. The immune responses (humoral and cellular) against the virus are unable to prevent a second virus infection of the host (because of virus variability induced by the previous factor and other causes), so that some cells could be co-infected by two viruses: this will also allow genetic recombination between the two viral RNAs existing each in two copies. The result is a "mosaic" virus in which many sequences from the two original viruses are entangled, starting from "hot spots" of recombination. This is particularly visible in Africa, probably because of repeated exposure to infection in many patients. The mosaic viruses, because of their selective advantage, then disseminate in the infected population. The original subtypes called A B C D E G... defined by the sequence of their envelope gene are thus replaced by A/G, B/C etc... depending on the geographic location.

Moreover, two other factors have been more recently identified: In the lymphocytes are expressed a family of genes coding for enzymes able to convert guanosine into adenosine in the viral DNA, fouling the viral genetic code (APOBEC3G). However, the virus has evolved a gene, Vif, which can more or less counteract this effect, rendering viable the viral DNA without completely avoiding mutations (16).

A last variability factor, whose importance has probably been overlooked, is oxidative stress (see below), a cause of RNA and DNA mutations (before integration of the proviral DNA): highly reactive molecules derived from oxygen can oxidise the bases, particularly guanine or deoxyguanine, thus modifying their coding capacity or inducing a wrong replacement in repair.

A combination of these factors could explain both the intrinsic variability of the virus in the host during the long evolution of infection, and also the increasing variability of the circulating strains as the epidemic is spreading in various populations.

We can at least act on this variability by decreasing the viral multiplication rate inside the host by antiretroviral treatment and also by neutralising the oxidative stress.

THE REMAINING PROBLEMS:

How HIV infection results in the destruction of the immune system
In the early years following the virus discovery, it was generally thought that the drop in CD4+ T cells was due to their direct infection by a cytopathic virus.

In fact, the viral isolates (like Bru) made in the early stage of the disease are not cytopathic; after binding to the CD4+ receptor of activated lymphocytes, they use a co-receptor (CCR5) which is the receptor for a chimiokine.

Only viruses isolated from patients at late stages of the disease are cytopathic (like Lai), and their direct infection of the remaining T lymphocytes (by using another chemokine co-receptor CXC4) could account for the final drop in these cells.

In fact, the number of activated CD4+ T lymphocytes (the ones which only allow full replication of the virus), is probably a limiting factor in the initial infection, after the first contact with dendritic cells and monocytes of genital or rectal mucosa. It is obvious that inflammation and co-infections (bacterial, viral) could increase the number of activated T lymphocytes and therefore could increase the risk of HIV infection.

Recently, the virus has been found associated with the Peyer patches existing around the small intestine which constitutes a major source of activated T lymphocytes.

At the onset of infection, the virus replication is high in all the lymphatic tissues, taking advantage of the delay in the reaction of the immune system (in time order, interferon, NK cells, CD8T cells, antibody response) and then decreases while persisting in some lymph nodes (Figure 6).

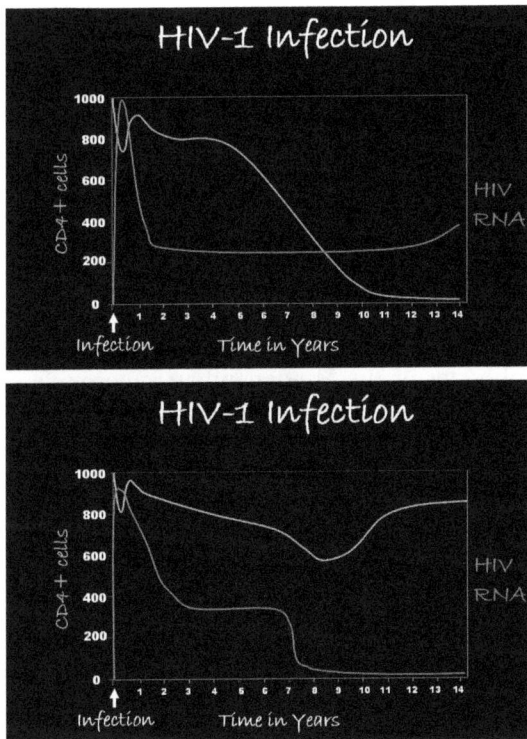

Figure 6. Evolution of HIV-1 infection in AIDS. Left: Untreated patients; Right: Patients treated by antiretroviral therapy at year 6.

This is the beginning of the chronic phase, which is generally asymptomatic, although lymphadenopathy is often present. It has been shown that the virus replication continues in the lymph nodes, despite the immune response. This starts declining, although there is a continuous renewal of T lymphocytes, both CD4+ and CD8+, which could last for years.

During this period, we have found two phenomena which could help explain the indirect destruction of the immune system:

One biological: apoptosis

One biochemical: oxidative stress

Apoptosis: my laboratory was the first to describe this programmed cell death in white blood cells cultured in a medium deprived of interleukin 2 (17). All the subsets, not only the CD4+ T cells, were affected when taken from the blood of asymptomatic HIV patients as well as from patients presenting with full blown AIDS: CD8+ T cells, NK cells, B lymphocytes, monocytes.

However, we found a good correlation between the drop in CD4+ T cells in patients and this in *in vitro* phenomenon (18) We surmised that in the *vivo* situation, cells were still alive but in pre-apoptosis.

Indeed, we could detect in infected patients a general phenomenon of immune activation (19), which has now become well recognised as a major factor of AIDS pathogeny.

At the biochemical level, we also showed that the lymphocyte population of asymptomatic patients (CD4+, CD8+, NK) displayed the biochemical signs of oxidative stress (excess of free radicals derived from oxygen): namely fast degradation of oxidised proteins, carbonylation of some of their amino acids (20). In the patients' blood, we could detect a similar hyper-oxidation of plasma lipids (21) and oxidisation of guanine.

What could be the origin of this strong oxidative stress? At least one HIV protein may contribute to it. It was shown by C. Flores, McCord and their collaborators that the Tat protein, among many functions, inhibits the expression in lymphocytes of the Mn-dependent superoxide dismutase gene (22). This enzyme is key to the transformation of the anion superoxide, highly oxidant into hydrogen peroxide. Tat has been shown to circulate in nanogram amounts in the blood of infected patients and to penetrate inside cytoplasm.

In addition, bacterial and viral co-infectors can also induce oxidative stress. We studied the possibility that a "cold" persisting bacterial infection could co-exist in HIV-infected patients.

These studies were initiated because we observed that *in vitro* co-infection of lymphocytes with some mycoplasma species (M. pirum, M. penetrans, M. fermentans) and HIV could greatly reinforce the cytopathic effect of the latter.

Moreover, these small bacteria lack catalase, an enzyme able to convert hydrogen peroxide into water. Therefore they also generate oxidative stress and, furthermore, are activators of lymphocytes (23).

In summary, the pathophysiology of AIDS is complex. HIV is the main cause, but could also be helped by accomplices and also have some indirect effects by wrongly activating the immune system through oxidative stress.

PROSPECTS FOR THE FUTURE

No cure. No vaccine, but maybe a cure by a vaccine.

The advent of HAART has transformed AIDS into a tolerable infection, but whatever the length of the treatment, the inhibitors used have not reached the level of a cure! As soon as this treatment is interrupted, virus multiplication resumes within a few weeks and the immune system declines again.

This observation led researchers to think that there is a reservoir of virus, to which the drugs have no access (24), probably because the virus stays in a latent form in some tissues.

Our project is to design quantitative tests to evaluate the size of this reservoir and to prevent it from giving rise to actively multiplying virus by boosting the immune system against the most conserved parts of viral proteins.

A schematic protocol of this therapeutic immunisation, aimed at achieving a functional eradication of HIV (25), could be the following:

1) First, antiretroviral therapy (HAART) for 3–6 months to reduce viral load in the plasma to undetectable levels and maintain it until the protocol has been terminated.

2) Then, treatment by antioxidants and immunostimulants such as an orally absorbable form of glutathione to reduce the oxidative stress induced by viral proteins and by HAART. Reduced glutathione is known to induce a shift from Th_2 to TH_1 responses, therefore reinforcing cell-mediated immunity. Its effect can be enhanced by some synthetic immunostimulants, which are now close to approval for clinical use by regulatory authorities.

3) After a two-week treatment by the former products, start specific immunisation against HIV proteins by a therapeutic vaccine. Trials with vaccine preparation made for a therapeutic use have already been carried out, with mixed results probably because the immune system of the patients was not sufficiently restored, or/and also due to the inadequacy of the immunogens. Our genetic engineering data indicate that the native HIV glycoprotein must be modified in order to make immunogenic the most conserved parts of the protein, including the pocket involved in HIV binding. This will result in a neutralisation capacity broad enough to cover potential escape mutants. I also advise adding during vaccine preparation two other proteins involved in immunosuppression, Tat and Nef, modified to become non functional while remaining immunogenic.

4) After this vaccination, interrupt HAART. If the protocol has been successful, there will be no virus rebound, as evidenced by a low viral load and an increase of the CD4+ T cell component. Regular monitoring of these two parameters will assess the durability of the immunisation. A strong cell-mediated immunity, in addition to the induction of neutralising antibodies, will permit the interruption of a cycle of new cell infections by newly formed viral particles. This control already exists spontaneously in a small number of HIV-infected patients, which show no immune depression even after many years.

This protocol is complex, but it will be less expensive and for the patient much more tolerable than life-long antiretroviral therapy.

The protocol can also be applied to patients in the early stages of HIV infection, perhaps with a better chance of success, as their immune system will have a better ability to respond.

If, in this optimistic scenario, HIV infection becomes a curable disease, the impact on the epidemic itself will be considerable: in developing countries, HIV infection represents a stigma for family and professional life. Many infected individuals do not want to be tested and to learn their status, and as consequence, they keep transmitting the virus to new partners. The prospect of being treated and cured immediately after the diagnosis of HIV infection will ease early testing and emergence of responsible behaviours.

Moreover, the success of a therapeutic vaccine will facilitate the design of an efficient preventive vaccine, based on the same viral components.

Meanwhile, it will be essential to make accessible the use of antiretroviral drugs to all patients who are eligible for them. This implies not only an international effort to lower the price of these drugs, which has already been partly achieved, but also a comparable effort to create adequate medical structures with trained doctors and research centres in developing countries. Our Foundation has chosen the mission to contribute to fulfilling these tasks in Africa.[*]

Figure 7. Centre Intégré de Recherche Bioclinique d'Abidjan (CIRBA), created in 1996.

[*] The World Foundation for AIDS Research and Prevention, in association with UNESCO and local governments, has created two Centres for AIDS Research and Prevention; The "Centre Intégré de Recherches Biocliniques d'Abidjan" – CIRBA (Figure 7) in Abidjan (Côte d'Ivoire) and the International "Chantal Biya" Reference and Research Centre for HIV-AIDS Prevention and Care-taking" (CIRBC) in Yaoundé (Cameroon).

REFERENCES

1. F. Barré-Sinoussi, J. C. Chermann, F. Rey, M. T. Nugeyre, S. Chamaret, J. Gruest, C. Dauguet, C. Axler-Blin, F. Vezinet-Brun, C. Rouzioux, W. Rozenbaum & L. Montagnier, "Isolation of a T-lymphotropic retrovirus from a patient at risk for acquired immune deficiency syndrome (AIDS)", *Science*, **220**, 868–871 (1983).

2. F. Brun-Vezinet, F. Barré-Sinoussi, A.G. Saimot, D. Christol, L. Montagnier, C. Rouzioux, D. Klatzmann, W. Rozenbaum, J.C. Gluckman & J.C. Chermann, "Detection of IgG antibodies to lymphadenopathy-associated virus in patients with AIDS or lymphadenopathy syndrome", *The Lancet*, 1253–1256 (1984).

3. D. Klatzmann, F. Barré-Sinoussi, M. T. Nugeyre, C. Dauguet, E. Vilmer, C. Griscelli, F. Brun-Vezinet, C. Rouzioux, J. C. Gluckman, J. C. Chermann & L. Montagnier, "Selective tropism of lymphadenopathy associated virus (LAV) for helper-inducer T-lymphocytes," *Science*, **225**, 59–63 (1984).

4. L. Montagnier, J. Gruest, S. Chamaret, C. Dauguet, C. Axler, D. Guetard, M. T. Nugeyre, F. Barré-Sinoussi, J. C. Chermann, J. B. Brunet, D. Klatzmann & J. C. Gluckman, "Adaptation of Lymphadenopathy Associated Virus (LAV) to replication in EBV-transformed B lymphoblastoid cell lines," *Science*, **225**, 63–66 (1984).

5. S. Wain-Hobson, J. P. Vartanian, M. Henry, N. Chenciner, R. Cheynier, S. Delassus, L. Pedroza Martins, M. Sala, M. T. Nugeyre, D. Guetard, D. Klatzmann, J. C. Gluckman, W. Rozenbaum, F. Barré-Sinoussi & L. Montagnier, "LAV revisited: origins of the early HIV-1 isolates from Institut Pasteur", *Science*, **252**, 961–965 (1991).

6. L. Montagnier, J. C. Chermann, F. Barré-Sinoussi, S. Chamaret, J. Gruest, M. T. Nugeyre, F. Rey, C. Dauguet, C. Axler-Blin, F. Vezinet-Brun, C. Rouzioux, G. A. Saimot, W. Rozenbaum, J.C . Gluckman, D. Klatzmann, E. Vilmer, C. Griscelli, C. Foyer-Gazengel & J. B. Brunet in *Human T cell leukemia/lymphoma viruses* (R. C. Gallo, M. E. Essex & L. Gross, eds.), "A new human T-lymphotropic retrovirus: characterization and possible role in lymphadenopathy and acquired immune deficiency syndromes", Cold Spring Harbor Laboratory, New York, 363–379 (1984).

7. M. Popovic, M. G. Samgadharan, E. Read, R. C. Gallo, , "Detection, isolation and continuous production of cytopathic retroviruses (HTLV-III) from patients with AIDS and pre-AIDS", *Science*, 1984; **224**: 497–500.

8. J. A. Levy, A. D. Hoffman, S. M. Kramer, J. A. Lanois, J. M. Shimabukuro & L. S. Oskiro, "Isolation of lymphocytopathic retroviruses from San Francisco patients with AIDS", *Science*, 1984, **225**, 840–842.

9. M. Alizon, P. Sonigo, F. Barré-Sinoussi, J. C. Chermann, P. Tiollais, L. Montagnier, S. Wain-Hobson, "Molecular cloning of lymphadenopathy-associated virus", *Nature*, **312**, no. 20/27, 757–760 (December 1984): printed version, manuscript.

10. S. Wain-Hobson, P. Sonigo, O. Danos, S. Cole, M. Alizon, "Nucleotide sequence of AIDS virus, LAV", *Cell*, 1985; **40**, 9–17.

11. L. Ratner, W. Haseltine, R. Patarca, K. J. Livak, B. Starcich, S. F. Josephs, E. R. Doran, A. Rafalski, E. A. Whitchorn, K. Baumeister, L. Ivanoff, S. R. Petteway, M. L. Pearson, J. A. Lautenbergen, T. S. Papas, J. Ghrayeb, N. T. Chang, R. Gallo, C. & F. Wong-Staal, "Complete nucleotide sequence of the AIDS virus, HTLV-III," *Nature* (Lond.), 1985, **313**, 277–284.

12. F. Clavel, D. Guetard, F. Brun-Vezinet, S. Chamaret, M. A. Rey, M. O. Santos-Ferreira, A. G. Laurent, C. Daughet, C. Katlama, C. Rouzioux, D. Klatzmann, J. L. Champalimaud, & L. Montagnier, "Isolation of a new human retrovirus from West African patients with AIDS", *Science*, 1986, **233**, 343–346.

13. S. Berthier, R. Chamaret, J. Fauchet, N. Fonlupt, M. Genetet, M. Gueguen, A. Pommereuil, Ruffault & L. Montagnier, "Transmissibility of human immunodeficiency virus in haemophilic and non-haemophilic children living in a private school in France", *The Lancet*, 598–601 (13 September 1986).

14. M. Seligmann, D. A. Warrel, J.-P. Aboulker, C. Carbon, J. H. Darbyshire, J. Dormont, E. Eschwege, D. J. Girling, D. R. James, J.-P. Levy, P. T. A. Peto, D. Schwarz, A. B.

Stone, I. V. D. Weller, R. Withnall, K. Gelmon, E. Lafon, A. M. Swart, V. R. Aber, A. G. Babiker, S. Lhoro, A. J. Nunn & M. Vray, "Concorde: MCR/ANRS randomised double-blind controlled trial of immediate and deferred zidovudine in symptom-free HIV infection", *The Lancet,* **343**, 871–881, 1994.

15. M. Zuniga. A. Whiteside. A. Ghaziani. J.G. Bartlett. Preface By L. Montagnier/R. Gallo., *A Decade of HAART. The Development and Global Impact of Highly Active Antiretroviral Therapy,* J. Oxford University Press (2008).

16. A. M. Sheehy, N. C. Gaddis, J. D. Choi, M. H. Malim (2002)", "Isolation of a human gene that inhibits HIV-1 infection and is suppressed by the viral Vif protein", *Nature* **418**, 646–650.

17. M. L. Gougeon, R. Olivier, S. Garcia, D. Guetard, T. Dragic, C. Dauguet & L. Montagnier, "Mise en évidence d'un processus d'engagement vers la mort cellulaire par apoptose dans les lymphocytes de patients infectés par le VIH", *C. R. Acad. Sci. Paris,* t. **312**, Série III, 529–537 (1991).

18. M. L.Gougeon, S. Garcia, J. Heeney, R. Tschopp, H. Lecoeur, D. Guetard, V. Rame, C. Dauguet & L. Montagnier, "Programmed cell-death in AIDS-related HIV and SIV infections", *AIDS Res. and Hum. Retrov.* **9**, 553–563 (1993).

19. M. L. Gougeon, H. Lecoeur, A. Dulioust, M. G. Enouf, M. Crouvoisier, C. Goujard, T. Debord & L. Montagnier, "Programmed cell death in peripheral lymphocytes from HIV-infected persons: Increased susceptibility to apoptosis of CD4 and CD8 T cells correlates with lymphocyte activation and with disease progression", *The J. of Immunol.,* **156**, 3509–3520 (1996).

20. G. Piedimonte, D. Guetard, M. Magnani, D. Corsi, I. Picerno, P. Spataro, L. Kramer, M. Montroni, G. Silvestri, J.F. Torres-Roca & L. Montagnier, "Oxidative protein damage and degradation in lymphocytes from patients infected with human immunodeficiency virus", *J. of Infect. Dis.,* **176**, 655–664 (1997).

21. O. Lopez, D. Bonnefont-Rousselot, M. Mollereau, R. Olivier, L. Montagnier, J. Emerit, M. Gentilini & J. Delattre, "Increased plasma thiobarbituric acid-reactive substances (TBARS) before opportunistic infection symptoms in HIV-infected individuals", *Clin. Chim. Acta,* **247**, 181–187 (1996).

22. M. Sevea, A. Faviera, M. Osmana, D. Hernandez, G. Vaitaitis, N. C. Flores, J. M. McCord, S. C. Flores, "The Human Immunodeficiency Virus-1 Tat Protein Increases Cell Proliferation, Alters Sensitivity to Zinc Chelator-Induced Apoptisis, and Changes Sp1 DNA Binding in HeLa Cells*1", *Biochemistry & Biophysics* Vol. **361**, 165–172 (1999).

23. Y. Sasaki, A. Blanchard, H.L. Watson, S. Garcia, A. Dulioust, L. Montagnier & M.L. Gougeon, "*In vitro* influence of Mycoplasma penetrans on activation of peripheral T lymphocytes from healthy donors or human immunodeficiency virus-infected individuals", *Infect. Immun.* **63**, 4277–4283 (1995).

24. D. D. Richman, D. M. Margolis, M. Delaney, Warner C. Grenne, D. Hazuda, R. J. Pomerantz, "The Challenge of Finding a Cure for HIV Infection", *Science* **323**, 1304–1307 (2009).

25. L. Montagnier, "Toward functional eradication of HIV infection", *Future HIV Ther* (2007) Vol. 1, 3–4.

Portrait photo of Luc Montagnier by photographer Ulla Montan.

Physiology or Medicine 2009

Elizabeth H. Blackburn, Carol W. Greider and Jack W. Szostak

"for the discovery of how chromosomes are protected by telomeres and the enzyme telomerase"

THE NOBEL PRIZE IN PHYSIOLOGY OR MEDICINE

Speech by Professor Rune Toftgård of the Nobel Assembly at Karolinska Institutet. Translation of the Swedish text.

Your Majesties, Your Royal Highnesses, Ladies and Gentlemen,

Every human being, with all the different cell types and tissues, develops from a single fertilised egg cell by means of repeated cell divisions. The structural map found inside our genetic material is copied in its entirety before each cell division. Each cell needs to carry the complete map inside it.

By the mid-20th century, it was clear that genetic material is made up of DNA strands inside the cell nucleus. These DNA strands are packed into chromosomes, numbering 46 in humans. Two past Nobel Laureates, Hermann Muller and Barbara McClintock, long ago observed that the ends of the chromosomes, which Muller named the telomeres, were special and particularly stable. One unanswered question was: In what way were these telomeres different, and what was their actual function?

Another question: How can telomeres be fully copied? Based on the knowledge that was available in the late 1970s, the DNA in chromosomes should become shorter each time a cell divided. But this is not what actually happens.

The first question – about the function of the telomeres – was answered after Elizabeth Blackburn and Jack Szostak met at a scientific conference in 1980 and began their research collaboration. In a pioneering experiment, they demonstrated that telomere DNA from one organism, a unicellular ciliated protozoan called *Tetrahymena thermophila*, would protect and stabilise chromosomes in an entirely different organism, yeast. The natural chromosomes in yeast cells were also shown to contain similar DNA sequences with the same function. Today we know that the protective function of telomeres is strongly conserved in the evolutionary chain and is present in all higher organisms, including us humans.

The second question – how telomere DNA could be formed and avoid becoming shorter each time a cell divided – was answered in an elegant way by Elizabeth Blackburn and Carol Greider. They discovered an enzyme that can produce telomere DNA. The very first proof that this enzyme existed came as a fantastic Christmas present, on Christmas Day 1984. The enzyme was given the name telomerase. It turned out to be unique in its structure and to consist of an enzymatically active protein component plus an RNA component that serves as a template for the formation of new telomere DNA.

Knowledge of telomeres and telomerase has led to important medical insights in many different fields.

One group of rare congenital diseases is caused by mutations that impair the functioning of telomerase. These diseases are characterised by bone marrow defects that result in reduced formation of new blood cells and can now be diagnosed with certainty.

If telomeres are not preserved, eventually cells cannot survive. Telomere shortening is thus one of several factors that affect the ageing process. In contrast, cancer cells can divide endlessly and nearly all of them have elevated telomerase activity. There is consequently hope that new drugs which target telomerase can be developed to fight cancer, and a number of clinical trials are under way.

The discovery of the vital role of telomeres in preserving chromosomal and genetic stability was made through research driven by curiosity and aided by simple model organisms. The findings of this research have provided fundamental insights into human biology and disease mechanisms.

Professors Blackburn, Greider and Szostak,

Your discoveries have revealed how the conserved function of telomere repeat sequences can maintain chromosomal and genetic stability and how synthesis of telomere DNA is achieved by the enzyme telomerase. In this way you have solved a long-standing and fundamental problem in biology, provided insight into disease mechanisms and raised hope that new therapies can be developed by targeting telomerase.

Today, the one hundredth Nobel Prize in Physiology or Medicine is being awarded. On behalf of the Nobel Assembly at Karolinska Institutet, it is my great privilege to convey to you our warmest congratulations. I now ask you to step forward to receive your Nobel Prizes from the hands of His Majesty the King.

Elizabeth Blackburn

ELIZABETH H. BLACKBURN

CHILDHOOD

I was born in the small city of Hobart in Tasmania, Australia, in 1948. My parents were family physicians. My grandfather and great grandfather on my mother's side were geologists. My great-grandfather on my father's side, before coming to Australia as a minister of the Church of England, had lived for some time in Hawaii, where he had collected Coleoptera (beetles). He continued his collecting in Australia, eventually selling his collection to the British Museum of Natural History. My uncle and aunt (my father's sister and my mother's brother) were also both family physicians, who moved to England, married there and permanently settled there to practice medicine and raise their families.

I was the second child of eventually seven siblings. I spent my first 4 years living in the tiny town of Snug, by the sea near Hobart. Curious about animals, I would pick up ants in our backyard and jellyfish on the beach. Then my family moved to Launceston, a town in northern Tasmania. Our first house, at 120 Abbott Street, was a one-storied, verandahed house of typical Australian suburban architecture. I started kindergarten at a girls' school, Broadland House Girls Grammar School in Launceston (Figure 1).

Figure 1. Elizabeth Blackburn (right) and her sister Katherine ready for Elizabeth's first day at school in Launceston, Tasmania. Circa 1953.

I kept tadpoles in rapidly-smelly-becoming glass jars in a back living room at home. When I was a preteen we moved to a larger house called Elphin House, which had a good-sized garden (Figure 2). Over the years we had many pets: at one stage I enumerated the family menagerie of the moment as consisting of budgerigars and canaries in an aviary in one corner of the garden, goldfish in a garden pond, chickens and pullets (for eggs and the occasional roast fowl) in a hen coop and henhouse, rabbits and guinea pigs in cages, and cats and a dog, who lived all over the house and garden. I was fond of all these animals, and of animals and nature in general.

Figure 2. All seven Blackburn family siblings in the garden at their home, 3 Olive Street, Launceston, Tasmania. From left to right, back row: Andrew, Elizabeth, Katherine, John, Barbara; front row: Caroline, Margaret. Circa 1965.

Perhaps arising from a fascination with animals, biology seemed the most interesting of sciences to me as a child. I was captivated by both the visual impact of science through science books written for young people, and an idea of the romance and nobility of the scientific quest. This latter was especially engendered by the biography of Marie Curie, written by her daughter, which I read and reread as a child. By the time I was in my late teens it was clear to me that I wanted to do science. I was educated at Broadland House Girls Grammar School, and received a generally excellent education. However, physics was not offered, so I took physics classes offered in the evenings at the local public high school. Latin and Greek were not taught at my girls' school either, a gap in my education that I rather regret later in life. But my school did provide an excellent piano teacher, Helen Roxburgh, by whom I was taught all throughout my school years in Launceston. I loved playing the piano, and even at one time wistfully hoped that I might become a musician. Fortunately I was also quite realistic about this, because I recognized that I was competent rather than greatly talented at piano playing, so I went in the direction of science.

My family moved, after some family disruptions, to the city of Melbourne, Australia, in time for me to complete my last year of high school at University High School. There I gained the confidence that I needed to apply for the undergraduate science degree at the University of Melbourne.

UNIVERSITY EDUCATION

I chose biochemistry as my major and graduated after 4 years with an Honours degree in Biochemistry. During that time, I had come to love biochemistry research, although I was just getting my feet wet in laboratory research.

The Chair of the Biochemistry Department, Frank Hird, then offered me a position as a Master's student in his research laboratory, where they investigated the biochemistry of amino acid metabolism. My undergraduate Honours thesis research advisors, Theo Dopheide and the late Barrie Davidson, had advised and encouraged me to do my Ph.D. abroad. Barrie in particular had urged me to consider going to the MRC Laboratory of Molecular Biology (LMB) in Cambridge, England, where he had done post-doctoral research. But in order to be accepted as a Cambridge Ph.D. student in biology, those from outside Britain were required to have done a year of research. The Master's degree with Frank Hird, studying the metabolism of glutamine in the rat liver, would constitute this required year.

Frank Hird taught his laboratory group members the joy and aesthetics of research. He said he thought each experiment should have the beauty and simplicity of a Mozart sonata. His laboratory group, dominated by his strong personality, was cohesive and we would sometimes drive to the hilly areas out-side Melbourne, all piled into his car, Mozart playing loudly on the car radio, to have an outdoor lunch picnic among trees and wildflowers.

While I was still in Frank Hird's lab, Fred Sanger visited Melbourne. Frank Hird had done research in England on amino acids with Fred Sanger shortly after the Second World War, providing an introduction to Fred in which Frank encouraged me to tell Fred of my hope to study for my Ph. D. in Cambridge at the LMB. It was arranged that I would join Fred's lab in the LMB and I was admitted as a Cambridge Ph. D. student.

The adventure of setting off to England, away from home and family, was a huge step, but I felt ready. My aunt and uncle and their family in Cambridge, who lived close to the LMB, became my anchor of a family away from home. I loved the LMB, the science being done there, the atmosphere of being at the epicenter of molecular biology, the intensity of the scientists and the constant discussions about science. It was a world of complete immersion. For my Ph.D. thesis research, I carried out sequencing of regions of bacterio-phage phiX 174, a small single stranded DNA bacteriophage. I transcribed fragments of the phage DNA into RNA and then used the methods that Fred had pioneered for piecing together RNA sequences. We combined the sequences derived by this method with the DNA sequencing that had been done by John Sedat and Ed Ziff and Francis Galibert, members of Fred's lab. All the sequences jibed. The first sequence of a 48 nucleotide fragment of this tiny bacteriophage DNA genome was a great excitement. I took it to show to my mathematically-talented Cambridge cousin, who was then about 12 years old, to see if any patterns emerged to his mathematically-inclined eye. He pointed out the repeats, but it was premature to think of analyzing DNA sequence patterns!

TO THE UNITED STATES

The world of discovering DNA sequences was opening up and I was entranced by its possibilities. I had planned to do a postdoctoral fellowship, beginning in 1975, with Howard Goodman and his close associate Herb Boyer of UCSF, a mutual decision made after an interview–cum-conversation Herb and I had walking in the garden of a monastery in Belgium at which we were attending a scientific conference, I still as a graduate student. But then love intervened: John Sedat and I decided to marry, and as John was going to Yale, I decided to see if I could change my postdoctoral research plans (for which I had obtained an Anna Fuller Fellowship to work at UCSF) to a laboratory at Yale. Howard Goodman wrote me a kind and understanding letter upon my letting him know the reasons for my change of plan, and I began inquiries into possibilities of a laboratory for my postdoctoral training at Yale.

Thus it was that love brought me to a most fortunate and influential choice: Joe Gall's lab at Yale. After a few hiccups engendered by misplaced international mail and other factors, at the beginning of 1975 Joe accepted me as a postdoctoral fellow in his lab, to which I was allowed to transfer my Anna Fuller Fellowship. I immediately began to work on finding ways to accomplish the sequencing of the DNA found at the terminal regions of the abundant, short, linear ribosomal gene-carrying "minichromosomes" that Joe and his colleagues, in parallel with Jan Engberg of Denmark, had discovered in the somatic nucleus of the ciliated protozoan *Tetrahymena thermophila* (which was at that time called *Tetrahymena pyriformis*, shortly thereafter to be renamed *Tetrahymena thermophila*)

TO THE UNIVERSITY OF CALIFORNIA

After finishing my postdoctoral training in Joe Gall's lab at the end of 1977, John Sedat and I, having married in 1975, moved to San Francisco, California. There John had accepted a position as Assistant Professor at the University of California San Francisco (UCSF). I had applied for several positions as an Assistant Professor in a variety of Universities and had been rejected from many of them, a discouraging experience. I had applied for such a position in the Department of Molecular Biology at the University of California Berkeley, but had not yet heard whether I was in the running for it. In the meantime, UCSF offered me a research track position and space in the Department of Biochemistry in the Genetics unit headed by Herb Boyer. My first NIH grant was the source of funding for my salary and research expenses. I had written this grant application with the encouragement of UCSF, in order to pursue my research on *Tetrahymena* telomeres and their associated proteins. This work grew out of that I had done in Joe Gall's laboratory at Yale. My grant was funded by the NIH General Medicine Institute. Unsure of my chances at obtaining funding, I had sent the same grant application to the National Institutes of Health, the National Science Foundation and the American Cancer Society, hoping for funding from any one of these. Reflecting the more informal scientific habits of the basic sciences community in those days, some time later one of the grant reviewers told me that he had been so intrigued by my photograph of the autoradiogram, showing that telomeric DNA in *Tetrahymena* was mysteriously packaged as something other than nucleosomes, that he had kept the photograph.

Then UC Berkeley offered me an associate professor position in the Molecular Biology Department, which I immediately accepted. Once again, I transferred my funding from UCSF, this time to my own laboratory, at UC Berkeley.

Because research was, and still is, such a central part of my life, my autobiography would be incomplete without describing my research experiences. Thus, to convey a fuller flavor of them, here I describe some of the events of my early scientific research on the molecular nature of ends of chromosomes.

EARLY WORK ON THE DNA AT THE ENDS OF EUKARYOTIC CHROMOSOMES

Very soon after arriving in Joe Gall's laboratory at Yale in early 1975, I started to apply methods for obtaining terminal DNA sequences to the *Tetrahymena* rDNA molecules. I had learned a collection of methods in Fred Sanger's laboratory in Cambridge, England, where I had just completed my Ph. D. Eager to sequence the end regions of these minichromosomes, with Joe's encouragement I set out right away early in 1975 to use end-labeling techniques on them. I incorporated ^{32}P isotope-radiolabeled deoxynucleosides residues into the *Tetrahymena* rDNA molecules using commercially available DNA polymerases for in vitro DNA repair enzymatic reactions. The results were immediately promising. First, it became clear that the end regions of the rDNA were being selectively labeled by certain combinations of ^{32}P isotope radiolabeled nucleoside triphosphate substrates. And by June 1975, I had become tremendously excited: I had obtained my first autoradiogram of the two-dimensional separation of the ^{32}P labeled depurination products. A strong signal of a run of 4 cytosine (C) residues was apparent. Furthermore, each such C4 sequence was flanked by a purine residue (that is, an adenosine (A) or a guanosine (G) residue; this initial data did not show which). The way the depurination reaction worked was the following: Ken Burton, a New Zealander, had shown that a chemical reaction could be done that cleaves the DNA backbone on both sides of every purine nucleotide but leaves intact any runs of pyrimidine nucleotides (such as C residues) that are uninterrupted by purine nucleotides. This so-called depurination method, when applied to a complex sequence DNA like a whole bacteriophage genome, was further made useful by Vic Ling, when he was a postdoctoral fellow in Fred Sanger's lab in Cambridge, England. Vic had shown that the resulting short pyrimidine tracts (mono-, di- tri-nucleotides, etc) would yield a pattern of products like a grid when a 2D fractionation method was used (see 2009 Nobel Lecture by Elizabeth Blackburn, in this volume). The most frequent products, on a random basis, of course are mono and dinucleotides, with the longer tri- and tetranucleotide tracts being less and less abundant on a random basis. Thus this strong C4 spot was interesting and informative.

It was also clear to me that the rDNA molecules were not simply lambda phage-like DNA. First, the terminal fragments were heterogeneous. In contrast, for any one type of lambdoid phage, in every viral particle the DNA molecule is just the same – a perfect carbon copy – as in every other viral particle. Second, on a per DNA molecule basis, much more incorporation of radiolabeled nucleotide precursor substrates occurred than expected if rDNA molecule ends were like those of the sticky DNA ends of lambda phages. It was also fortunate that the C4 repeat had such a regular nature and that it happened to have four Cs in a row. Even three C's would have been striking, but harder to unravel. As it was, the C4 spot consistently stood out like a beacon through my repeated productions of 2D fractionations ("homochromograms" as we called them in Fred Sanger's lab).

Next, I needed to validate independently that what I was radiolabeling *in vitro* validly reflected the rDNA sequence, and also try to get a closer estimate of the number of C4 runs per rDNA molecule. I therefore decided I needed to ^{32}P isotope-label the rDNA *in vivo*. The vast majority of the ^{32}P isotope (chemically in the form of inorganic phosphate ions) taken into the cells would be incorporated into other molecules, including the much more abundant cellular RNA, with very little ending up in DNA, and even less in the rDNA (in *Tetrahymena*, only a percent or so at most of the DNA is rDNA). Therefore, with some trepidation I asked Joe Gall for permission to order sufficient ^{32}P phosphate to be added to the cell growth medium for labeling the rDNA. This meant handling 2 milliCuries at once, which Joe's lab had not done before. But I knew that this was the only way currently available to get a sufficient amount of ^{32}P into rDNA to detect the C4 spot in an autoradiogram, above background. Possibly with some trepidation on his side too, Joe agreed. I worked in the "hot room", a room set aside for doing work with radioactive isotopes. On October 9, 1975 the ^{32}P was shipped to Yale. My laboratory notebook from that time reads: "^{32}P stored in refrig. until use – assayed for 10/14 . 3 pm 10/14. Zeroed [the cell culture] on 1% PPS medium blank.... Added 2 mCi 32p as h3 PO4 in water in 1 ml from syringe." For the next preparation I raised it to 5 milliCuries. Curious about this very radioactive departure from the more usual activities of the lab, my Gall labmates periodically looked in through the window set in the door of the hot room as I worked.

By October 22, 1975, I had the ^{32}P rDNA purified. I triumphantly wrote down in my notebook that day my plans for this precious sample:

"1) Depurination
2) denaturing gel after EcoRi treatment
3) 1.4% agarose gel"

One by one, I inflicted various nucleases on the terminal region of the rDNA. I found it could be selectively radiolabeled using one triphosphate ^{32}P labeled at a time. Then I carried out the battery of analyses possible at the time that would allow me to piece together the nucleotide sequence: I digested the radiolabeled end regions with Endonuclease IV nuclease or micrococcal nuclease, and performed depyrimidations and nucleotide nearest-neighbor analyses, and spleen phosphodiesteriase digestions (Figure 3).

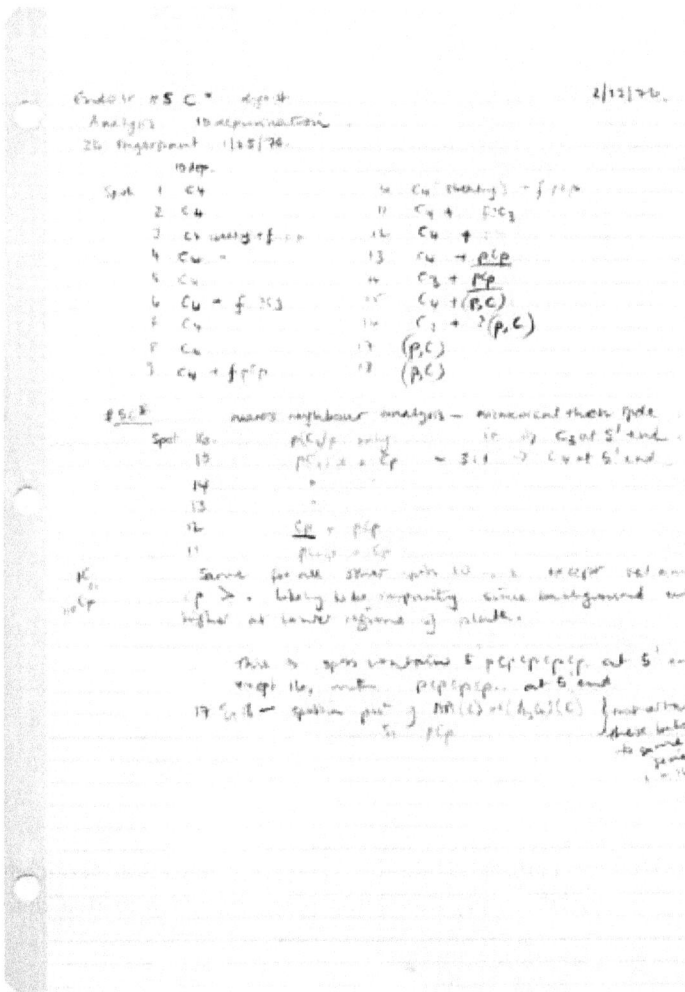

Figure 3. Lab Notes 02/12/1976.

Piecing together the rDNA end sequence was a matter of careful puzzle-solving. At an intermediate point, my notebooks of the time show that I had to consider two possibilities – CCCCAG and CCCCAA repeats (Figure 4). But it was apparent that there were a large number of repeated copies.

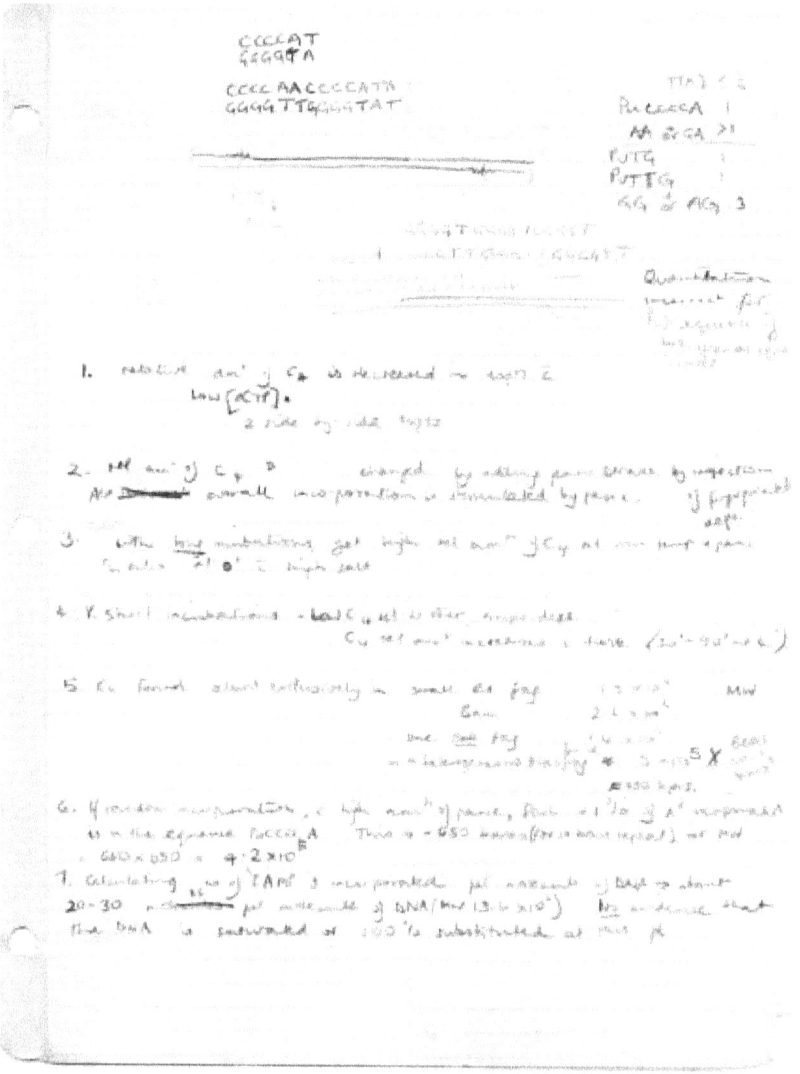

Figure 4. Lab Notes.

But by April 8, 1976, I was confident that the correct sequence was deduced, because the entry in my laboratory note-book page headed with that date reads:

"Sequence data from:

1) Depurination
2) Endo IV – deps [my abbreviation for depurinations]
   – nearest neighbours
3) partial micrococcal
   – deps
   – nearest neighbours
partial spdes [my abbreviation for spleen phosphodiesterase digestions]

4) Gel analysis of Endo IV digest of restriction fragments t ends of rDNA
   → AACCCC repeated
   → (AACCCCAACCCCAACCCC)etc."

Another laboratory note-book page, dated August 17, 1976, shows I was already referring, in a routine way, to (CCCCAA)n sequence – by then in the course of experiments designed to see whether this same repeated sequence was also present in the other (much longer) chromosomal DNAs of *Tetrahymena*.

I put together a picture which tried to take into account all of my many observations. The deduced sequence consisted of a tandem array of CCCCAA repeats. One experiment done in 1979, radiolabeling the rDNA using just ^{32}P-labeled dCTP, and unlabeled dATP, and separating the products on a denaturing gel electrophoresis, showed this visually as a beautiful ladder of tiger stripes extending up the gel. The size of every band in this regular ladder was 6 bases more than the band below it! This strikingly characteristic visible pattern of bands presaged the pattern that would later become important for our discovery of telomerase enzyme activity, as described in this volume in the 2009 Nobel Lecture by my co-awardee, Carol Greider.

In the months following April 1976, much of my effort was also devoted to trying to understand the arrangement of the strand discontinuities along the tract of CCCAA repeat DNA. For this, I performed a great many experiments following the kinetics and specificity of radiolabeling the RNA end regions, using multiple different enzymes and protocols and analyses. This also was a matter of piecing things together – there was no template for me to work from as this was all uncharted territory.

In the late 1970s and early 1980s, I did a variety of radiolabeling experiments trying to divine the structure right at the termini of telomeres of both ciliated protozoans and yeast linear plasmids. I could put together a composite but still incomplete picture. Some of the features, I realize in retrospect, might be attributed to the terminal G strand sequence at the very ends of the telomeres assuming G-G paired or G-quartet structures. But other features are not so readily explainable. Why I was able to label the strands with DNA polymerase or a kinase to get the patterns of labeled strands and nucleotides I did has still not been completely fitted into a coherent view of the molecular structure of DNA ends. Various kinds of in vitro radio-labeling experiments had suggested that, in both *Tetrahymena* rDNA and the macronuclear DNAs of hypotrichous ciliates, there is a short overhang of the G-rich strand consisting of a few telomeric repeat sequences. Currently the view is that in mammalian telomeres there is a long protruding G-rich strand. Yet this does not take into account the clear evidence for the short C strand repeat oligonucleotides that I discovered can be readily melted off the telomeric DNA. This I found for both the *Tetrahymena* rDNA minichromosome molecules and linear plasmids purified from yeast. These tiny telomeric sequence oligonucleotides could be radiolabeled and clearly identified by two dimensional fractionations. However their significance is still unknown. To this day, aspects of the structure at the very terminal region of the telomeric DNA are enigmatic; the very ends of chromosomes remain as challenges.

TELOMERE PROTEINS: EARLIEST ATTEMPTS AND FAILURES

In my early work, my molecular views of telomeres were first focused on the DNA; not only because DNA was uppermost in my mind, but for several years DNA was also the only component of the telomeres that was identified. This was not for want of trying. I thought that DNA would not be the entire story of chromosome ends and, by extension from work emerging about chromatin in general in the 1970s, that it was likely that the telomeric DNA repeats tract would be packaged with proteins. The 1970s had seen great interest in chromatin, and the discovery of nucleosomes as the basic packaging unit of eukaryotic DNA. Telomeric sequences in *Tetrahymena* looked very intriguing to me in that regard, and as soon as I had identified the telomeric DNA I wanted to get my hands on whatever packaged it. Therefore, while still a postdoctoral fellow in Joe Gall's laboratory, I performed micrococcal nuclease treatment on isolated Tetrahymena nuclei. I found that the $CCCCAA_n$ tracts of the telomeres were protected in chromatin as a heterogeneous class of DNA fragments very different from that expected for nucleosomal packaging.

Soon after moving from Joe Gall's lab at Yale, while still temporarily at the University of California San Francisco (UCSF) in an independent research position (before I moved to the University of California Berkeley as an Assistant Professor), on March 1, 1978, I wrote to Joe Gall: "I am getting quite excited about getting a $CCCCAA_n$-binding protein complex from the *Tetrahymena* macronuclei, so I've been busily making rDNA, the $CCCCAA_n$ probe, and macronuclear micrococcal nuclease digests. Results so far are that I've found a simple salt fractionation that enriches for $CCCCAA_n$ sequence plus putative protein(s). The plan at the moment is to purify this some more so I can get some structural characteristics of any such complex, i.e. S value, and some identification of protein(s) in terms of 1-D and 2-D gel eletrophoretic properties…The other aspect of course is to fish for something that will stick to a $CCCCAA_n$ column. "

there exists the possibility that a protein is tightly or covalently linked to the rDNA. A 5'-covalent linkage of the linear DNA of adenovirus to a protein has been found (93, 94). Three kinds of experiments will be carried out to search for any similar protein in Tetrahymena rDNA:

a) rDNA will be extracted from Tetrahymena as described (54), but the pronase treatment will be omitted. The rDNA will be examined by electron microscopy for a terminal protein (93). b) Tetrahymena cells will be labelled in vivo with ^{3}H amino acids (69), after starvation and refeeding to induce preferential rDNA synthesis (55). This would be expected to induce concomitant synthesis of rDNA-specific proteins, including any putative proteins that are bound to the termini of these molecules. rDNA will be isolated with and without protease treatment, and its internal and end restriction fragments assayed for ^{3}H protein peptide residues. c) The experiment shown in Figure 2 will be repeated as described in section A. After phenol-chloroform extraction of the DNA, part of the sample will be treated with proteinase K . This will remove any protein covalently bound to the DNA that may not have been dissociated. A consequent change in the mobility, on gel electrophoresis, of the DNA that hybridized with repeating CCCCAA DNA would indicate a tightly linked protein.

4) Search for a chromosomal protein that specifically binds repeating CCCCAA sequence DNA

Properties expected: As described earlier, it has been shown that DNA containing the repeating hexanucleotide CCCCAA is protected from limited nuclease digestion of whole macronuclei in a manner quite different from the bulk of nuclear DNA. Thus it is predicted that a DNA-binding, structural protein (or proteins), specific for this sequence, is responsible for the observed protection. Histones may not be involved in this binding, since they would have to be in a different arrangement from their normal form in nucleosome cores (95) to account for the observed difference in the protection of repeated CCCCAA sequence DNA. Furthermore, it is reasonable to expect that the proteins associated with sequences at the termini of DNA molecules would have special features. If a covalently-bonded protein is found at the rDNA terminus, a single protein is still unlikely to be large enough to account for the length of the repeating hexanucleotide DNA fragment that is protected from nuclease digestion.

Bacterial systems could provide a model for such a DNA sequence-specific binding protein. The lac and λ repressors, and RNA polymerase, are neutral to slightly acidic proteins (see 90 for references). Sequence specificity of binding is conferred by heterologous subunit interactions in the case of RNA polymerase, and the actively-binding forms of the repressors are oligomeric. Symmetry of the multimeric forms of repressors, Class II restriction endonucleases and perhaps in the prealbumin thyroid hormone receptor protein (113) is reflected in symmetry of the DNA sequence recognized (for review see 112). It is of interest that the repeating hexanucleotide sequence CCCCAA is not palindromic. However, because the sequence is tandemly repeated, it is likely that more than one protein molecule is bound, and such a protein may be found in oligomeric forms.

Non-histone chromosomal proteins of rat liver nuclei have been shown to have some sequence-specificity in their binding to DNA (97). Therefore, the protein that binds specifically to repeated CCCCAA DNA will initially be sought in

21

Figure 5. Scanned page from grant application, 1977.

By 1980 I had done experiments to show that telomeric tracts of DNA in Tetrahymena were encapsulated in a protective sheath of protein that did not include nucleosomes. The vast majority of chromosomal DNA is packaged as nucleosomes: DNA-protein complexes. Each nucleosome is a flattened ball made up of histone proteins, around which the DNA is wrapped twice. The very basic (positively charged) histone proteins neutralize the negative charges of the phosphate chemical groups arrayed along the phosphodiester backbone of DNA and allow chromosomal DNA to become very closely packed and compactly folded in the nucleus. Nucleosomes in artificially stretched-out chromosomes are like beads on a string, although mostly in the nucleus they are closely packed into shorter thicker fibers. If one clips up chromatin using an enzyme, micrococcal nuclease, that cuts across the two strands of the linker DNA between neighboring nucleosomes, after getting rid of the histones, one can see that there are nucleosome-sized fragments of DNA left – a fragment of about 142 base-pairs is protected by the histone core of the nucleosome, once the DNA linkers have been trimmed away. This kind of nuclease clipping behavior is a hallmark of a nucleosome. In contrast to nucleosomal regions of chromosomes, special regions of DNA, for example promoters that must bind transcription initiation factors that control transcription, have proteins other than the histones on them. The telomeric repeat tract turned out to be such a non-nucleosomal region. We found that if we clipped up chromatin using an enzyme that cuts the linker between neighboring nucleosomes, it cut up the bulk of the DNA into nucleosome-sized pieces but left the telomeric DNA tract as a single protected chunk. The resulting complex of the telomeric DNA tract plus its bound cargo of protective proteins behaved very differently, by various tests, from standard nucleosomal chromatin, and therefore we concluded that it had no histones or nucleosomes.

By 1977, it was known from work of Rekosh *et al.* that adenovirus DNA has a covalently bonded terminal protein, presumably for viral genome replication. Thus, in 1979, Marsha Budarf, a postdoctoral fellow in my laboratory at UC Berkeley, began using used radioactive iodine procedures (the Bolton-Hunter reagent) to see if we could find any comparable protein at the ends of rDNA. Although Marsha found a covalently attached protein (that in hindsight may have been topoisomerase I) enriched toward the end of the rRNA transcribed region, it was not enriched in the terminal parts of the rDNA molecules. She was unable to detect any other covalently attached protein elsewhere on the rDNA. Any evidence for a protein on the bulk of the rDNA molecule ends, such as their behavior in gel electrophoresis and the appearance of the rDNA molecules under the electron microscope, was conspicuously lacking. This made me feel all the more confident that there was no covalently attached protein at the very ends of this minichoromosome. But what other proteins were at telomeres?

My lab was the first to try to identify these protective proteins. We used biochemical fractionations of *Tetrahymena* nuclear extracts. My 1979 notebooks record that, together with my technician San-San Chiou in the Department of Molecular Biology at UC Berkeley, over and over I made attempts to purify the telomeric proteins from nucleoli. Nucleoli are the tiny bodies within the *Tetrahymena* nucleus that harbor the actively transcribed rDNA minichromosomes. Fractionations after fractionations, mostly using sucrose gradients, were patiently performed by San-San. Then we scaled up the preparations – I purchased a huge industrial-sized Waring blendor that loomed like a leviathan on the laboratory bench. *Tetrahymena* cells were blended in order to disrupt them just enough to shake their nucleoli free from the rest of the nuclear contents. At one time my note-book laconically reported: "Respun only one-third of total.....Waring blender broke."

All these early efforts were to no avail. In retrospect, the experimental approach had been reasonable – to purify nucleoli, as being the most enriched form of telomeric chromatin known, then to digest them with micrococcal nuclease into fragments, the end ones containing the telomeric DNA terminal tracts and their bound proteins. Then, I would further fractionate these away from the rest of the chromatin by selective precipitation in potassium chloride solutions, or fractionate them by size on sucrose gradients. The goal was to see what protein(s) would co-purify, through these multiple fractionation steps, with the telomeric repeat tract DNA, which I followed through the multiple steps by its hybridization signal. But we were only able to obtain limited amounts of chromatin and binding factors, and we tried without success to get enough to identify any factors that might be specific to the rDNA ends. Looking back, I see that we were fighting against the numbers game – our detection methods were too frail, our preparation scale-ups too modest. Therefore, it was yeast genetics and approaches done by others that turned out to provide the next great leaps forward in understanding telomeric proteins. That I failed in this by my early attempts using *Tetrahymena* made me all the more determined, if anything, to use other approaches to try to understand the nature and biological significance of those strange-seeming repeated sequences at the ends of chromosomes.

I also recall that our failure to find telomeric proteins taught a lesson that became useful when it came to our work on *Tetrahymena* telomerase. As Carol Greider's Nobel lecture describes, at one point the value of scaling up the telomerase activity preparations became evident to her. Thus, when Carol proposed the purchase of a very large glass column for preparative gel filtration chromatography, I was very willing to make this expensive-seeming purchase, ruefully recalling the past history of my too-pusillanimous scale-ups of *Tetrahymena* chromatin preparations.

Figure 6. Mission Bay Laboratory Group BBQ, August 2007.

TO THE UNIVERSITY OF CALIFORNIA SAN FRANCISCO

I became a Full Professor at UC Berkeley in 1986 (after 8 years on the faculty of UC Berkeley), and in the same year a mother (our son Benjamin David was born in December 1986). By around 1989, I decided that as the long drive to Berkeley each day from our home in San Francisco made it difficult to pursue both science and our family life optimally, it was time to begin investigating alternatives. I settled upon a professorship at UCSF, and the move of my laboratory to UCSF's Department of Microbiology and Immunology was accomplished in mid-1990. I have remained on the faculty of UCSF ever since. There, I have had the great good fortune to be able to keep delving into the nature and mechanisms of telomeres and telomerase. Together with colleagues in and out of UCSF and with my many talented students and postdoctoral fellows and technicians in my laboratory (Figure 6), I have been able to address the wondrous biological systems comprised of telomeres and telomerase. A fanciful depiction evoking both telomere dynamics and telomere researchers is shown in Figure 7. This painting, done by the artist Julie Newdoll in 2008, elicits the idea of a telomere as an ancient Sumarian temple-like hive, tended by a swarm of ancient Sumarian Bee-goddesses against a background of clay tablets inscribed with DNA sequencing gel-like bands.

Figure 7. Julie Newdoll, *Bee Goddesses.*

OUT OF THE LABORATORY

In the 1990s my research's implications for humans began to intrigue me, but with scientific research, faculty and Department Chair duties, family and many associated commitments, I had little time to indulge in delving into the philosophical and policy questions that can arise as science opens new possibilities. I served as President of the American Society for Cell Biology in 1998 and become more cognizant of the world of national science policy. Thus it was that in late 2001, the request to consider becoming a member of a newly created U.S. Federal Commission, the President's Council on

Bioethics, had a certain appeal. I felt that my knowledge of the relevant fields of science, and long experience in the world of research, would be useful contributions to the Council, a body that, as a Federal Commission, would be advisory on some matters of national science policy. A further appeal was the coincidence with my growing thinking about these issues. I reasoned that if I joined this Council, it would be an opportunity to contemplate some of these dimensions of research's ramifications, and the possible reverberations of my own area of research.

Time for quiet contemplation of these and related questions in the abstract was not forthcoming. I understood from the beginning that the Bioethics Council would be occupied with publicly debated topics including human somatic cell nuclear transfer and embryonic stem cell research, as well as other topics less clearly defined at the outset of the council's deliberations. I thought I should agree to serve on this Council because, as a seasoned scientist (particularly in cell and molecular biology), I might be able to offer perspectives that would be helpful in advising national scientific policy. I knew the topics upon which this Council, appointed by the George W. Bush administration, would advise would be politically charged ones. For this reason especially, I felt that a strong base of scientific fact and evidence would be particularly important, and useful advice in this vein was something that I could in fact offer to this advisory body.

I publicly made clear my views on some of the council's recommendations, views that did not generally accord with those of the White House or with those of the Council's Chair. After two years, I was informed by the Personnel Office of the George W. Bush White House that I would no longer be on this Council. This dismissal from the Council received quite a lot of public attention at the time. In the course of it, I was overwhelmed by the great many letters and communications I received. Almost without exception positive and supportive, they came from all over the United States and even from as far afield as a musician in London. His somewhat (to me) unexpected concern for science policy brought home to me how widespread is the wish among the public that science policy be informed by good scientific evidence. This entire episode was a broadening education. It reinforced my love of the searches for truth to which so many in research and academia aspire.

PEOPLE WHO HAVE HAD IMPORTANT INFLUENCES ON MY LIFE AS A SCIENTIST

I am indebted to so many individuals that I can only describe a few of them here. Growing up, three of my schoolteachers in particular encouraged my interests in biology and chemistry and mathematics, not least by letting me know that they believed in my abilities to succeed in these areas – Nan Hughes, Jenny Phipps and Len Stuttard.

As I embarked on research in biological science, my teachers, advisors and mentors – notably Frank Hird in Australia, Fred Sanger in Cambridge,

England, and Joe Gall in the U.S.A. – not only imparted their scientific knowledge, visions and wisdom, but also their examples of how to be a scientist. In particular, a photograph of Joe Gall from 1999, although taken several years after I had been in his lab, captures in a succinct visual way some of Joe's characteristics that influenced me when I was a member of his lab group (Figure 8). I took the photograph during a conference he was attending in Prague in the summer of 1999. During the conference a partial eclipse of the sun took place, and all the conference participants rushed out of the lecture hall to witness its progress. Joe is seen in the photograph demonstrating that it could be seen very simply and safely: All one had to do was hold a flat sheet of paper under a leafy bush so that the light, diffracted through the leaves onto the paper, caused to appear on the sheet of paper images of the "bite" being taken out of the disc of the sun by the moon passing in front of it. I recall that most of the conference participants had never seen this applied optics demonstration before. This photo evokes at once Joe Gall's desire and ability to teach – by his use of a very striking demonstration to teach something new to the conference participants – and, not least, one glimpse of his wide knowledge encompassing optics and natural science in general.

Figure 8. Joseph G. Gall. Prague, 1999.

Like so many who are fascinated by chromosome behavior, I owe much to Barbara McClintock for her scientific findings. But in addition, Barbara McClintock also gave me a memorable lesson: in a conversation I had with her in 1977, during which I had told her about my unexpected findings with the rDNA end sequences, she urged me to trust my intuition about my scientific research results. This advice was surprising to me then, because intuitive thinking was not something that at the time I allowed myself to admit might be a valid aspect of being a biology researcher. I think her advice recognizes an important and sometimes overlooked aspect of the intellectual processes that underlie scientific research, and for me it had a liberating aspect to it. For this, also, I am very grateful to Barbara McClintock.

Figure 9. Elizabeth and her son, Ben, at the piano. Circa 1990.

My husband, John Sedat, himself an accomplished scientist, has always urged me to dig deeper into myself and find the reserves of strength I might not have tapped – his encouragement in this way has helped me through years of doing science. Our son Ben (Figure 9) inspired me to try to find ways of combining family and science, something that I have tried to convey to young scientists making their careers. Finally, my parents were both family physicians. From them I imbibed a sense of the importance of serving people kindly and as well as one can. I continue to believe that bioethics, done well and underpinned by the best available scientific evidence, can be an important part of our consideration, as a society, of the impact on people of scientific research in the biological sciences and medicine.

TELOMERES AND TELOMERASE: THE MEANS TO THE END

Nobel Lecture, December 7, 2009

by

ELIZABETH H. BLACKBURN

Department of Biochemistry and Biophysics, The University of California San Francisco, San Francisco, CA 94158, U.S.A.

INTRODUCTION

DNA carries coding and noncoding sequences. Noncoding DNA both regulates and ensures the continued inheritance of DNA's coding information. In eukaryotes, by protecting the chromosome ends and thereby the chromosomes themselves, telomeric DNA is a class of noncoding DNA that ensures the stable inheritance of the genetic material. Research begun in the 1930s on the cytogenetics of telomeres was followed by a molecular understanding of telomeric DNA and its maintenance, which began in the 1970s and continues apace today. This fundamental, question-driven basic research has led into realms of human health and disease that have turned out to inform medicine in new ways.

BEGINNING THE ENDS

"You corn kernels, … may you succeed, may you be accurate."
Popul Vuh

Tracing the beginnings of the interwoven stories of science can be arbitrary, as beginnings are so often lost in the mists of time. For me, arguably the story of telomeres and telomerase began thousands of years ago, in the cornfields of the Maya Highlands of Central America. Today, under the brilliant, shifting sunlight of the Central American highlands, lush corn plants cover every inch of sloping land wherever they can gain a foothold. There, over millennia, agricultural breeding generated corn (maize) crops from the ancestral plant teocinte. Estimates place the early cultivation of corn in the Central American highlands to around 7,000 years ago, and while early maize cobs dated from then were tiny, over millennia they progressively got bigger and bigger. Agricultural breeding is a process of consciously selecting the "best" plants. It was known that "like begets like", so that if one used the kernels from the biggest ears of corn in the planting for next year, a better crop would result. Intensely cultivated areas were carved out of the Central American rainforests and devoted to the production of corn. Maize came to occupy a central position in the agriculture and culture of the ancient Maya, and the Mayans had a maize goddess. Their ancient Council book, the Popul Vuh, includes many references to maize. The Popul Vuh even evokes genetic principles: "You corn kernels,... ...may you succeed, may you be accurate" (Popul Vuh).

Figure 1. A form of the Mayan corn god.

As maize became important for human food worldwide, modern agricultural research on maize breeding continued the corn breeding begun thousands of years ago in the Central American highlands.

THE TELOMERE CONCEPT
"This is the beginning of the end."
Charles Maurice de Talleyrand 1754–1838 (announcing Napoleon's defeat at Borodino).

Perhaps another, more modern beginning to the story of telomere research is the discovery of X-rays by Roentgen. Hermann Muller, working on the fruit fly *Drosophila*, showed that X-rays could be used to produce mutations and chromosome fragmentation. By the end of the 1920s it was understood that the hereditary material was in chromosomes: Mendel's work, begun on heritable traits, had been integrated with findings showing that the inheritance

patterns of genetic traits (or genes) corresponded with the regular movements of chromosomes in cells in meiosis and mitosis. By the 1930s, in the
United States maize breeding research using such genetic principles came
to be undertaken under government sponsorship, in agricultural research
stations. And in one such station, the Missouri Agricultural Research Station,
the geneticist and cytologist Barbara McClintock worked with maize, using
the methods she had developed for examination of individual chromosomes.
Genes were arrayed along chromosomes, as Muller's work showed. But in the
1930s there was no particular interest in what was at the ends of those arrays
of genes, until it was noticed that the ends had some distinct properties (for
a brief review, see [1]) In the early 1930s McClintock concluded that "the
natural ends" of chromosomes (McClintock's 1931 phrase; [2]) were
functionally different from experimentally-induced or accidental chromosomal breaks. A "stickiness" of broken ends of chromosomes (causing
chromosomal fusions) was one of their defining features, while in contrast
telomeres, the natural ends of chromosomes, had no such stickiness. This
recognition arose from McClintock's research on broken ends of chromosomes and their behavior [3]. Independently, Muller reached the same
conclusion from his fruit fly work, and in 1938 named these ends
"telomeres". [4] (Reviewed in [1]) .

DIVING INTO POND WATER

*"Now this is not the end. It is not even the beginning of the end. But it is, perhaps,
the end of the beginning."*
Sir Winston Churchill, Speech in November 1942

On reading the insightful early cytogenetic work of McClintock and Muller
from the 1930s and 1940s, it is sometimes hard to remember that their
deductions of the fundamental cytogenetic properties of the natural ends
of chromosomes preceded any knowledge that the genetic material is DNA.
The molecular mechanisms underlying the telomeric properties were
completely unknown when, in the mid-1970s, I first began research using
DNA purified from the ciliated protozoan *Tetrahymena*, as described below.
From these molecular analyses emerged the nature of the specialized DNA-
protein complex that comprises the telomere, distinguishing it from an
accidental DNA break.

First, the sequence and structural features of telomeric DNA had to be
understood. By the early to mid-1970s, viral and bacteriophage DNAs,
and in some cases their ends, had been studied both biochemically and
genetically. But what was the end of a cellular DNA in a eukaryotic nucleus – a
chromosomal end – like? What was most daunting to a molecular biologist
interested in that question in the 1970s, at the time just before the advent of
DNA cloning methodologies, was the sheer length of typical chromosomal
DNAs.

By the early 1970s Kavanoff and Zimm had carefully isolated chromosomal DNA in as intact a form as possible from fruit fly cells. [5] The kind of

molecular weight range they deduced corresponded to long DNA molecules extending from one end of the chromosomes to the other. Thus the typically long chromosomal DNAs from a cellular nucleus were thousands of times longer than phage DNAs. This presented an enormous technical hurdle with respect to analyzing their telomeric regions. Answering the question of the molecular nature of telomeres meant going into pond water.

The specific pond water denizen in question was a single-celled ciliated protozoan, *Tetrahymena thermophila.* In the early 1970s Joe Gall, at Yale University, had discovered that *Tetrahymena* harbors a class of abundant, homogenous, short, linear chromosomes ("minichromosomes"). These were the key to my being able to analyze telomeric DNA directly. I first encountered *Tetrahymena* when I joined Joe Gall's lab as a postdoctoral fellow at Yale. Although single-celled, a ciliated protozoan such as *Tetrahymena thermophila* contains two different types of nucleus. As Grell describes it [6] "the majority of ciliates... have generative nuclei capable of unlimited reproduction as well as somatic nuclei which perish sooner or later to be re-formed by descendents of the generative nuclei." Thus *Tetrahymena* economically combines both "soma" and "germline" into one cell. The abundant "minichromosomes" that Joe Gall discovered resided in the somatic nucleus. These linear DNA molecules bore the genes encoding ribosomal RNAs (rDNA), their high abundance ensuring sufficient expression for the large *Tetrahymena* cell. Kathleen Karrer, then a graduate student in Joe Gall's lab, had just discovered that the rDNA molecules consisted of two equal halves in a palindromic arrangement. This made them even more attractive: each telomeric end region would be the same as the other end region! The abundance and relative shortness (only 20,000 base pairs) of these molecules would, I reasoned, make it feasible to apply methods on them like those that had been used by Ray Wu and colleagues, for example, to sequence nucleotides at the ends of bacteriophage DNA in the early 1970s.

It was with these rDNA molecules that, very soon after arriving at Yale in early 1975, I started to use methods for determining the DNA sequences at the rDNA ends. I describe my experiments that succeeded in piecing together the telomeric sequence in more detail in the autobiography in this volume. Briefly, initially, upon fractionation of depurination products of radiolabeled rDNA end regions, the strong CCCC sequence spot seen was particularly informative (Figure 1).

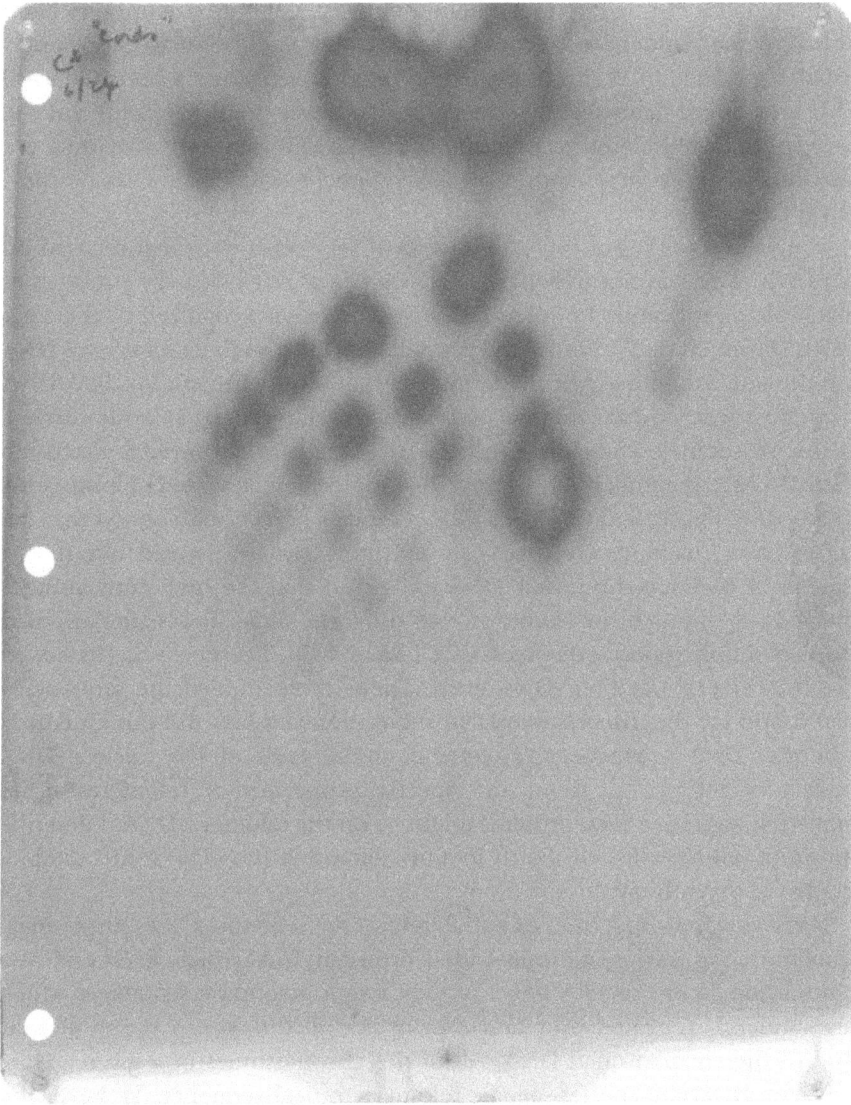

Figure 2. First autoradiogram showing a prominent CCCC sequence (the bottom right spot).

By August 1976 I was confident that the *Tetrahymena* rDNA molecules ended in (CCCCAA)n sequence, and I was looking to see whether this same repeated sequence was also present in the other (much longer) chromosomal DNAs of *Tetrahymena*, and exploring the molecular structures of the DNA end regions. The results describing tandemly repeated CCCCAA sequences at the rDNA ends were published in 1978. [7] One experiment done in 1979, radiolabeling the rDNA using just ^{32}P-labeled dCTP, and unlabeled dATP, and separating the products by denaturing gel electrophoresis, showed a beautiful ladder of tiger stripes extending up the gel. The size of every band in this regular ladder was 6 bases more than the band below it – a strikingly visual confirmation of the repeated hexameric sequence I had deduced! This arrestingly characteristic pattern of bands was the first example of the pattern that would later become important for our discovery of telomerase enzyme activity, as described in the 2009 Nobel Lecture by my co-awardee, Carol Greider.

In my early work, our molecular views of telomeres were first focused on the DNA. This was not only because DNA was so central to the problem of incomplete replication of linear DNAs, as had been recognized by the early 1970s (reviewed in Blackburn and Szostak. [8], but also for several years DNA was the only component of the telomeres that was identified. By 1980, DNA sequences were known for the ends of a few different eukaryotic nuclear DNAs. By then we had shown that the DNA ends of *Tetrahymena* macronuclear DNAs in general, not only the rDNA molecules, consisted of long arrays of the same simple 6-nt CCCCAA (C_4A_2) repeats [9]. Others showed that the ciliate Oxytricha and its relatives had very short tracts of 8-nucleotide (C_4A_4) repeats at their macronuclear DNA ends, and that the high copy-number linear rDNA minichromosomes in two different slime molds similarly had tracts of simple repeat sequences, CCCTAA and $C_{1-8}T$ respectively (Reviewed in [10, 11] [12]) at their DNA ends. These all resembled the sequence I had found for the *Tetrahymena* rDNA molecules. But how did this emerging common DNA sequence arrangement at the ends of the nuclear DNA molecules inform us about the special properties of telomeres? One approach was to see what proteins might be on the telomeric DNA. I describe my early unsuccessful efforts to identify telomeric proteins in my autobiography in this volume.

Now we know that the essential telomeric sequences are surprisingly similar among phylogenetically widely divergent eukaryotes. Each end of a chromosome consists of a block of very simple telomeric sequences which are tandemly repeated over and over again, all the way to the molecular end of the chromosomal DNA. All of the chromosomes in a given organism have the same species-specific telomeric repeat sequence. However the same telomeric repeat sequence crops up in very diverse eukaryotes. For example, human telomeres consist of AGGGTT repeats, tandemly repeated for thousands of nucleotides at the ends of all of our chromosomes. The same repeated AGGGTT sequence is the telomeric sequence of the mold, *Neurospora*, the slime mold, *Physarum*, and the trypanosome protozoan parasites. (This makes telomeric DNA sequences possibly the world's worst sequences for deducing phylogenetic relationships!) Telomeric DNA generally has a strand composition asymmetry, resulting in a G-rich and a C-rich strand. It is the G rich strand that is always oriented in the 5' to 3' direction toward the end of the chromosome. This overall structural conservation of telomeres suggests that this general arrangement and composition of DNA strands is of fundamental importance for telomere function.

THE LINES OF EVIDENCE THAT LED TO THE CONCEPT THAT TELOMERASE ACTIVITY EXISTED

That telomeric DNA had certain molecular behaviors indicative of dynamic properties in vivo had emerged by the early 1980s. Four main lines of such molecular evidence were instrumental in spurring me to hunt for a new type of enzymatic activity that might synthesize telomeric DNA and elongate telomeres. This evidence took the form of molecular observations on telomeric DNA not readily explainable by any of the knowledge then about DNA replication or recombination.

First, the telomeric CCCCAA repeat tracts (which we eventually ended up referring to by their sequence on the complementary DNA strand, GGGGTT repeats) in the ciliates *Tetrahymena* and *Glaucoma* were heterogeneous in length; that is, the DNA molecules in the population carried different numbers of repeats.[7, 13] Perfect DNA replication of parental DNA to make two daughter DNAs was not predicted to produce such heterogeneity.

Second, during development of the somatic macronucleus in different *Tetrahymena* strains, telomeric GGGGTT repeat tracts were found to become joined, by then-mysterious means, to various sequences in the rDNA minichromosomes; that is, new telomeres were forming on macronuclear chromosomes. Meng-Chao Yao, continuing work he had started as a Ph. D. student in Martin Gorovsky's laboratory and then as a postdoctoral fellow in Joe Gall's lab (at the same time I was there), had observed this first for *Tetrahymena* rDNA telomeres.[14] But a single TTGGGGTT sequence already present at this position in the precursor DNA sequence had made it conceivable that this sequence could itself somehow be a seed sequence for repeated unequal recombination events to generate multiple repeats, for example. However, then my lab at Berkeley made similar observations for other rDNAs and non-rDNA telomeres of the somatic nucleus, with the difference that in these cases the telomeric DNA sequences were found to be joined to sequences where there was no initial GGGGTT repeat at all. [15] Thus in 1982 I wrote about these observations: "...the sequences common to the macronuclear DNA termini must be acquired by these subchromosomal segments during their formation. Two types of routes can be envisaged: Telomeric sequences are transposed or recombined onto the developing macronuclear DNA termini, or the simple, repeating telomeric sequences are synthesized de novo onto these termini by specific synthetic machinery" [15]. Simultaneously, David Prescott's group in Colorado had made the observation that T_4G_4 repeats similarly appeared to become joined onto the ends of the short chromosomes of the macronucleus in a hypotrichous ciliate. [16]

Third, as described in detail in Jack Szostak's 2009 Nobel Lecture, we had discovered that yeast telomeric sequence DNA (irregular TG_{1-3} repeats, which Janice Shampay, a graduate student in my lab at UC Berkeley, first sequenced as part of our collaboration with Jack) was added directly to the ends of *Tetrahymena* GGGGTT repeat telomeres maintained in yeast.

[17, 18] In this collaboration, we showed that a telomere from the ciliated protozoan *Tetrahymena*, consisting of GGGGTT repeats, was able to function as a telomere in the yeast S. cerevisiae in a particular sense: specifically, the *Tetrahymena* telomeric sequences, when put into a yeast cell on the ends of a linerized plasmid DNA molecule, could stabilize the plasmid, such that now it was maintained indefinitely as an extrachromosomal, linear DNA molecule through many rounds of replication, mitosis and even meiosis. But in addition, something very interesting always happened to the introduced foreign (*Tetrahymena*) telomere in yeast. We found that yeast telomeric repeats, which Janice Shampay sequenced and found to be TG_{1-3} repeats, were added to the distal end of the foreign, GGGGTT-repeat telomere after it had been maintained in dividing yeast cells. Other observations made soon after, in the course of following up these findings, further highlighted the dynamism of telomeric DNA in cells. For example, by the early 1980's Janice Shampay had also observed that if we introduced such a high-copy-number plasmid into yeast (thereby adding of the order of 100 extra telomeres into these cells), the telomeres of the chromosomes themselves, whose average length had heretofore remained steady for 300 generations of previous mitotic passaging, now underwent a slow shortening over the ensuing cell divisions (EHB and Janice Shampay, unpublished results).

Fourth, Piet Borst and his collaborators published an intriguing observation in 1983. They were monitoring the inheritance of a gene of trypanosomes (which cause sleeping-sickness) encoding a variant surface antigen. These antigens play important roles in the parasite's ability to evade the host's immune response. The gene they had found was located on a telomeric restriction fragment. In the course of passaging the trypanosome cells, this gene's restriction fragment steadily became progressively longer, implying that the telomeric DNA tract was growing.[19]

Finally, from yet another independent direction, a cytogenetic observation by Barbara McClintock reinforced my nascent notion that some undiscovered kind of developmentally-controlled cellular enzymatic activity might act on telomeres. In essence, first in a conversation, and later in a 1983 letter, McClintock described how, long ago, she had identified a maize mutant that had lost the normal capacity of maize to heal broken chromosome ends that specifically exists very early in plant development – just after fertilization. In the letter McClintock wrote to me in 1983, she explains this (Figure 3):

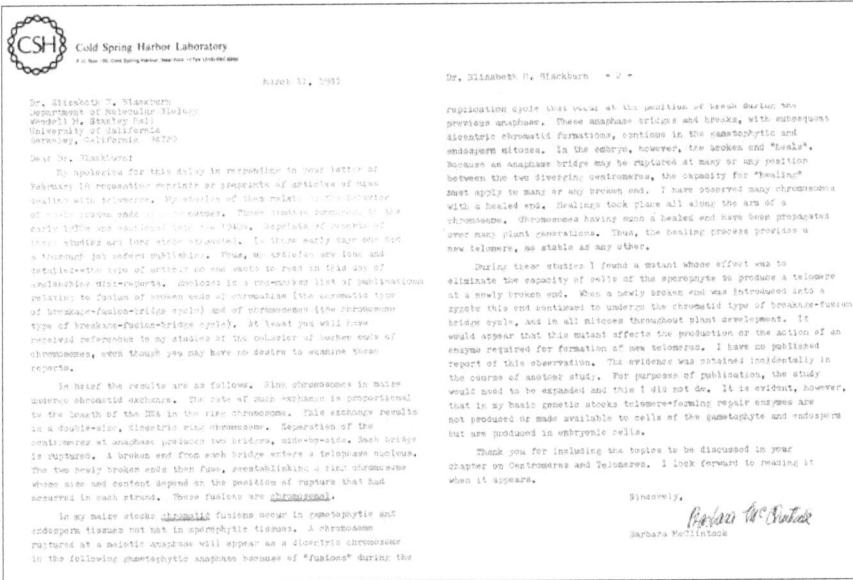

Figure 3. McClintock letter to EHB 1983.

Finding a mutant implied that there is a gene associated with the ability to heal – a gene that could be mutated to nonfunctionality. I was struck by the implication that in zygotes, a fully functional telomere ("healed end" in Mc Clintock's terminology) was generated from a broken chromosome end not just by chance but rather, by an active, developmentally controlled process; a process, furthermore, occurring just after fertilization – the developmental stage equivalent to when ciliate chromosomes become broken (albeit deliberately in their case) and telomeric DNA is efficiently added to the freshly produced DNA ends.

This remarkable information was another one of the reasons that I decided to look for telomerase. The capacity of ciliates to form de novo telomeres just after fertilization (equivalent to the zygote) was just too striking a parallel to ignore.

TETRAHYMENA CELLS BY A BIOCHEMICAL APPROACH

"If your knees aren't green by the end of the day, you ought to seriously re-examine your life."
– Bill Watterson (American Author of the comic strip Calvin & Hobbes, b. 1958)

Tetrahymena cells provided an attractive system to use to hunt for this putative telomeric DNA-adding enzymatic activity. My choice of approach was to prepare extracts from *Tetrahymena* cells. As Joe Gall had pointed out when I proposed sequencing DNA end regions, they could be grown in large quantities relatively inexpensively. Furthermore, their developmental time course could be synchronized, thanks to the efforts of several laboratories, notably those of David Nanney, Peter Bruns, Ed Orias, Sally Allen and their colleagues. Hence, one could make cellular extracts from a large population of cells all undergoing macronuclear development and, concomitantly, I hypothesized, the putative telomere addition reactions. This, therefore, seemed to me likely to be a developmental stage when any such activity would be in high demand by the cell and therefore, I reasoned, would allow the best chance of its being detectable. The big question was the choice of substrates to use. What DNA would be best to use to prime any telomeric DNA addition, and what nucleotide building-block precursors would be required? Would both strands of the telomeric DNA have to be synthesized together in a coupled reaction, or perhaps even in a reaction that had to be coupled to the production of the freshly-cut DNA ends? Would it be preferable to provide DNAs that resembled the freshly-cut ends in the developing macronucleus, or pre-existing telomeric repeat tracts that could be further elongated: both reactions would be expected to be performed in *Tetrahymena* cells at this stage. To make sure I did not miss any of these possibilities, I added a mixture of all four deoxynucleoside triphosphates and all four ribonucleoside triphosphates, an energy-generating (ATP-generating) enzyme system, and a mixture of cloned DNA fragments, purified from bacterial cells, that would present to any enzymes in the *Tetrahymena* extracts both telomeric and nontelomeric DNA termini. I prepared cell extracts from cells at this developmental stage, adapting a method that my graduate student Peter Challoner had in turn adapted from one used by Tom Cech and collaborators to examine rDNA gene expression (which led Tom Cech to the discovery of self-splicing RNA). Peter had found that incubating such extracts (for a different experimental goal) allowed him to detect changes in the DNA restriction fragments that he had added to his extracts. The changes had even hinted at some form of alteration specific to the telomeric DNA ends. As described elsewhere (Appendix, [1]) in early 1984 I was able to see increasing amounts of telomeric GGGGTT-hybridizing repeat sequences were somehow generated during the course of the reactions. The hunt was on!

THE DISCOVERY OF TELOMERASE

"… to make an end is to make a beginning."
– T.S. Eliot 1888–1965, Four Quartets: "Little Gidding"

The next immediate need was to greatly simplify and refine the reaction conditions, in order to unravel what was actually occurring during the reactions that were being carried out by enzymes apparently present in the *Tetrahymena* cell extracts. In 1984 Carol Greider joined my lab at UC Berkeley as a Ph.D. student and was immediately interested in doing just this. This work, which led us to discover telomerase activity, is described in detail in Carol Greider's 2009 Nobel Lecture in this volume, so I summarize only briefly some points here.

July 15, 1985

Dr. Benjamin Lewin, Editor
Cell Editorial Offices
292 Main Street
Cambridge, MA 02142

Dear Ben,

Please find enclosed a manuscript entitled "Identification of a specific telomere terminal transferase activity in Tetrahymena extracts" by Carol W. Greider and Elizabeth H. Blackburn, which we are submitting for publication in Cell. We report here the discovery of a novel telomere elongation activity which we believe forms the basis of telomere replication and stabilization in vivo, so we think it should be of general interest to Cell readers.

Thank you for your consideration.

Sincerely yours,

Elizabeth Blackburn

Carol Greider

EB/CG:cv

Figure 4. Letter to Editor of *Cell*, 1985.

We discovered that short fragments of DNA, synthesized chemically as DNA oligonucleotides and therefore available in high concentrations, would get telomeric GGGGTT repeats added to their 3' ends when incubated with *Tetrahymena* extracts. This enzyme reaction was more efficient when the extracts were made from cells at the developmental stage when new telomeres are added during macronuclear development. Fortunately, the reaction, at least in the test tube, was not obligatorily or mechanistically coupled to synthesis of the complementary DNA strand or to DNA cleavage. [20]

Ribonuclease treatment abolished this telomeric DNA repeat addition capability of the extract. Hence the enzyme activity needed RNA. Protease treatment also destroyed the enzyme reaction, implicating required protein component(s) as well as RNA.[21]

The essential RNA component of telomerase was identified and found to contain a sequence, 5' CAACCCCAA 3'. This sequence is complementary to one and a half repeats of the 6 nucleotide repeat sequence that was synthesized *in vitro*. Starting with this powerful hint, we found that all the properties of the synthesis reaction *in vitro*, including its particular DNA and nucleoside triphosphate precursor requirements, added up to a coherent model by which telomere synthesis by telomerase (as we eventually named the enzymatic activity) is templated by repeated rounds of copying of this short template sequence in the telomerase RNA. The synthesis is aided by alignment of the 3' end of the DNA primer on the template, thereby positioning the primer for addition of the next templated repeat.[22]

We thus had discovered that *Tetrahymena* telomerase added tandem repeats of the *Tetrahymena* telomeric DNA sequence, TTGGGG, onto the 3' end of a variety of G-rich telomeric DNA sequence oligonucleotide primers, independently of an exogenously added nucleic acid template. The cellular activity that carried out this reaction was both RNAse- and protease-sensitive. Similar experiments were then derived for telomerase activities from the ciliates *Euplotes* and *Oxytricha*, – by Dorothy Shippen-Lenz in my lab and by Alan Zahler in David Prescott's lab, respectively, and from human cells by Gregg Morin in Joan Steitz's lab at Yale. Each telomerase synthesized its own species-specific sequence – GGGGTTTT repeats (hypotrichous ciliates) or AGGGTT repeats (human cells). These other telomerase activities were very like the *Tetrahymena* telomerase in their primer recognition and other characteristics, including ribonuclease-sensitivity, arguing for the generality of this enzyme activity among eukaryotes. When Dorothy Shippen identified and sequenced the RNA moiety of the telomerase of the ciliate *Euplotes*, it gratifyingly contained the sequence 5' CAAAACCCCAAAA 3'. Experiments indicated that this sequence was indeed the templating domain for synthesis of GGGGTTTT repeats, the telomeric sequence of *Euplotes*. Together, these findings established telomerase as a widespread, reverse transcriptase, unusual in being of cellular origin and in carrying its own internal RNA template for repeated DNA synthesis.

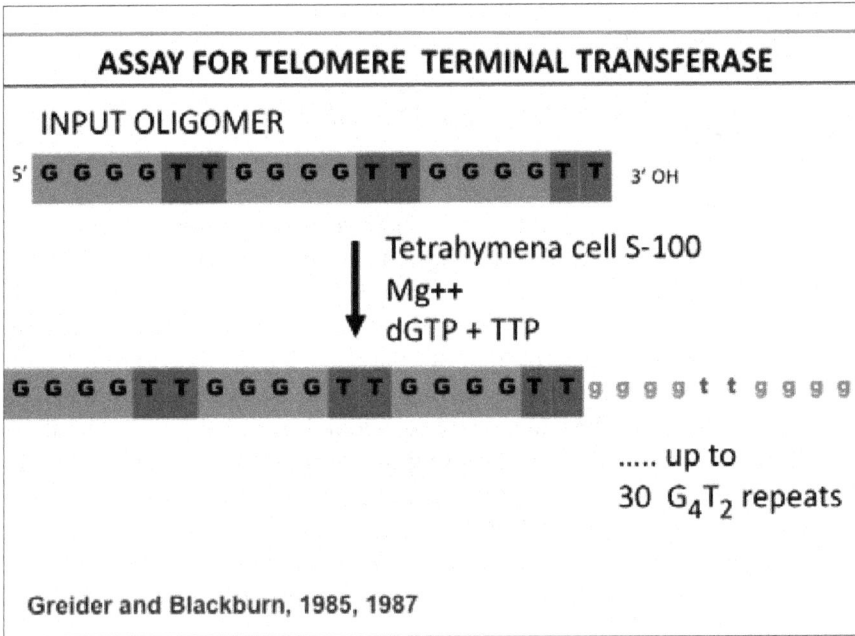

Figure 5. The original telomerase assay. Greider and Blackburn, 1985, 1987).

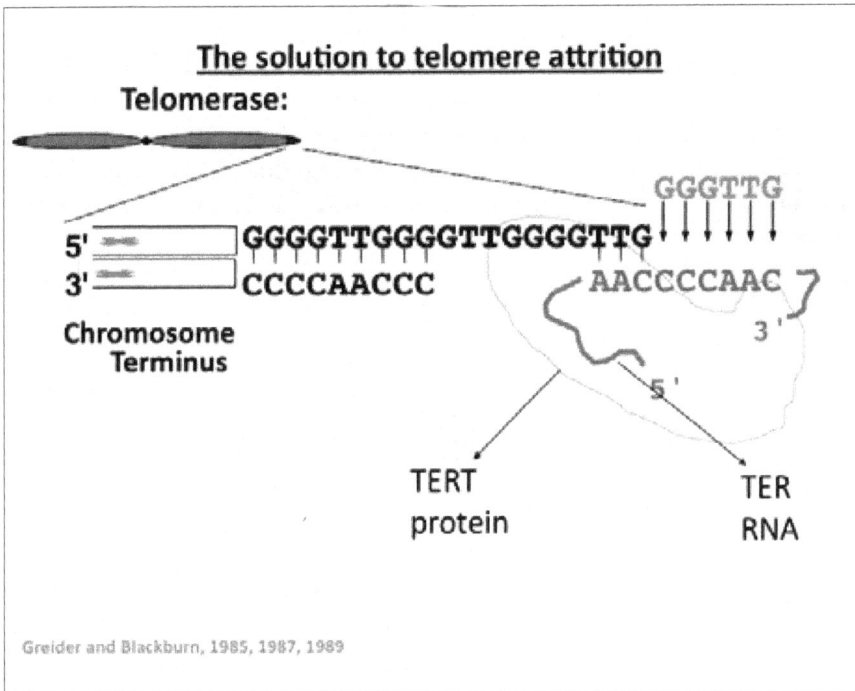

Figure 6. A model for the action of the *Tetrahymena* telomerase. Greider and Blackburn, 1985, 1987, 1989).

DEMONSTRATION OF THE REVERSE TRANSCRIPTASE ACTION OF
TELOMERASE *IN VIVO*
*"They didn't have to walk around to see what was under the sky; they just stayed
where they were. [And] as they looked, their knowledge became intense."*
– POPUL VUH, P. 165.

We now had good evidence that the telomerase enzyme could synthesize the
G-rich strand of telomeric DNA *in vitro*. Proving that the 5' CAACCCCAA
3' sequence in the *Tetrahymena* telomerase RNA gene was the template for
telomere synthesis *in vivo* required site-directed mutagenesis of this sequence in the telomerase RNA gene (which we called the TER gene). Again,
Tetrahymena provided the first key to being able to do these experiments. Its
telomerase RNA gene had been recently cloned by Carol Greider, and we
had devised a system in my lab for overexpression of such mutated genes in
Tetrahymena cells. We inserted the engineered telomerase RNA genes into a
self-replicating vector devised by Guo-Liang Yu, a graduate student in my lab.
Guo-Liang then introduced them into *Tetrahymena* cells by microinjection of
the DNA molecules. He analyzed and sequenced the telomeres in the cells
expressing the mutated telomerase RNAs. These experiments, done with
help from Guo-Liang's labmates Laura Attardi and John Bradley, established
the *in vivo* role of telomerase in three ways. As predicted from the sequence
of the telomerase RNA and from the in vitro experiments described in Carol
Greider's Nobel Lecture, altered telomere repeats specified by the mutant
gene appeared in the telomeres that he cloned out of the transformant cells.
This proved that telomerase was the cellular reverse transcriptase enzyme
that synthesizes telomeres in cells by copying its own internal RNA template
within the TER moiety of the enzyme complex. The cells rapidly showed
abnormal nuclei indicative of failure to segregate their DNA properly. This
indicated that the correct DNA sequence was necessary for proper nuclear
behavior [23].

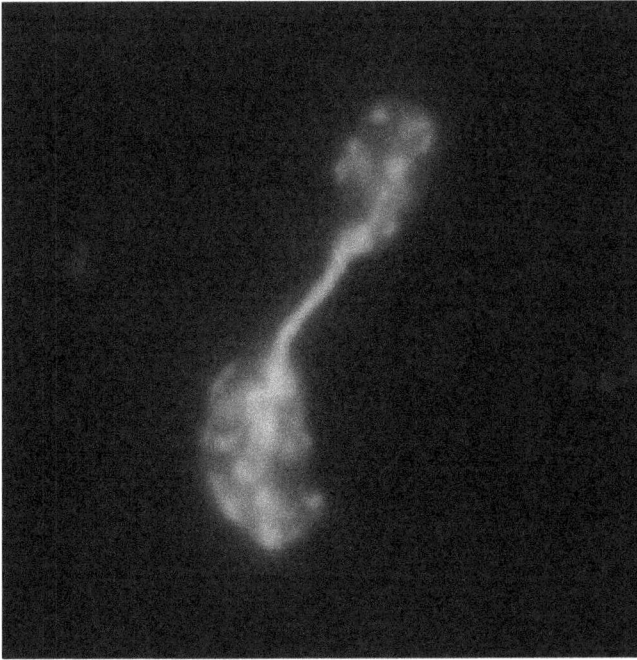

Figure 7. A *Tetrahymena* cell expressing a telomerase RNA with mutated template attempts to divide.

DEMONSTRATION OF THE NEED FOR TELOMERASE FOR CELL GROWTH

"Like as the waves make towards the pebbled shore,
So do our minutes hasten to their end."
– WILLIAM SHAKESPEARE, 1564–1616, SONNET 60.

In addition to proving the templating role of telomerase RNA in *Tetrahymena*, we obtained a bonus result from these experiments. *Tetrahymena* cells are normally effectively immortal. With one particular template mutation, Guo-Liang found none of the predicted sequence DNA was added onto telomeric ends. Instead, the cells continued to grow for only about 20 to 25 more cell divisions. During that time their telomeres progressively shortened. The cells then ceased to divide. This result was the first demonstration that interference with normal telomerase function itself could limit cellular lifespan. It established that continuing action of telomerase was necessary for replicative immortality of cells [23].

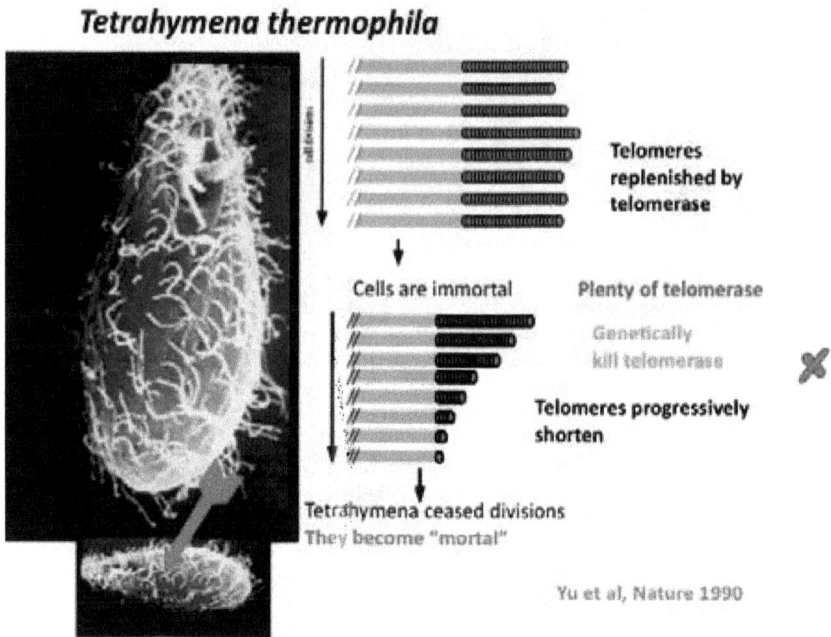

Figure 8. Tetrahymena cells' dependence on telomerase. (Yu *et al., Nature,* 1990).

The action of telomerase thus could explain how replication of the 5' ends of the chromosomal DNA can be completed, without the loss of terminal sequences that would result from normal semi-conservative DNA replication mechanisms: continuous addition of telomeric DNA to the chromosomal ends by telomerase could counterbalance this predicted terminal DNA attrition.

TELOMERES AS PROTEIN-DNA COMPLEXES
"Having well polished the whole bow, he added a golden tip."
– Homer ("Smyrns of Chios"), The Iliad (bk. IV, III).

As recounted in my autobiography in this volume, soon after identifying the telomeric sequence, I found that in *Tetrahymena* chromatin, telomeric DNA tracts were protected by bound protein(s), distinct from nucleosomes. I tried to identify the proteins on *Tetrahymena* telomeres, but did not succeed in this. It was others' work, initially using yeast molecular genetic approaches, that unlocked the door to telomeric proteins. Now an extensive list of proteins associated with telomeres, from various eukaryotes, is known. Many of these have been characterized to varying extents with respect to biochemistry, structure, occupancy levels on telomeres and circumstances that lead to measurable changes in these occupancy levels. Many functions are deduced, often by looking at the consequences to cells of mutating or deleting the protein. But despite the extensive work that has built up the current molecular knowledge of the telomeric DNA-protein complex, the actual picture of a telomere is in some ways still a rather ghostly and partial image. This is almost certainly because telomeres are highly dynamic.

TELOMERES AS A DYNAMIC HOMEOSTATIC SYSTEM

"Stability is not immobility."
– Klemens von Metternich, Austrian statesman, 1773–1859.

During the 1990s, the view of a telomere that emerged was that of a self-regulating entity, normally resilient to change and buffered from it by a variety of molecular mechanisms. Mike McEachern, a postdoctoral fellow then in my laboratory, proposed a model for telomere dynamics based on his experiments with telomeres in *Kluyveromyces lactis,* a budding yeast. The essence of the model is that, first, the rate of shortening of a telomere in the absence of telomerase stays constant as that telomere shortens. But the probability of lengthening it by telomerase actually changes depending on the telomere length – the shorter the telomere, the more likely it is to be lengthened by telomerase action. Mike deduced this largely from a series of experiments in which he altered the telomerase RNA template sequence to direct the synthesis of various repeats. Anat Krauskopf, then concurrently a postdoctoral fellow with Mike in my lab, extended these findings: Telomere length regulation became altered in a way that tracked with altered binding of a yeast telomeric protein to the mutated telomeric sequence. Together, Mike's and Anat's experiments showed that a major contributor to such negative regulation of telomerase action on telomeres is the telomere protein-DNA complex structure itself. Thus the telomere itself was a like a gatekeeper, regulating access of telomerase onto the telomere, even in the presence of excess telomerase in the cells. Over subsequent years much more has been learned about the details of the proteins involved, but this general model has stood the test of time.

A general and important corollary concept is that telomeres can exist in two states: capped or uncapped. Capped telomeres signal the cells to keep on proliferating, all other things being well. But uncapped telomeres in the cell signal the cell; if uncapping is persistent, it signals the cell to arrest its divisions. Mike McEachern and Anat Krauskopf showed that one of the most striking properties of a telomere is how resilient it can be to molecular insults of a variety of types, and then, like the last straw, just one more molecular change is sufficient for the telomere to collapse catastrophically into disaster. Thus it emerged that cells have evolved elaborate and overlapping, redundant or mutually reinforcing mechanisms to ensure that their telomeres stay functional. [24]

Currently, the combined picture from the results from many researchers is that the telomere in a cell is a highly dynamic structure. Rather than being a rock-stable complex, it is perhaps reminiscent of a swarm of bees: the size and shape of the swarm overall appears the same, but in reality its composition is constantly changing as the bees (the telomeric proteins) of the swarm constantly come off it and are replaced by other bees.

SIMILAR MOLECULAR MACHINERIES: DIFFERENT LIFE HISTORIES
"Have regard to the end." [Lat., Finem respice (or Respice finem).]
– Chilo of Sparta (Chilon).

The structure and the function of telomeres are highly evolutionarily conserved among eukaryotes. This conservation underlies why in the early 1980s Jack Szostak and I were able to successfully propagate *Tetrahymena* telomeres in the distantly related organism baker's yeast. As described above, and in detail in Jack Szostak's 2009 Nobel Lecture, we found that yeast telomeric sequences were added to the introduced *Tetrahymena* telomeric ends on a plasmid, thereby stabilizing the plasmid and allowing it to replicate indefinitely, in linear form, as an extrachromosomal plasmid. Similar conservation applies to the telomerase mechanism for telomere maintenance: throughout the eukaryotes telomerase, a specialized ribonucleoprotein reverse transcriptase, is used to maintain the ends of eukaryotic chromosomes, with relatively rare exceptions. Telomerase RNA and core protein of telomerase, TERT, each retain well-recognizable conserved features in even the most distantly related eukaryotes.

In the face of this widespread conservation of telomeres and telomerase, extending down to the deep roots of eukaryotic evolution, a fascinating finding is the great variety of telomere maintenance stories that play out during the lives of different eukaryotes. Among mammals alone, even under favorable living conditions species clearly differ in their maximal possible lifespans, implying that maximum lifespan has considerable genetic determination. Humans can have a life expectancy of about eighty years, and laboratory mice about two years. Thus, it is reasonable to contemplate the possibility that the rate-limiting steps causing aging and eventual death may differ between these two species, despite the common underlying cellular and molecular mechanisms they share. Even within the mammals, the qualitative and quantitative contributions of telomere maintenance to cellular proliferative lifespans seem to differ widely [25]. And, extending further out from mammals to invertebrates, despite much conservation of fundamental molecular and cellular mechanisms, it is possible that those mechanisms that contribute to their aging and death from old age may be divergent from those that are quantitatively important or rate-limiting for aging and lifespan in humans. All of these considerations have raised the question of whether telomere maintenance is a quantitatively important determinant of normal human aging and lifespan.

TELOMERASE IN HUMAN HEALTH AND DISEASE
a) Telomerase in cancer cells
"We ought to consider the end in everything." [Fr., En toute chose il faut considerer la fin.]
– Jean de la Fontaine, Fables (III, 5).

One special and notable context in which telomerase plays a prominent role in humans is in human cancer cells. Hyperactive telomerase in the cancer cells is a prominent characteristic of the great majority of most types of malignant human tumors. In this setting of the cancer cell – which, importantly, has undergone multiple other genetic and epigenetic changes in its progression to tumorigenicity – telomerase plays cancer-promoting roles. Most clearly, it promotes cellular immortality by providing cancer tell telomeres with the means for continuous replenishment. The high level of telomerase that characterizes human cancer cells thus is a rational target for anti-cancer therapies.

b) Telomere maintenance and human life histories
"The end crowns all,
And that old common arbitrator, Time,
Will one day end it."
– William Shakespeare 1564–1616, The History of Troilus and Cressida (Hector act IV, v).

As described above, abrogating telomerase in otherwise effectively "immortal" single-celled species causes progressive telomere shortening over several cell generations followed by cessation of cell division ("senescence"). This naturally led to the question of whether the same progressive process operates to cause human aging and limit human lifespan. Rare genetic mutations in telomerase component genes leading to reduced telomerase levels and telomere shortening in humans clearly have adverse disease-causing effects and can prevent the affected individual from attaining an old age. Yet until recently, for the vast majority of people, who by definition are not "mutants", the contributions of insufficient telomere maintenance to aging and lifespan limitation was less clear. What does one observe in the human population in general?

Large amounts of epidemiological molecular data on humans and their *in vivo* telomere maintenance have now accumulated. From our present knowledge of telomeres, the picture that has emerged is that telomere maintenance is linked to human aging and diseases of aging. First, telomere shortness in white blood cells is linked to a large and impressive list of the major diseases of aging: in multiple cohorts, often involving hundreds to thousands of individuals, short telomere length has been found to be associated with risk of, and incidence of, cardiovascular disease, stroke, vascular dementia, osteoporosis and obesity and risks for diabetes and certain cancers. Longer mean white blood cell telomere length is not consistently linked to longer lifespan, but longer telomere length has been linked to more years of healthy life, in a cohort of people in their seventies. [26]

Second, telomerase activity is not only present in many normal human somatic cells but also, importantly, quantifiable ([27–30]) in adult (including elderly) humans; even in resting white blood cells, as well as in stem and proliferating progenitor cell types, telomerase is active. This means that telomere shortening in normal cell populations has the possibility of being counterbalanced, or even reversed, throughout life. While cross sectional studies show a slow loss of telomeric length across populations of humans in general, the datapoints are noticeably scattered: it is not uncommon for an 80-year old's telomeres to be as long as those in a 30-year old. What might account for such scatter? In fact, lengthening of telomeres in white blood cell populations is now found to be much more common than expected from the previous models of inexorable telomere shortening throughout life. However such models had been based almost solely on cross sectional studies and on the presumption of lack of telomerase in the normal cells of adult humans.

What determines and regulates the variation in long-term telomere maintenance in people? While genetic influences have been detected, non-genetic factors are also coming to the fore as significant influences on telomere length maintenance in human white blood cells. In summary, in humans, telomere maintenance status results from the integrated influences of many factors, genetic and non-genetic. The non-genetic influences include modifiable factors; notably, psychological stress, behavioral and even nutritional factors.

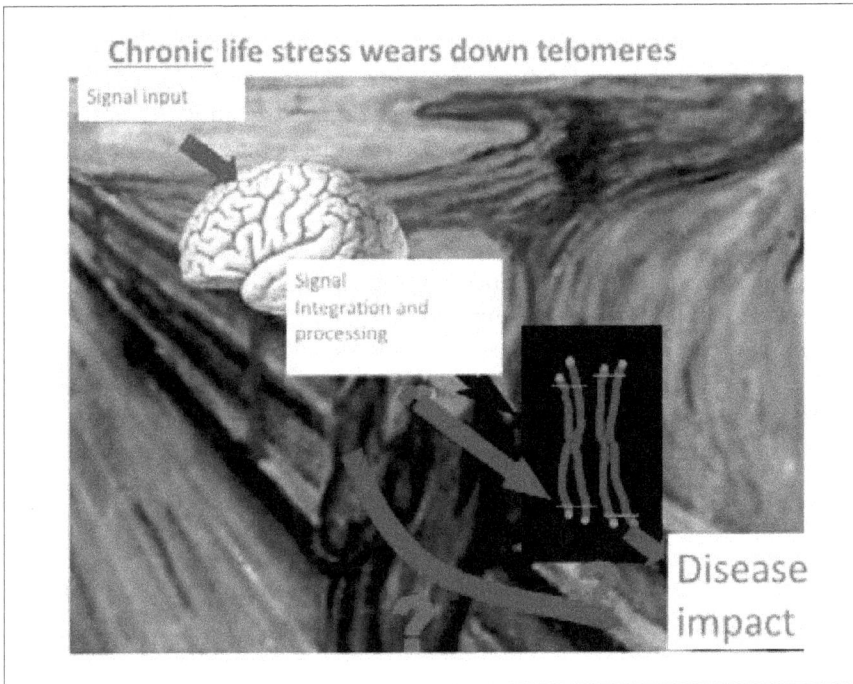

Figure 9. Input of stress on telomeres and its disease impact.

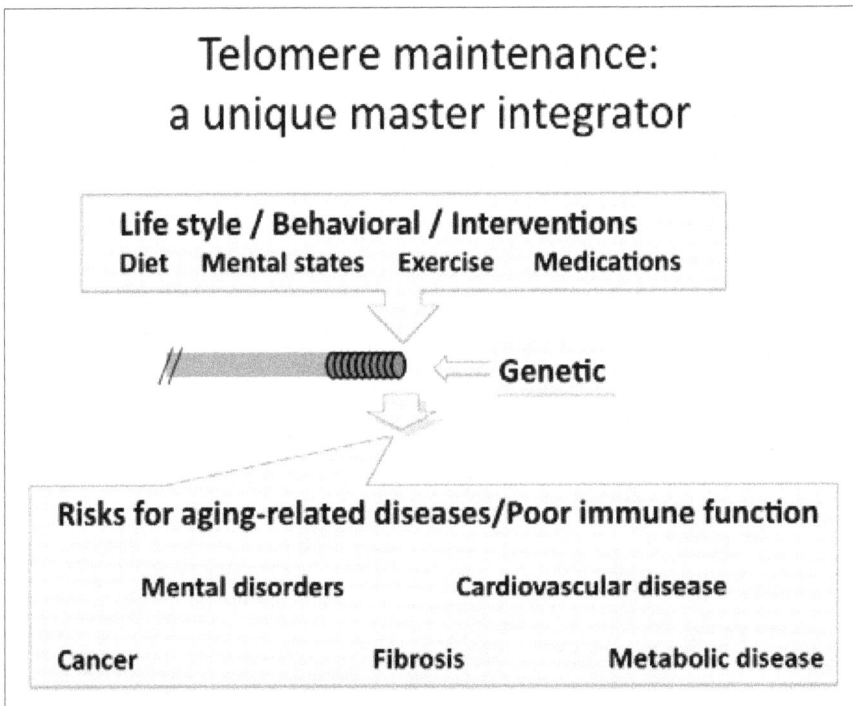

Figure 10. The telomere as an integrator of many factors.

The major conditions and diseases or disease risks occurring with human aging have now been associated with shortness of blood cell telomeres: prominently, cardiovascular disease, cancers, diabetes and impaired immune system function in various forms. Thus, telomere shortness is not specifically associated with any one disease. Rather, this seeming non-specificity of telomere maintenance may instead be more usefully considered as reflecting – perhaps causing – aging more fundamentally. Telomere maintenance status may be a truer integrative measure of actual "biological age" than chronological age. Furthermore, human life conditions impact on telomere maintenance in humans. Perhaps telomere monitoring will become as common as regular weighing as an integrative indicator of health. Certainly, these findings and implications are taking the field of telomere and telomerase biology into realms far from the single-celled pond microorganisms in which I began this work.

ACKNOWLEDGEMENTS

I am indebted to my many valuable colleagues with whom I have been blessed over the years, without whom I would have done much less.

REFERENCES

1. de Lange, T., V. Lundblad, and E. Blackburn, eds. *Telomeres*. Second Edition, 2006, Cold Spring Harbor Press: Cold Spring Harbor, NY. 1–19.
2. McClintock, B., *Cytological observations of deficiencies involving known genes, translocations and an inversion in Zea mays*. Agricultural Experiment Research Station Bulletin, University of Missouri College of Agriculture, 1931. 163: p. 4–30.
3. McClintock, B., "A correlation of ring-shaped chromosomes with variegation in Zea Mays," *Proc Natl Acad Sci U S A*, 1932. 18: p. 677–681.
4. Muller, G., "The remaking of chromosomes," *The Collecting Net*, 1938. 8: p. 182–195.
5. Kavenoff, R., "Chromosome-sized DNA molecules from *Drosophila*," *Chromosoma*, 1973. 41(1): p. 1–27.
6. Grell, K., *Protozoology*. 1973, Berlin: Springer-Verlag.
7. Blackburn, E. and J.C. Gall, "A tandemly repeated sequence at the termini of the extrachromosomal ribosomal RNA genes in *Tetrahymena*," *J Mol Biol*, 1978. 120(1): p. 33–53.
8. Blackburn, E. and J. Szostak, "The molecular structure of centomeres and telomeres," *Ann Rev Biochem*, 1984. 53: p. 163–194.
9. Yao, M., E. Blackburn, and J. Gall, "Tandemly repeated C-C-C-C-A-A hexanucleotide of *Tetrahymena* rDNA is present elsewhere in the genome and may be related to the alteration of the somatic genome," *J Cell Biol*, 1981. 90(2): p. 515–520.
10. Blackburn, E., "Telomeres and their synthesis," *Perspectives in Science*, 1990. 249(4968): p. 489–490.
11. Johnson, E., "A family of inverted repeat sequences and specific single-strand gaps at the termini of the *Physarum* rDNA palindrome," *Cell*, 1980. 22(3): p. 875–86.
12. Emery, H. and A. Weiner, "An irregular satellite sequence is found at the termini of the linear extrachromosomal rDNA in *Dictyostelium discoideum*," *Cell*, 1981. 26 (3 pt 1): p. 411–9.
13. Katzen, G., G. Cann, and E. Blackburn, "Sequence-specific fragmentation of macronuclear DNA in a holotrichous ciliate," *Cell*, 1981. 24(2): p. 313–20.
14. King, B. and M. Yao, "Tandemly repeated hexanucleotide at *Tetrahymena* rDNA free end is generated from a single copy during development," *Cell*, 1982. 31(1): p. 177–82.
15. Blackburn, E., *et al.*, "DNA termini in ciliate macronuclei," in *Cold Spring Harbor Symp Quant Biol*. 1983: Cold Spring Harbor Press.
16. Boswell, R., L. Klobutcher, and D. Prescott, "Inverted terminal repeats are added to genes during macronuclear development in Oxytricha nova," *Proc Natl Acad Sci U S A*, 1982. 79(10): p. 3255–9.
17. Szostak, J. and E. Blackburn, "Cloning yeast telomeres on linear plasmid vectors," *Cell*, 1982. 29(1): p. 245–55.
18. Shampay, J., J. Szostak, and E. Blackburn, "DNA sequences of telomeres maintained in yeast," *Nature*, 1984. 310(5973): p. 154–157.
19. Bernards, A., *et al.*, "Growth of chromosome ends in multiplying trypanosomes," *Nature*, 1983. 303(5918): p. 592–7.
20. Greider, C. and E. Blackburn, "Identification of a specific telomere terminal transferase activity in *Tetrahymena* extracts," *Cell*, 1985. 43(2 Pt 1): p. 405–413.
21. Greider, C. and E. Blackburn, "The telomere terminal transferase of *Tetrahymena* is a ribonucleoprotein enzyme with two distinct primer specificity," *Cell*, 1987. 51(6): p. 887–898.
22. Greider, C. and E. Blackburn, "A telomeric sequence in the RNA of *Tetrahymena* telomerase required for telomere repeat synthesis," *Nature*, 1989. 337(6205): p. 331–37.
23. Yu, G.-L., *et al.*, "*In vivo* alteration of telomere sequences and senescence caused by mutated *Tetrahymena* telomerase RNAs," *Nature*, 1990. 344(6262): p. 126–132.

24. Blackburn, E., "Switching and signaling at the telomere," *Cell,* 2001. 106(6): p. 661–73.

25. Gomes, N., J. Shay, and W. Wright, "Telomeres and telomerase. Inter-species comparisons of genetic, mechanistic and functional aging changes," in *The Comparative Biology of Aging,* N. Wolf, Editor. 2010, Springer Netherlands. p. 227–258.

26. Njajou, O., *et al.,* "Association between telomere length, specific causes of death, and years of healthy life in health, aging, and body composition, a population-based cohort study," *J Gerontol A Biol Sci Med Sci,* 2009. 64(8): p. 860–4.

27. Epel, E., *et al.,* "Accelerated telomere shortening in response to life stress," *Proc Natl Acad Sci U S A,* 2004. 101(49): p. 17312–15.

28. Epel, E., *et al., Cell aging in relation to stress arousal and cardiovascular disease risk factors.* Psychoneuroendocrinology, 2006. 31(3): p. 277–87.

29. Lin, J., *et al., Analyses and comparisons of telomerase activity and telomere length in human T and B cells: insights for epidemiology of telomere maintenance.* J Immunol Methods, 2010. 352(1–2): p. 71–80.

30. Epel, E., *et al., Dynamics of telomerase activity in response to acute psychological stress.* Brain Behav Immun, 2009.

Portrait photo of Professor Blackburn by photographer Ulla Montan.

CAROL W. GREIDER

I was born in San Diego, California in 1961. My brother Mark was born in January of the previous year. My father Kenneth Greider was a physicist who had recently graduated with a Ph.D. from University of California at Berkeley. My mother Jean Foley Greider also had received her Ph.D. from UC Berkeley in Botany. My father worked in high-energy nuclear physics and my mother was a mycologist and a geneticist. After both parents completed postdoctoral fellowships in San Diego in 1962, my father took a faculty position in the Physics Department at Yale and so the family moved to New Haven, Connecticut. My mother took a postdoctoral position at Yale in the laboratory of Norman Giles, where she worked on Aspergillus as well as other fungal species. A few years later in 1965, my father took a faculty position in the Physics Department at UC Davis and so the family moved back to California. My mother first took a teaching position at a Sacramento community college and then later at American River College in nearby Sacramento.

DAVIS

Mark and I grew up in Davis, where we could walk to school. My parents built a house in a development in West Davis shortly after we moved to Davis. The street was conveniently located about a four block walk from the West Davis Elementary School (Grades K-4) and half a block from the new West Davis Intermediate School (Grades 5 and 6). Mark and I would walk to school together as kids, and later bike to high school, year-round. It gave us a sense of independence to come and go. The idea of parents driving their kids to school was one I had never heard of until moving to the east coast and becoming a parent myself. This early responsibility was something that shaped my sense of independence. For me school was something that was a kid's responsibility. Parents were not really involved.

In December of 1967 my mother died when I was in first grade and Mark was in second. In retrospect, this event played a major role in my learning to do things on my own. Mark and I continued to get ourselves to school and to go on with our lives as best we could. School was not easy for me. I was put in remedial spelling classes because I could not sound words out. I remember a special teacher coming into the classroom every week to take me out for special spelling lessons. I was very embarrassed to be singled out and removed from class. As a kid, I thought of myself as "stupid" because I needed remedial help. It was not until much later that I figured out that I was dyslexic and that my trouble with spelling and sounding out words did not mean I was stupid, but early impressions stuck with me and colored my world for a time.

HEIDELBERG

In 1971 my father was invited take his sabbatical at the Max Planck Institute for Nuclear Physics in Heidelberg, Germany. We moved to Germany for the year and Mark and I went to the *Englisches Institut,* a private school. Despite its name, it was a typical German Gymnasium and all of the instruction was in German. So for the first six months or so, we learned German by immersion. Mark and I took the city bus to school each day so we quickly learned to navigate the public transportation system, as well as navigate our way around a new school and new language and new culture. In Davis we had been used to getting to school on our own, so we welcomed this independence and developed an appreciation for how things were done in a very different culture.

I remember my grades were particularly poor in this school and especially so in the English class. The English teacher would give a dictation and we were supposed to write down what she said in English. It seemed too simple and pointless to me, but when I got my graded notebooks back, the scores were usually D's or F's because every other word was misspelled. Looking back over those notebooks later, I saw the pattern of backwards words and letters and gross misspellings that led me to suspect I was dyslexic. The other confusing thing about school for me was the "religion" class. You had to declare if you were Catholic or Protestant (as if those were the only choices) and then each group had their own class. Back home, my father was music director for the Unitarian Church, but as kids, we rarely went to church. It was too hard to translate what Unitarian meant to the Germans, so my father asked the school to excuse me from this religion class, and instead have a free period to do homework. This is how I met my friend Jiska, who was one of the few Jewish kids in the school and who was also excused from religion class. In my friendship with Jiska, both of us different from the rest, I began to develop an appreciation for people who were not like the others and who stood a bit outside the mainstream. This understanding of and affinity for people outside the mainstream served me well later in life. In high school and college I never felt the need to be part of a popular group, but rather sought out friends for their personal qualities. This appreciation may have also shaped many choices later in life; for example, working on the unusual organism *Tetrahymena.*

I spent a lot of time on my own in Heidelberg, playing down by the stream near our house or hiking the hill to the top of Boxberg. I took the bus into town on my own and learned to dress and speak like a German. There was a large American army base in town and I did not want to be mistaken as an army kid. I liked being more unusual: an American kid who understands German culture. By the middle of the year, I had learned German and became fluent in speaking and reading, but like all other written tasks, the writing and grammar eluded me. Mark and I had some German-American friends a few stories up in our apartment complex and we made up games like tapping out a code on the radiators and sending notes on string outside the kitchen windows to communicate. These games irritated the other apartment residents and resulted in the building manager coming to talk to my father. We were typical kids in that fashion, breaking some rules, where we could, but not going too far.

DAVIS – PART II

When we returned from Germany I went into 6th grade, which was a transition year, the last in intermediate school before junior high school. I spent much of it readjusting to being back and making new friends. Unlike many scientists I know, I was not a kid who knew from early on that I wanted to be a scientist. I think one important thing I learned in my early years was to focus intently on the task at hand, such as learning German when we were in Heidelberg, to the exclusion of other things going on around me. This survival skill served me very well in later years. Focusing on certain goals and ignoring obstacles came naturally to me.

In junior high school I learned that I could be good at school. I remember liking the freedom to choose classes and the pleasure of learning and doing well. My perseverance and love of reading had somehow allowed me to overcome many disadvantages of dyslexia, and I read a lot of books for pleasure. I found I had to memorize words to spell them, as sounding them out did not work for me. This coping mechanism proved also to have an upside; memorization in biology and history was easy for me. My father encouraged us to do well in school and to do it for ourselves. He said that we should want to do well because it would "open doors" for us. He emphasized that being able to choose what you want to do in life is so important, and doing well early on will allow more possibilities in the future. I also discovered the pleasure of the outside reward of getting all A's in classes, it made me feel good and I got positive feedback from people outside the family.

In high school I focused on doing well in my classes and finding a supportive group of friends. In junior high I had been attracted to outsiders, perhaps from my experience in Germany. But the outsider group I found myself with in junior high was not as interested in school as I was. I took the opportunity of the change in schools from Emerson Junior High to Davis Senior High as an opportunity to find a new group of friends. I met Lori Lopez and Resi Zapfel at an American Field Service (AFS) Club meeting in the first weeks of high school, and they quickly became friends. Resi was an exchange student from Austria and Lori was the AFS club president. Lori's family and Resi's host family the Robertsons became like a second family to me. I liked the foreign students in the ASF group, and the American kids who were a part of this group were not interested in mainstream popularity. I affiliated myself with the AFS student group throughout high school and was even president of the club my senior year. I did not focus particularly on science in high school, or join any science-related groups, although I continued to do well in all of my classes; I considered it a challenge to get all A's. I never considered myself one of the smart kids, they seemed confident and driven. I just enjoyed learning and especially spending time with friends.

After my junior year at Davis High School, I knew I needed to think about where to go to college. I had done well in biology in school and was particularly captivated by my 12th grade biology class, where we learned a lot of physiology from a very motivated science teacher who had a Ph.D. I loved learning new material and being challenged, so I decided to major in biology in college. Many of my fellow high school graduates intended to go to nearby schools, either UC Davis or UC Berkeley. I did not want to go to either. I wanted to do something different from the norm, get out and have new experiences. My friend Alyssa Ingalls, whom I had known since 6th grade, was taking a trip to visit several University of California schools with her family Liz and Bob Young. I was happy to be invited along on this school tour. We visited UC Santa Cruz, UC Los Angeles and UC Santa Barbara.

I had a contact at UCSB, Beatrice Sweeney, who was a professor there and who had known and worked with my mother at Yale. My father put me in touch with Beazy, as she was known, and Alyssa and her family and I got a tour of the campus from her. Beazy was a cell biologist by profession but a naturalist at heart. She took us for a walk on the beach near her house and told us fascinating stories about the biology of all the marine animals and plants that we walked by. I was captivated by her and by the beautiful UCSB campus. I decided I wanted to study Marine Ecology at UCSB.

SANTA BARBARA

Beazy was on the faculty of the College of Creative Studies, a small college that is part of UCSB. The College of Creative Studies was founded by Marvin Mudrick, a professor in the English Department, to foster independent learning and interaction between disciplines. The requirements to get into CCS were significantly higher than to get into UCSB. My grades were very good, but my Scholestic Aptitude Test (SAT) scores were not. I never spent time practicing to take standardized tests and the dyslexia made them hard for me. I was very happy when CCS accepted me. So off I went in the fall to Santa Barbara.

The most important thing about UCSB and CCS was that Beazy [Professor Sweeney] encouraged me to begin working in a lab my freshman year. I was scared that I needed more time to adjust to college, but she said to start as soon as possible. I did a project first with Adrian Wenner studying sand crab populations in Santa Barbara. Though I thought I wanted to be a Marine Ecologist, this experience did not captivate my attention. The science was mostly statistics, which I did not understand or relate to. Beazy kept in close contact with me and saw I needed a different experience. So I then worked with a postdoc in Beazy's lab studying the movements of chloroplasts during dark/light circadian cycles in *Pyrocystis,* a dinoflagellate. I enjoyed the work in the lab. I liked coming in to do my own experiments and was challenged when Beazy said I had to come up with a way to plot and describe my experiment on my own, with no set form. The simultaneous pain and joy of trying to create something that made sense to describe my observations was exhilarating.

I enjoyed watching cells and describing the circadian rhythms, but after a while I felt the work was too descriptive. So next Beezy took me to work in Les Wilson's lab on microtubule dynamics. I am not sure if it was the topic or the personalities in the lab, though it was likely both, that captivated me. I worked first with Kevin Sullivan and later with David Asai studying microtubule associated proteins. The work in the lab was focused on understanding molecules and how they interact and behave. We would do experiments to examine how fast microtubules would assemble from the tubulin building blocks, then make a change to the tubulin preparation and see how that affected the results. Being able to manipulate molecules and understand the mechanics of how things worked fit my way of thinking. In addition, talking with both Kevin and David and the others in the lab was fun. People knew each other well, were playful, and would tease each other a lot. There were inside jokes and an easy way of laughing about experiments as well as everyday life that was infectious. I worked with Kevin for my sophomore year studying the assembly kinetics of chick brain microtubules under different conditions. The experience that Beezy and the CCS program provided me, to try out several different laboratory experiences, was instrumental in my finding how much I loved mechanistic thinking and biochemical experiments. By comparing several different labs, it became clear to me when I was having fun and when I was not. I saw that laboratory work was about people and interactions as well as about science. It could be playful and was appealing as a potential path I could enjoy.

GÖTTINGEN

My junior year in college I spent as a student at the University of Göttingen in Germany. Ever since my early experience in Heidelberg and visits to see Resi Zapfel (now Schmall) in Austria, I wanted to experience what it was like to live as a student in a foreign country. I took the opportunity to go to Germany for a year on the University of California's Education Abroad Program (EAP). Before I left, Kevin Sullivan and Les Wilson encouraged me to continue lab work in Germany. They contacted Klaus Weber who had a lab at the Max Planck Institute for Biophysical Chemistry, and he agreed I could work there. I would split my time between classes at the University, such as Biochemistry and Genetics, and time at the Max Planck working on intermediate filaments in the Weber lab. In addition to lab work, I also became close friends with a number of Americans in Göttingen who were also on exchange programs.

At the beginning of the second semester I was looking for biology courses in the course catalogue and found one on chromosomes that looked interesting. When I showed up in the assigned room at the right time it turned out it was the regular lab meeting for Professor Ulrich Grossbach. Professor

Grossbach had listed his lab meetings as a course so the graduate students could get credit. I was very embarrassed to walk into a private lab meeting, but the researchers in the group were all very nice and they asked me to stay. Michel Robert-Nicoud, a research associate in the group, took me under his wing and asked if I wanted to help in a study of polytene chromosomes of *Chironomus*, a diptera distantly related to *Drosophila*. I enjoyed learning how to do the preparations. It was satisfying to prepare the salivary glands just right and see the giant polytene chromosome under the microscope. I finished the work I had begun in the Weber lab and moved to work with Michel in the Grossbach lab.

Michel collaborated with Tom and Donna Jovin, who were also at the Max Planck Institute for Biophysical Chemistry, on an unusual left handed helical form of DNA called Z-DNA. Tom and Donna had studied the biophysics of sequences that could form this unusual DNA structure. To understand if Z-DNA is found in natural chromosomes and where it might be located, they developed antibodies to Z-DNA. They were collaborating with Michel Robert-Nicoud to locate the Z-DNA by staining the giant *Chironomus* polytene with their Z-DNA antibody. There were controversies about whether Z-DNA might be located in bands or interband regions of the chromosome. There was also discussion about whether the regions that stained with the antibody normally had Z-DNA or if the binding of the antibody itself induced Z-DNA where it might not normally be. There was a lot of excitement in the lab about this project and Donna Jovin was preparing to submit a paper on these findings. It was thrilling to know that my work staining chromosomes was of use for *real* experiments and not just as make-work, and might be part of a publication. This experience with *Chironomus* polytene chromosomes gave me an appreciation for the beauty of chromosomes. It may be that I gained an affection for chromosomes that I brought with me several years later when I first met Liz Blackburn.

SANTA BARBARA – PART II

When I returned to Santa Barbara for my senior year I wanted to go back to work in the Wilson lab. Kevin Sullivan was writing his thesis and planning to move to a postdoctoral position. Kevin suggested I work with David Asai who was a research associate in the Wilson lab. Kevin was very excited about his future studying the genes for tubulin, because he said genes and DNA were the most exciting work going on. Talking to Kevin and David helped me decide that I wanted to go to graduate school. I enjoyed the camaraderie in the lab and liked the challenge to think creatively. I worked hard my senior year, and CCS made it possible for me graduate in 4 years by their flexibility about transferring credit from my course work in Germany.

For graduate school entrance I took the Graduate Recorded Exam (GRE) exams and, as with the SATs, did not do well. I applied for admission to eight different graduate programs, but did not make it through the numerical cut off for grades + GRE's. I got many rejection letters. However, two schools did decide to interview me. I may have seemed like an interesting case to those people who actually read the applications, rather than pre-screening with a numerical cut-off. I had a 3.9 GPA and A+'s in O-Chem, P- Chem and pharmacology, a lot of lab experience, but poor GRE scores. California Institute of Technology interviewed me and each of the 10 professors with whom I talked asked me why my GREs were low. I talked science with all of them and also explained the dyslexia and poor scores on standardized tests. After the interview I was accepted to Cal Tech. UC Berkeley also accepted me and asked me to come for an interview. It was during that interview that I met Elizabeth (Liz) Blackburn. I felt her enthusiasm for chromosomes and telomeres was infectious. I wanted to talk to her more after the allotted interview time so I made plans to come back again the next week to talk in more depth about her telomere work. After that interview, I decided I wanted to go to Berkeley and work with Liz.

Both of my advisors at UCSB, Bea Sweeney and David Asai encouraged me to go to Cal Tech instead. David had done his Ph.D. there and felt it was a special place to he wanted me go there too; Beazy did not want me to go to Berkeley "just because my parents had gone there." Somehow my interest in potentially working with Liz was great enough to for me to go against the recommendations of two mentors. So I signed up as a Ph.D. student in the Department of Molecular Biology at Berkeley.

UC BERKELEY

When I got to Berkeley I had missed the week of orientation for new students, because I decided to attend the wedding of my friends Monica and Chris Morakis whom I had met in Göttingen. My first few weeks at Berkeley felt overwhelming. Although I had done biochemistry, I had not taken any molecular biology courses and had never worked with DNA. My classmates were an impressive bunch with a strong background in molecular biology and it seemed they were all clearly smarter and better prepared than I was. It was thrilling to be part of such an impressive group of interesting people, and soon we all became very close friends.

Although I had come to Berkeley to work with Liz Blackburn, all first year students had to do three laboratory "rotation projects" for 2–3 months each before decisions were made about which lab to join. My first rotation was with Richard Calendar studying phage P2 and P4 interactions. I was very fortunate that that year, two of us first year students were both assigned to Rich's lab at the same time. My fellow 'roton', Jeff Reynolds, was very smart, very friendly and it seemed he knew everything about DNA. So I could lean on Jeff and his knowledge to get me started at Berkeley. From those first days, Jeff became, and still is, one of my best friends.

My second rotation project was in Liz Blackburn's lab. There was a certain amount of anxiety among the first year students as we could not choose our rotation labs; assignments were made by the Department chair, Nickolas

Cozarelli. I was very happy to get assigned to Liz's lab because of my strong interest in working with her. For the rotation I worked on a project to clone telomeres from trypanosomes and the related species Leishmania. By the time I arrived in the lab, Liz and Jack Szostak had already shown that telomeres from *Tetrahymena* would function as telomeres in yeast. This was incredible because *Tetrahymena* and yeast are in different kingdoms phylogenetically. They had shown that when *Tetrahymena* telomeres were ligated to both ends of a plasmid, they allowed that plasmid to be grown as a linear chromosome in yeast. By removing one *Tetrahymena* telomere they were able to clone a functional yeast telomere. I was using this same technique to try to capture telomere fragments from Leishmania. I enjoyed the laboratory environment and by talking to people I got a sense of what projects I found most interesting; I was intrigued by the question of how telomeres get elongated.

In the second quarter, I also took a graduate course on chromosomes taught by Liz in which students were assigned papers that they then presented to the entire class. I was assigned the Szostak and Blackburn 1982 *Cell* paper that identified yeast telomeres. I was petrified, having never presented a paper in front of a large group before. I studied the paper inside and out. I was scared, but I was energized and got a thrill out of presenting that paper. I found it satisfying to convey the excitement I had about telomeres to my fellow students.

Janice Shampay, a student in Liz's lab had recently published an important follow-up paper to the *Cell* paper. They showed that *Tetrahymena* telomeres had yeast sequences added to them as the linear plasmid was maintained in yeast. The excitement grew with the idea that these telomere sequences must be somehow added to chromosome ends. A previous rotation student and friend of mine, Jim Bliska, had done his first rotation in Liz's lab. He had been testing ways to find an activity that might elongate telomeres. From what I knew about telomeres, I thought this project was exciting because it directly approached the heart of the biggest question: How are telomeres elongated?

I had to wait until after my third rotation before I could ask Liz about working with her, according to the graduate program rules. Toward the end of the 3rd rotation, I made an appointment to talk to Liz. As I went into her office I was both scared and excited. I asked her first if I could work in her lab, and second, whether I could work on the telomere elongation project. I was thrilled when she said "yes" to both. I think the conversation lasted all of a minute, but it was a very momentous minute for both of us.

THE BLACKBURN LAB

I joined Liz's lab in May of 1984 and I set out to see if I could find biochemical evidence for telomere elongation in *Tetrahymena*. Liz had first sequenced telomeres in *Tetrahymena* and she reasoned that this single celled ciliate would be a good source for a telomere elongation activity. Each cell has over 40,000 telomeres and perhaps more importantly, there is a stage of its life cycle where new telomeres are added onto fragmented chromosomes. I made extracts from *Tetrahymena* cells and examined whether artificial telomeres could be elongated by enzymes present in the extracts. These experiments are described in detail in the accompanying published lecture.

After about nine months of trying variations on experiments, we found our first strong evidence for telomere elongation. An 18 nucleotide telomere "seed" was elongated with a repeated sequence that was six bases long – precisely the length of the TTGGGG telomere repeat in *Tetrahymena*. Now we had a biochemical assay that we could use to determine if this was a new telomere elongation mechanism. We set out to critically examine whether the 6 base pattern we were seeing was indeed due to a new activity or perhaps instead was a well known polymerase fooling us. Liz and I worked very well together. We would talk most every day and each of us would assert what we thought should be done next. Often we agreed but sometimes we did not, and we would try to convince the other of our reasoning. I remember for one experiment we talked for a long time and neither of us would give up our stance. It was an impasse. The next day when I came in to the lab and we talked, we had both shifted sides. I decided to do her proposed experiment first. We both laughed that we had each convinced each other.

I learned many important lessons that first year after the initial telomerase discovery. Mostly, I learned the importance of questioning your own assumptions. We did not set out to prove we had a new enzyme, rather we imagined all the ways our own thinking could be deceiving us and allowing us to interpret our results in a way that favored our bias. I learned that getting the correct answer is more important than getting an answer you might hope for. I learned to step aside from myself and view my data through the eyes of a skeptic. We worked for a year before we convinced ourselves that the telomere terminal transferase was indeed a unique activity. The initial discovery was in December of 1984 and the paper was published in *Cell* in December 1985.

STANLEY HALL COLD ROOM – UC BERKELEY

We first called the activity we identified "telomere terminal transferase" because it transferred telomere sequences onto termini, but later shortened it to "Telomerase". My friend and fellow student Claire Wyman and I would joke around in the lab a lot. Claire pointed out telomere terminal transferase was too long and suggested various humorous names as alternatives. Names were further discussed later that night over a few beers and telomerase was one Claire had proposed initially as a joke. She thought it was funny, but Liz and I both liked it.

The next most exciting question was – where does the information for the TTGGGG repeat addition come from? I wondered if there might be an RNA component that specifies the TTGGGG sequences added. I set out to do an experiment to pre-treat the *Tetrahymena* extract with either DNase or RNase or nothing and see if that affected the activity. I remember that day Tom Cech was visiting Berkeley for a seminar. Tom had a long-standing interest in both telomeres and in RNA biology. Liz and I met with Tom in the morning and I told him about my idea of testing to see if the activity was RNase-sensitive. He agreed that was an interesting experiment. Throughout the day, as he was being walked around the department from appointment to appointment, Tom would stop by the lab and see how the experiment was going. I was flattered that he was so interested.

The RNase experiments indicated that activity was eliminated when RNA was degraded, implying there was an RNA component. Liz and I felt that the best way to really show that an RNA was involved was to find the actual RNA. So I went into the cold room to try to purify the enzyme. I read as many books on biochemical purifications as I could, and set out to purify telomerase. As a complete amateur, I spent an inordinate amount of time in the cold room setting up and running columns to purify telomerase. My friends would come to find me to go get coffee at Café Roma, and I would have on my puffy down jacket covered over with an extra-large white lab coat. They joked that I looked like the Pillsbury Doughboy.

The friends in Stanley Hall were a very close group. We would walk to get a latte at least once every day. We would talk science, tell jokes, tease each other and complain to each other about experiments that did not work. There were a lot of practical jokes that we played on each other. I was having trouble with experiments one afternoon and complained to Jeff that I was "bored". So late that night Jeff filled my umbrella with home made confetti with the word *bored* on each piece. The next day I was leaving genetics class, it was raining so I opened my umbrella and thousands of pieces of paper fell out. I knew I had to retaliate. The next day I got into the lab very early. I went to Jeff's lab bench; he had 40 bottles of different chemical reagents for his experiments lined up on the shelf above his bench. They were all glass bottles filled with clear liquid, I removed the labels that were taped on for every one of them (I marked each with a number underneath and kept a paper key). When Jeff came in to work in the morning, he started his experiment for the day, reached up for his TE buffer and found 40 identical unlabeled bottles. He was shocked at first, then, being clever, he saw the small numbers on the bottom and realized what I had done. He came into our lab and said "OK so where is the key?" I pretended to not know what he was talking about, but was glad when he admitted I had gotten him back. These kinds of jokes were common in Stanley Hall. Often they involved dry ice inside plastic tubes, which would burst and sounds like a bomb when placed in a metal garbage can.

We found every excuse imaginable to have parties at one of our graduate student houses. One party invitation flyer copied a Departmental memo that said "Emergency water outage-Party time" we decided this was a good reason for a time for a party at Jeff's house. Some of our parties involved making up skits for the "follies" where we would roast our professors and fellow students. We all worked very hard and we played hard too. The creativity was not just at the lab bench, but spilled over to our daily life together at work; being creative in all aspects of our lives in the lab and out was wonderful. Spending time with people who understood me and what I was doing and who loved to laugh and play was extremely rewarding.

After four years in graduate school, my thesis committee members encouraged me to finish up and look for a post-doctoral position. I remember Jasper Rine specifically telling me it would be good to finish the thesis in four years, because I had enough material and it looked good to finish quickly, so why not try? Mike Botchan, who was on my committee, strongly encouraged me to apply for postdoctoral fellowship positions at Cold Spring Harbor Laboratory, where he had been for a number of years before coming to Berkeley. So I sent letters inquiring about positions to four people at CSH, Bruce Stillman, Yasha Gluzman, Doug Hanahan and Mike Wigler, and was asked to go there for an interview.

COLD SPRING HARBOR LABORATORY

I think there may have been only eight or ten people in the audience for my interview talk at CSH. I gave a talk on telomerase activity in the James library. All four lab heads with whom I had applied to work were there, as well as Jim Watson whom I had never met before. It was a cold and rainy day, and afterwards Jim Watson wanted to take me to lunch. I was both excited and terrified at lunch and did not know what to say, but he was clearly interested in telomerase. Several days after the interview when I was back in Berkeley, I got a call from Bruce Stillman, he said that he would be happy to have me as a postdoc in the lab if I wanted to come, but that there was also an opportunity to have an independent position as a Cold Spring Harbor Fellow and work on whatever I wanted. I had not heard of or applied to an independent fellowship position so I was a bit surprised. I later found that Mike Botchan from Berkeley had quietly nominated me for this without my knowing. When Bruce called, I first said that I would just work with him as a postdoc; but then I thought it over for a week and realized there were so many interesting questions I still wanted to ask about telomerase that I would love to keep working on it. So I called Bruce back and told him that I would like to accept the independent Fellow position. So I filed my thesis in November of 1987 and continued to work on trying to identify and sequence the RNA that co-purified with telomerase for a few months. January 1, 1988 I started as a Cold Spring Harbor Fellow.

My main goal in my new lab at CSHL was to clone the gene for the telomerase RNA. I had already obtained several partial sequences through direct RNA sequencing using specific RNases. I made short oligonucleotides to the regions of RNA sequence and used them to probe genomic libraries from *Tetrahymena*. After searching through many libraries, I found one clone where the sequence matched the partial RNA sequence AND also contained CAACCCCAA, the complement to the TTGGGG telomere sequence. I was excited and told my friends in the building about the sequence. I was very surprised to hear later at lunch in Blackford Hall that many other people knew of the result. A few hours later Bruce Stillman stopped me on the street to say he heard I got a great result. News traveled fast at CSH and people really cared about what other people were doing. It was fun to again be with people who cared about each other and who kept up with what science people around them were doing; it was a very exciting time.

Having a clone with a telomere repeat was tantalizing, but how could I show that it was required for telomerase activity? I devised a series of experiments using antisense oligonucleotides and RNase H to show that this RNA was indeed required for telomerase activity. I wrote a draft of a paper and sent it to Liz since I had initiated the sequencing efforts while working in her lab. I presented my work in Bruce Stillman's lab meeting and he encouraged me to propose a model for how I thought the enzyme might work in the paper. This model, drawn crudely on a Macintosh SE, has stood the test of time. It turned out what I conceived of as a possible mechanism is indeed the way telomere repeats are made by copying the RNA template. I sent Liz the clone encoding the RNA component before our paper was published, and she and her student Gou-Liang Yu were able to express a telomerase RNA with a change in the template sequence in *Tetrahymena* and show that change was incorporated into the telomere repeats. This was definitive evidence for the templating model proposed.

Having success in cloning telomerase soon after arriving at CSH was a great start. I soon had my first graduate student, Lea Harrington, and was rapidly promoted through the different scientific staff positions to the position of "Rolling 5". We continued to pursue our curiosity about the function of telomerase and role of telomerase in cells that are discussed in more detail in the Nobel lecture. In 1993 I married Nathaniel Comfort, whom I met when he was the science writer in the Public Affairs office at CSHL. In 1996 our son Charles Comfort was born in Huntington, New York. Nathan completed his Ph.D. in history of science at the University at Stony Brook in 1997 and was offered a position on the faculty at George Washington University. I was concurrently offered a position in Tom Kelly's department of Molecular Biology and Genetics at Johns Hopkins University. So when Charles was one year old, we moved to Baltimore to start new lives.

JOHNS HOPKINS UNIVERSITY SCHOOL OF MEDICINE

I was very fortunate to come to Johns Hopkins to a very interactive and cohesive department. Although the institution as a whole is much bigger than CSH, the department of Molecular Biology and Genetics felt as small and homey as CSH. I was fortunate to have outstanding graduate students and postdocs come to work with me. I was able to branch out into both yeast genetics and mouse genetics and follow my interests in what happens to cells when they don't have telomerase. I enjoy having smart people around to talk to who are excited by the work on telomeres. The different directions the lab has gone have been driven not just by my own interests but by the interests of the students in the lab. Finding something new that nobody knew before is exhilarating, and discussing ideas with students and postdocs and helping them to pursue their most interesting questions leads to new insights.

Two years after I moved to Johns Hopkins, my daughter Gwendolyn was born. Having two kids and a full time job in the lab is a challenge, but having Charles and Gwendolyn is the best thing that has ever happened to me. My lab knows that I am a mom first, and the flexibility that academic science provides makes having a career and a family possible. I can go home when needed, or to a school play in the middle of the day, then come back and finish my work-day; or work from home on the computer. The main thing is to find the time to get things done, it is not the hours at work but the overall productivity that counts. Having flexibility takes a huge amount of pressure off.

In 2002 Tom Kelly, the Department Director (the Hopkins name for Chairman) told me he was leaving Hopkins to take a position at Memorial Sloan-Kettering Cancer Center in New York. I kept the note that my assistant left on my desk that day that said "Tom Kelly wanted to talk to me" – and marked it as a Black Day; I actually cried when he told me he was leaving. Tom was an ideal director for the department. He cared about everybody and worked hard to help create the collegial environment that attracted so many top scientists. After a two-year search process, I was appointed as the Daniel Nathans Professor and Department Director for Molecular Biology and Genetics. I am extremely honored to hold the Daniel Nathans Chair, as it was Dan who created the department and established the interactive environment that Tom Kelly helped build. Dan Nathans, who died in 1999, personified thoughtfulness, caring and above all integrity, traits that we all strive to show the way that Dan did.

I know that I cannot fill the shoes of either Tom Kelly or Dan Nathans, but I try to bring my own style of leadership to the department. Being director is made easy by the terrific faculty in the department; the science is outstanding and everybody talks and cares about the other faculty. The flat structure of the department that was established early on makes it clear that everyone has a voice. Decisions are made by discussion and consensus and not in a top-down fashion. I have been able to learn about leadership in a hands-on fashion and the faculty have all helped me tremendously in that.

MENTORS, FRIENDS AND LESSONS

One of the lessons I have learned in the different stages of my career is that science is not done alone. It is through talking with others and sharing that progress is made. Work done today, of course, builds on the past work of many others, but in addition, experiments are often suggested by friends and colleagues either directly or indirectly. The ideas generated are not always the result of one person's thoughts but of the interaction between people; new ideas quickly become part of collective consciousness. This is how science moves forward and we generate new knowledge.

I am grateful to the many scientists who have influenced and helped me in my journey from Davis to Baltimore: Bea Sweeney, Michel Robert, Kevin

Sullivan, David Asai, Les Wilson, Elizabeth Blackburn, Jasper Rine, Mike Botchan, Bruce Stillman, Rich Roberts, Dan Nathans, Tom Kelly. These colleagues and many others have helped me move from one stage to the next and taught me many essential lessons along the way. I would not have been able to do the science that I have done without the students, postdocs and wonderful technicians who brought their energy and great ideas to the lab. Finally, the close friends I have made in Davis, Berkeley, CSHL and Baltimore are my constant support group I value them above all else.

TELOMERASE DISCOVERY: THE EXCITEMENT OF PUTTING TOGETHER PIECES OF THE PUZZLE

Nobel Lecture, December 7, 2009

by

CAROL W. GREIDER

Daniel Nathans Professor and Director, Department of Molecular Biology and Genetics Johns Hopkins University School of Medicine, 725 N. Wolfe Street, Baltimore MD 21205, U.S.A.

The story of telomerase discovery is a story of the thrill of putting pieces of a puzzle together to find something new. This story represents a paradigm for curiosity-driven research and, like many other stories of fundamental discovery, shows that important clinical insights can come from unlikely places. In this paper I describe the process of scientific discovery – at times frustrating, at times misleading and perplexing, but yet also at times wonderfully exciting. The willingness to keep an open mind, to enter uncharted waters and try something new, along with patience and determination, came together to tell us something new about biology. Fundamentally this story shows how curiosity and an interest in solving interesting problems can lead to a lifetime of exciting discoveries.

IDENTIFYING THE PUZZLE: TELOMERE SEQUENCES DEFINED

Telomeres posed a puzzle for biologists for a many years: it was clear that a chromosome end differed from a chromosome break – but how? In the 1930s, Herman Muller and Barbara McClintock showed that natural chromosome ends have special properties that afforded their protection; they coined the word *telomere* to describe these natural chromosome ends, deriving the word from the Greek *telo* = end and *mere* = part (McClintock, 1939, 1941; Muller, 1938). The puzzle of how these ends functioned remained unsolved for many years, until their molecular structure was characterized.

In 1978, Blackburn and Gall identified the first telomere sequence using the ciliate *Tetrahymena*, which contains 40,000 telomeres. They found that the chromosome end was made up of tandem, consecutive repeats of the simple sequence CCCCAA (Blackburn and Gall, 1978). Their discovery of this simple repeated sequence turned out to be the key to understanding telomere function.

The identification of tandem repeats in the telomeres of *Tetrahymena* was followed by the identification of similar repeats in the telomeres of other organisms, including *Oxytricha, Physarum*, yeast, and trypanosomes (Bernards *et al.*, 1983; Boswell *et al.*, 1982; Johnson, 1980; Szostak and Blackburn, 1982). This conservation of telomere sequence across a wide variety of organisms suggested that telomere function might also be similar in these species. This conservation also suggested that telomeres and their function – of maintaining chromosome ends, and hence protecting them from destruction – arose early in evolution, such that all species with linear chromosomes evolved with the same ancestral mechanism of maintenance, albeit with some variation.

CURIOUS FACTS ABOUT TELOMERES: SOME PIECES OF THE PUZZLE

Liz Blackburn and others were interested to know how the simple repeats functioned as telomeres to confer protection upon a chromosome end. Soon after the telomere sequence was first identified, several curious facts about telomeres were uncovered. First, telomeres were heterogeneous in length: in a population of cells, different chromosome ends possessed telomeres comprising different numbers of repeats (Blackburn and Gall, 1978). How was this established? When a population of telomeres was digested using restriction enzymes it generated so-called "fuzzy bands" when the digest was separated using agarose gel electrophoresis (Figure 1). Why was this? The variable-length telomeres had generated heterogeneous fragment lengths, which all migrated to slightly different positions on the agarose gel, yielding blurred (or 'fuzzy') bands, rather than the sharper, more distinct bands that are typical of restriction fragments which are all of a similar size.

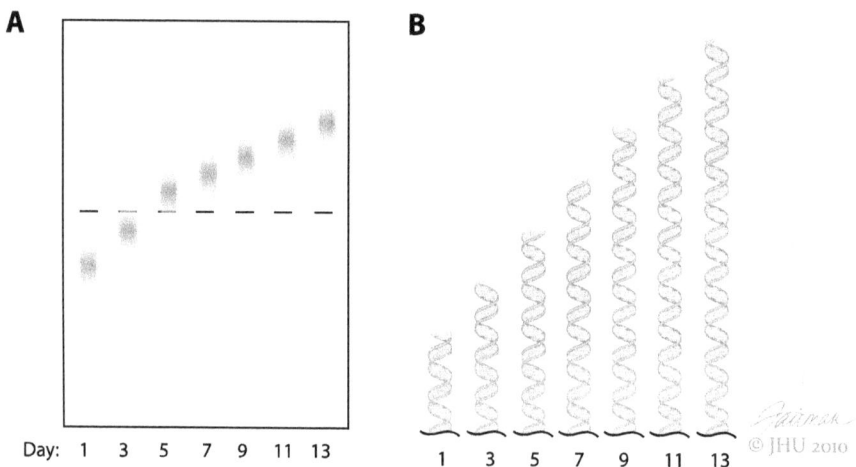

Figure 1. **Telomere elongation in Trypanosomes and *Tetrahymena*.**
Keeping cells in continuous log phase growth results in progressive elongation of the telomeres. A) Diagram of a gel showing the heterogeneous length telomeres from *Tetrahymena*. Each lane represents increased numbers of cell divisions. B) Diagram of the interpretation of telomere elongation seen in gel in part A.

A second curious fact was that the telomeres in trypanosomes grew longer as the cells were grown in culture (Bernards et al., 1983). The same surprising result was found when similar experiments were performed with *Tetrahymena*: when kept in continuous log phase growth, the telomere fragments grew progressively (Larson *et al.,* 1987) (Figure 1). This elongation was unexpected. In fact, it was the opposite of what had been predicted on the basis of published models for telomere replication. In 1974, James Watson suggested that a linear chromosome should get shorter as cells divided, based on the mechanism by which DNA polymerase replicates DNA during cell division (Watson, 1972). A similar idea was also proposed by Alexi Olovinkov, who suggested that chromosome ends might shorten after rounds of DNA replication (Greider, 1998; Olovnikov, 1973). Why, then, were the telomeres getting longer? And why were they fuzzy?

COLLECTING MORE PIECES OF THE PUZZLE: TELOMERE SEQUENCE ADDITION

When Liz Blackburn and Jack Szostak met for the first time at a conference, they were both interested in DNA ends. They knew about the curious structure of telomeres and their elongation, and they saw a way to use these puzzle pieces to perform a long-shot experiment: they decided to test whether *Tetrahymena* telomeres could function as telomeres in yeast. The experiment was a long shot because these two species are very distantly related, having diverged from a common ancestor millions of years ago. Despite being a long-shot it was an experiment they could do, so they forged ahead.

Blackburn and Szostak took a circular yeast plasmid and cut it once with a restriction enzyme to make it linear. Jack knew from his earlier experiments in yeast that this linear DNA would be rapidly degraded if put into yeast cells. However, they wondered whether the addition of *Tetrahymena* telomeres to this linear DNA would cause the ends to be protected – as they are in *Tetrahymena* – so preventing the normal rapid degradation of the DNA when transformed into (that is, added to) yeast cells. Liz purified the *Tetrahymena* telomeres and Jack ligated them onto the linearized yeast plasmid (Figure 2). When this construct was transformed into yeast, it was stable: it replicated and was maintained as a linear chromosome fragment.

This maintenance of the linear plasmid was a stunning result: it indicated that the *Tetrahymena* telomeres functioned and protected DNA ends in yeast, a very distantly-related organism (Szostak and Blackburn, 1982). Emboldened by this success, Liz and Jack took one step further forward; they devised a method to identify the naturally occurring yeast telomere sequence. Instead of attaching *Tetrahymena* telomeres to the linear plasmid, they set out to find the parts of the yeast genome that protect and maintain the linear plasmid, and which therefore contain the yeast telomere. To do this, they removed one end of the linear plasmid to which *Tetrahymena* telomeres had been added, and ligated random genomic fragments of yeast

DNA to this free end. Those random fragments that harbored a telomere allowed stable maintenance of the linear plasmid.

In doing these experiments, Liz and Jack noticed another curious fact: the *Tetrahymena* telomeres on the plasmid maintained in yeast were longer than they had started out. They were curious to know why. So, together with Janice Shampay, they set out to sequence the cloned yeast telomeres and the *Tetrahymena* telomere end. When they sequenced the yeast telomere they found it to have the same kind of tandemly-repeated sequences as had been found in all other organisms so far. More strikingly, determining the sequence of the *Tetrahymena* telomeres from the linear plasmid showed why they were longer: the initial *Tetrahymena* telomeres had been extended by terminal addition of yeast-specific sequences. (Shampay et al., 1984) (Figure 2)

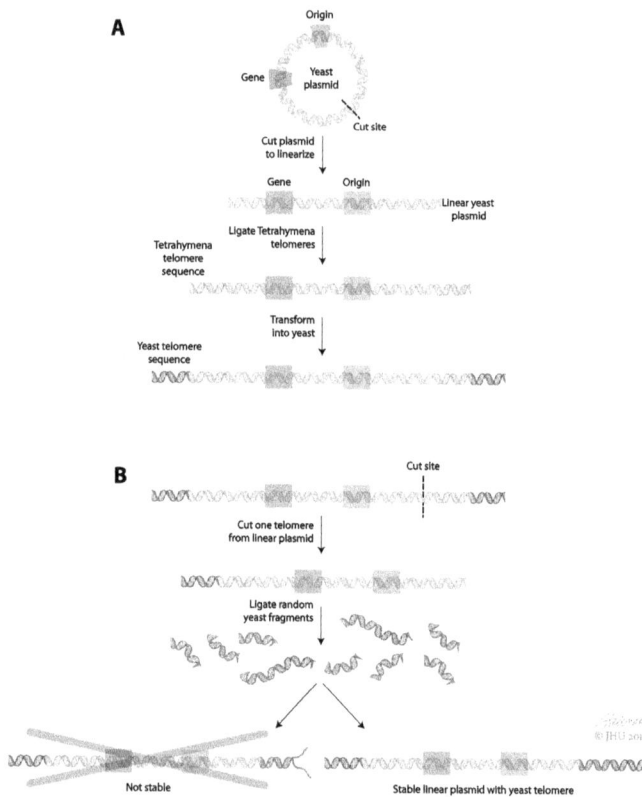

Figure 2. **Evidence for telomere elongation.**
A) Tetrahymena telomeres function in yeast and are elongated. A circular yeast plasmid was linearized at a unique cut site and Tetrahymena telomeres were added. This linear DNA was transformed into yeast and the linear plasmid was stable. During propagation in yeast, the terminal DNA was elongated by addition of yeast telomeres sequences. B) Cloning of a yeast telomere. The linear plasmid shown in part A was cut to remove one end and random pieces of yeast genomic DNA were ligated onto the end. Those products that had internal yeast DNA added were not stable – they would be degraded by nucleases. However, when a piece of DNA that contained a telomere was ligated, this plasmid was stable and was propagated as a linear plasmid. Yeast telomere fragments were first identified using this strategy.

This last puzzle piece was a very important one. Models of telomere maintenance and elongation had been proposed, which involved telomere-telomere recombination (Figure 3a). Even though Jack's expertise lay in understanding these recombination models, both he and Liz recognized that the addition of yeast-specific sequences onto the *Tetrahymena* ends could not really be explained by those models. Instead, the data suggested to Liz and Jack the existence of an active elongation mechanism in yeast, whereby DNA was generated *de novo* rather than being the result of a recombination event (Figure 3b). They wrote in their paper, "We propose that terminal transferase-like enzymes are responsible for extending the 3' G+T rich strand of yeast telomeres." By continually following the clues (and keeping an open mind) they had come to the conclusion that no known enzyme could do the sequence addition, and so proposed instead that there must be an unknown enzyme that adds telomere sequences.

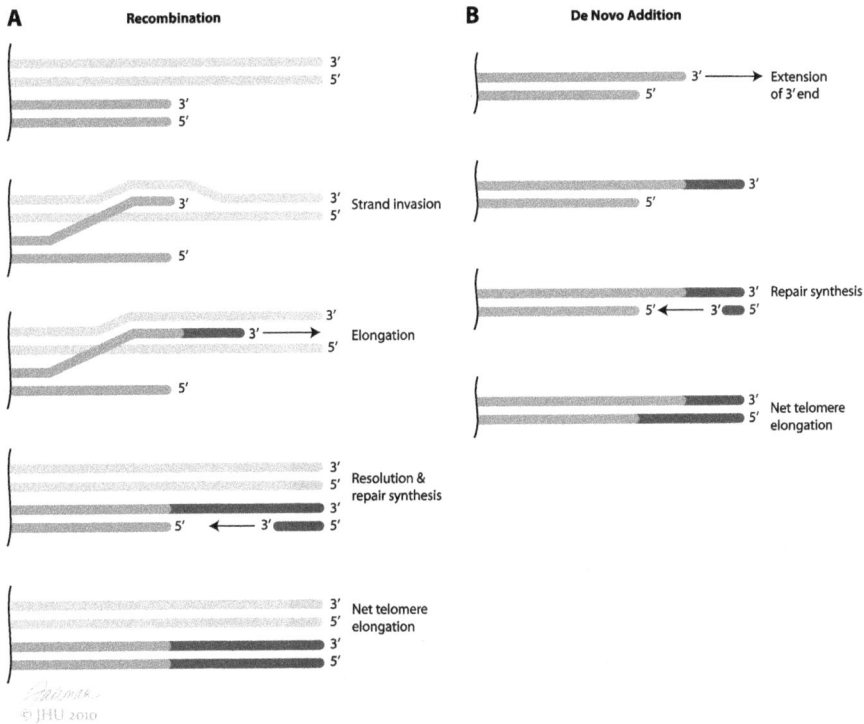

Figure 3. **Two models for telomere elongation.**
A) Recombination can elongate telomeres. Since telomeric repeats are homologous on all chromosome ends a short telomere may copy off of a long telomere in a gene-conversion type of recombination. The 3' end of the short telomere base pairs with the longer telomere and polymerase extends the end of their strand. The other strand can be copied by conventional polymerase activity. This results in the net elongation of the short telomere, while the long telomere remains long. B) Alternative model for telomere elongation proposes that de novo elongation lengthens telomeres. The discovery of telomerase indicated that this mechanism is the predominant mechanisms for telomere elongation in most species.

LOOKING FOR TELOMERE ELONGATION: DEFINING THE EDGES OF THE PUZZLE

When I joined Liz's lab in May of 1984 I set out to look for this unknown enzyme. We decided to use biochemistry to see if we could identify an enzyme that might elongate telomeres. There was no established protocol for finding an unknown enzyme, so we had fun and made one up. In fact, more precisely, we continually made up new protocols. It was like biochemical improvisation: we started with one concept of an assay that might allow detection of addition onto telomeres, but kept modifying the assay after each set of experiments. We changed the reaction conditions, the substrates, and even method of detection. After nine months we found something (!) and so we had another piece of the puzzle. But how did our nine-month search for this puzzle piece unfold?

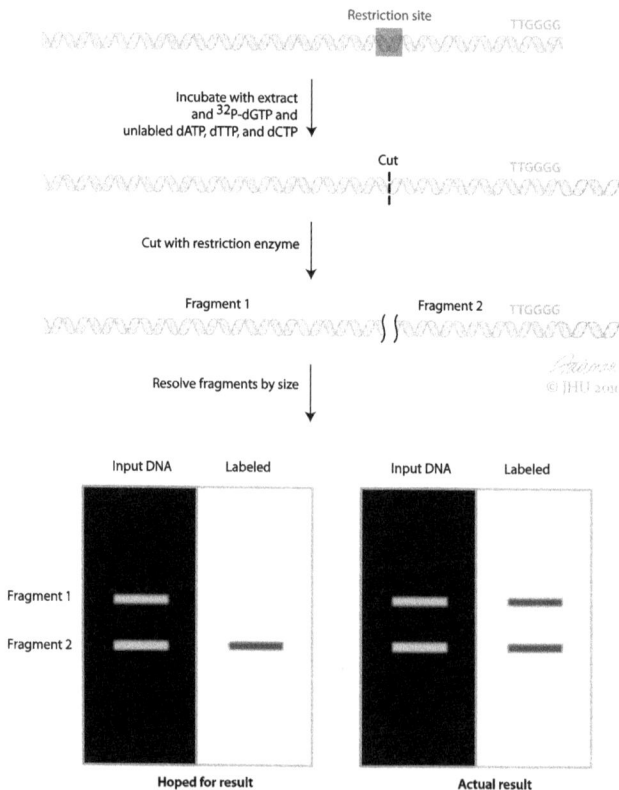

Figure 4. **Initial assay for telomere-specific end labeling**
A restriction fragment was purified from a plasmid. The fragment had telomeric DNA on the right end and non-telomeric DNA on the left end. There was a restriction enzyme cut site that would allow two different sized pieces (Fragment 1 and Fragment 2) to be generated. After incubation with ^{32}P dGTP and ^{32}P dCTP and unlabeled dATP and dTTP the fragment was purified and cut with the restriction enzyme. The products were resolved on an agarose gel and then the gel was exposed to X-ray film. The hoped for result was that the Fragment 2 with the telomeric end would be preferentially labeled (Left bottom panel). The actual results showed that both ends were equally labeled (Right bottom panel), prompting a revision of the assay.

First we needed an assay – a way to detect if telomere elongation was happening. The first assay we tried explored whether a piece of DNA that included a telomere would incorporate DNA precursors more readily than a piece of DNA containing non-telomeric sequences. The idea was that if there *was* an enzyme that actively elongated telomeres, we might be able to detect it through its activity in association with telomere DNA. For this assay, we developed a substrate that was meant to mimic a telomere in the cell: a linear DNA fragment that contained a telomeric sequence at one end but not at the other (Figure 4). I incubated this linear fragment of DNA with the nucleotides dA, dC, dG and dT – the building blocks of DNA – the dC and dG had been labeled with the radioactive isotope ^{32}P. We reasoned that an elongation enzyme might preferentially elongate the end of the DNA fragment that contained the telomeres; if it did, we expected to see more of the ^{32}P label incorporated into the telomeric end than the end lacking a telomere.

I incubated the linear DNA substrate in an extract made from *Tetrahymena* nuclei. The extract was prepared in a manner that we hoped would allow all of the enzymes normally present in the nuclei to be active. We also added radiolabeled dCTP and dGTP and unlabeled dATP and dCTP to serve as DNA precursors. After incubation for an hour, I purified the linear fragment from the extract to examine what had happened to the ends. Following its purification, I cut the DNA fragment to generate two unequal sized fragments to distinguish between the telomeric and non-telomeric ends. (The smaller of the two pieces contained the telomere end.) We could then separate and identify the two different-sized fragments on an agarose gel.

After separating the two pieces by size, we exposed the gel to X-ray film, because the fragments that had incorporated the ^{32}P-labeled precursor would generate a dark band on the film and we could thus identify them. We then looked to see if there was more radioactive label incorporated in the fragment that had the telomere end than the non-telomere end (Figure 4). Unfortunately, no matter how we prepared the extract, both fragments always incorporated similar amounts of label. We knew we needed a different way to approach this problem.

A PUZZLE-SOLVING STRATEGY: GETTING THE ASSAY RIGHT

The most productive way to solve a puzzle is to attack it with the right strategy. Since we did not know precisely what we were looking for, we tested a number of different approaches to see which plan might be successful. After each experiment, we thought of new changes to make to the next experiment. Two of the many changes we tried proved important: using sequencing gels to resolve the DNA fragments after incubation, and using synthetic oligonucleotides as telomere substrates instead of large restriction fragments. How did these two changes affect the outcome of the experiments?

After our initial attempts that I've described above, we sat and puzzled about the fact that both the telomere end and the non-telomere end showed incorporation of the radioactive label. We realized that exonucleases, which were expected to be present in the extract, were likely generating some single-stranded DNA within the linear DNA fragment; repair polymerases would then fill out the single strand region, causing radioactive label to be incorporated into both ends of the fragment.

We thought hard about a way to get around this problem. First, we changed the approach: rather than looking for increased label incorporation, we decided to look for changes in the size of the fragment. If there was an enzyme that extended telomeres not only should the telomeres become labeled with the radioactive precursors, but *also* the size of the fragment should increase as the telomere is extended. (The repair polymerases in the extracts, which could cause both telomere and non-telomere ends to be labeled would not be capable of generating DNA that was *longer* than the fragments added at the start of the assay.)

To see a change in size of a fragment we needed to have a short fragment: a small change in size of a large fragment would be too hard to detect, but a small change in size of a very small fragment would be noticeable. We repeated the experiment as described above but, when cutting the fragment, made the cut very close to the telomere end to generate a telomere fragment just 34 base pairs long. The non-telomere and telomere fragments were then separated on a gel that was usually used for DNA sequence analysis and which could distinguish between fragments that differed in length by just a single base.

I worked from May through December trying different variations of this experiment, staring hard at the sequencing gels but never quite convincing myself there was much of a change in fragment size. So, in December of 1984, we decided to make another change to the assay: we changed the substrate that we were adding to the reaction. Instead of a long linear DNA fragment, I tested a synthetic 18 residue oligonucleotide $(TTGGGG)_4$ as the substrate. Eric Henderson, a postdoctoral fellow in the lab, was studying the unusual structure of DNA oligonucleotides made of these G-rich telomere sequences, and offered some of his synthetic oligonucleotide, which I decided to use instead of the DNA restriction fragment to see if it might be elongated.

A

Single-stranded telomeric
oligonucleotide primer

B

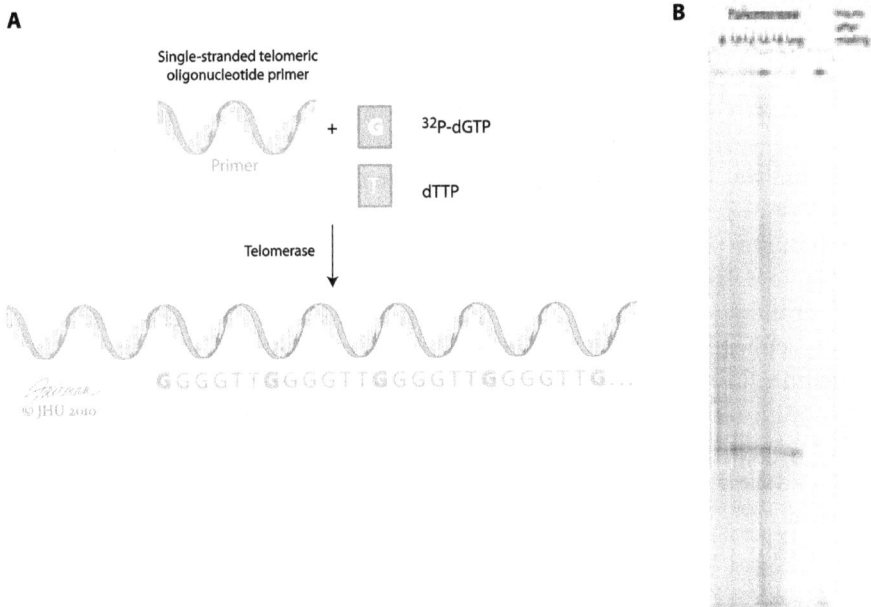

Figure 5. **Telomerase activity assay**
A) Diagram of primer elongation assay. A single stranded DNA oligonucleotide with
the sequence (TTGGGG)4 was added with 32P dGTP and unlabeled dTTP and incubated in Tetrahymena cell extracts that we hoped would have telomerase activity. The primer was extended by addition of a six base repeating pattern B) Telomerase
elongation of telomeric primer. Extracts from a time course of Tetrahymena development
were made at different numbers of hours of development (indicated at top by hours).
Extracts were also made from vegetatively growing cells (marked Veg). The banding
pattern represents the six base repeat of TTGGGG that is added to the primer. This
autoradiogram was the first experiment in which telomerase activity was seen on December
25, 1984.

So, I set out to examine the elongation of the telomeric oligonucleotide. I made cell extracts from *Tetrahymena* and incubated them with the
oligonucleotide and radiolabeled nucleotide precursors (Figure 5A). It
took over a week to set up the experiment, do the reactions, and then run
the sequencing gel. To maximize the signal generated by the radioactive
label, I exposed the gel to X-ray film for three days. When I went to the
lab to develop the X-ray film, I was thrilled to see a repeating pattern of
elongation products that extended up the gel. (Figure 5B). The
oligonucleotide substrate was being elongated to give products that varied in
size by six bases, giving the repeating pattern seen on the gel. This was the
first visualization of telomerase activity.

TESTING OURSELVES: DO THE PIECES REALLY FIT, OR ARE WE
FORCING THEM?

I talked to Liz the next day and showed her the gel. We were both talking
at the same time, trying to understand the meaning of the repeating pat-

tern. We knew that this could be the result of the enzymatic activity we were looking for, but we also wanted to be sure the pattern we were seeing was indeed generated by a novel enzyme. Were we truly seeing something new, or was our own wishful thinking coloring our interpretation of the result?

To be sure we were not forcing the interpretation of the elongation pattern, we set out to test various alternative explanations that might generate a repeating pattern like the one we were seeing. For example, we thought the TTGGGG primer might be annealing to double-stranded genomic DNA that might be present as a contaminant, such that a conventional polymerase could generate the TTGGGG repeat addition when replicating the DNA (Figure 6A). Alternatively, the primer might be self-annealing (that is, pairs of the primer might be sticking to one another), generating a double-stranded substrate for a conventional polymerase to copy (Figure 6B).

To address these and other concerns, we devised an ever-evolving set of control experiments to determine if the repeat addition was the result of the activity of a previously identified enzyme. For one control, we treated the samples with aphidicolin, which inhibits conventional DNA polymerases. More importantly, we used a CCCCAA primer, which would be expected to be elongated if simple copying of telomere repeats by DNA polymerase was occurring (rather than nucleotides being added *de novo*). The fact that the CCCCAA primer was not elongated ruled out the trivial explanation that the repeating pattern was coming from the copying of endogenous DNA.

These were exciting times. Once I could repeatedly see the primer elongation activity, it was fun to test various ideas about how it might be generated. I would come in to the lab every day, eager to test the next set of experiments and find something new. The final experiment that convinced both Liz and me that we had something new was when we did the converse of the experiment that Liz and Jack Szostak had done, which had been published in *Cell* in 1982. They had put *Tetrahymena* telomeres into yeast cells and shown that a yeast telomeric sequence was added to the ends. By contrast, we made a synthetic yeast sequence telomere oligonucleotide primer and put it in *Tetrahymena* extracts – and found that the *Tetrahymena* telomere repeats were added to the yeast telomere.

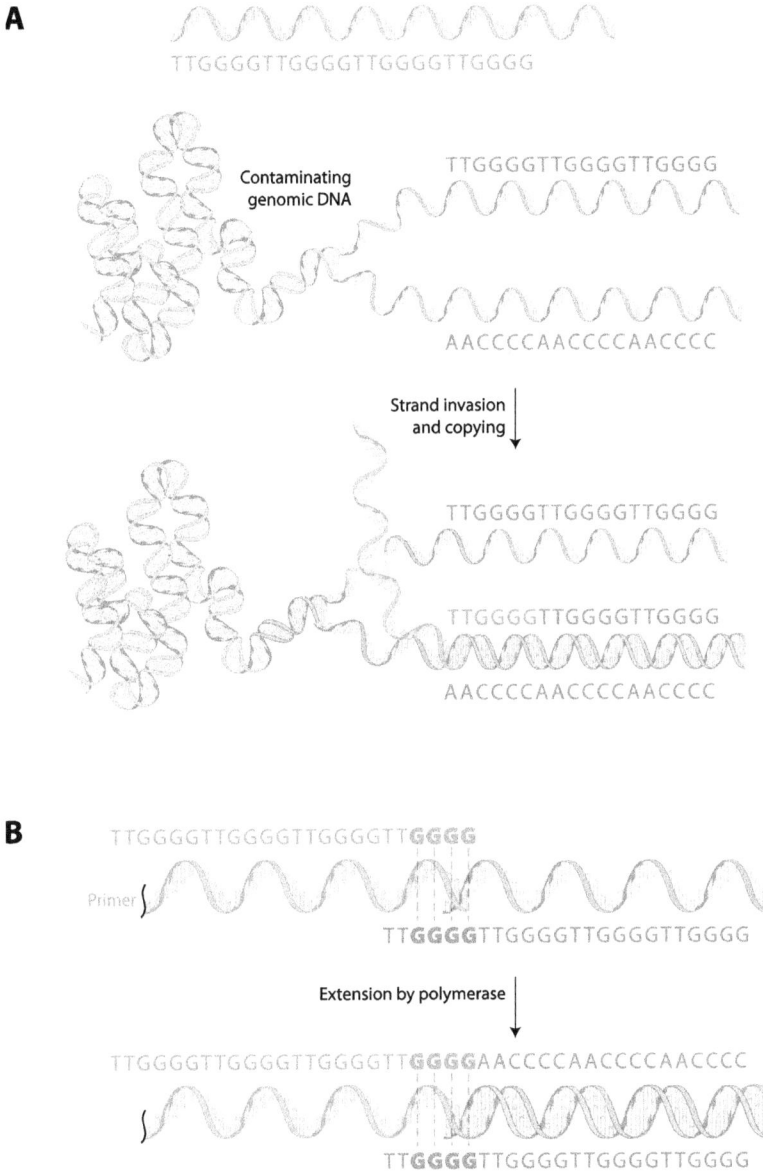

Figure 6. **Diagram of possible telomere elongation artifacts**
A) If the primer telomeric oligonucleotide were to anneal to contaminating DNA in the extract, any conventional DNA polymerase could copy the telomeric sequence and might generate the six base repeating pattern. B) An alternative artifact might come from two telomeric primers self-annealing thought G-G non-Watson-Crick base pairing. Extension of these "primer dimers" might also generate a 6 base repeated pattern.

How could we tell that *Tetrahymena* and not yeast telomere repeats were being added? The yeast sequence primer had three Gs at the end, while the *Tetrahymena* telomere sequence had four. The banding pattern of sequences added onto the yeast primer was one base longer than would have been the case if the yeast sequence had been repeated: this extra base was the extra

G required to complete the GGGG found in the *Tetrahymena* sequence. This result was quite stunning. The shift in banding pattern convinced us a new enzyme was in action: we could not imagine how conventional polymerases would elongate a yeast sequence with *Tetrahymena* repeats. We wrote up the paper, which was published in *Cell* in December of 1985 (Greider and Blackburn, 1985).

THE NEXT PART OF THE PUZZLE: SEQUENCE INFORMATION

The next big question was where the information for the addition of TTGGGG repeats was coming from. Liz and I talked through several different models. The one I wanted most to test was that there might be an RNA component that specifies the sequences added to the chromosome end. I set out to do an experiment to pre-treat the extract with either DNase (which would digest DNA) or RNase (which would digest RNA) or nothing (as a control) to see if this pre-treatment affected the activity of telomerase. On the day of this experiment, Tom Cech, who had a long-standing interest in both telomeres and RNA, was visiting Berkeley to give a seminar. Liz and I met with Tom in the morning and described my idea of seeing whether the activity would be sensitive to treatment with RNase. He agreed that was an interesting experiment.

Throughout the day, as Tom was being escorted around the department from appointment to appointment, he would stop by the lab and see how the experiment was going. We found that pre-treatment of the *Tetrahymena* extract with RNase did indeed block the elongation activity. Establishing that RNA was needed for elongation was a key clue: it allowed us to think about possible mechanisms by which the enzyme would specify the TTGGGG repeats that were added.

FOLLOWING THE CLUES: IS THERE A TEMPLATE?

The inactivation of telomerase by RNase treatment suggested a clear hypothesis: the TTGGGG repeats are made by copying from an RNA template, which is a component of the functioning telomerase. The most powerful way to test the validity of this hypothesis was to find the actual RNA template. To isolate a component of telomerase, we needed first to purify the enzyme from all of the other proteins present in the crude extract. I established a multistep purification protocol for telomerase, using conventional column chromatography (Figure 7A). I would separate the extract into fractions and test each fraction for activity (to identify the fraction containing telomerase). I would then take the active fractions and subject them to another, different separation step. I used size exclusion, ion exchange, dye binding, and heparin binding columns to successively purify telomerase (Figure 7B). I then examined the active fraction to look for an RNA that was always present when telomerase was active. I purified the RNAs from active fractions, and labeled them with ^{32}P (Figure 7C); we then narrowed down the likely candidates by determining which RNAs reproducibly co-purified with telomerase.

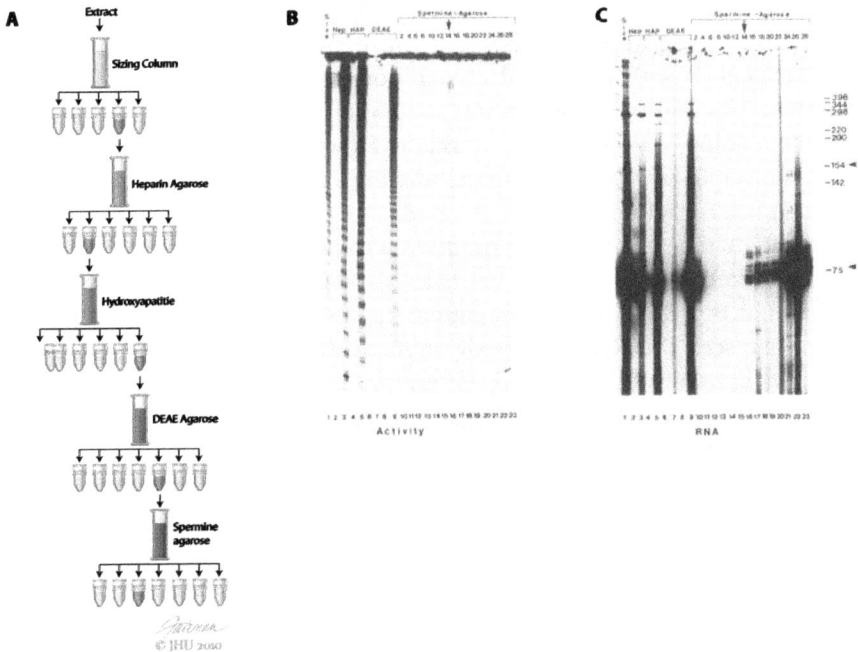

Figure 7. **Purification of telomerase and identification of the co-purifying RNA**
Purification scheme for telomerase. A) Extracts were fractionated first on a sizing column, the fractions were all assayed for activity and those that had activity were again fractionated on a Heparin Agarose column. Further purification continued in the same manner, after each column the active fractions were loaded on a subsequent column including Hydroxyapatite, DEAE Agarose and finally Spermine Agarose B) Gel showing the active fractions for each of the columns diagrammed in part A. C) To identify the co-purifying RNAs, all RNAs from each of the active fractions was end labeled with ^{32}P and resolved by gel electrophoresis. The faint RNA band in lane 16 migrating near the 154 base marker (indicated on the far right) was later found to be the telomerase RNA. RNA gel Parts B and C are reprinted from the publication *Cell* 51, 887–898; Greider, C.W., and Blackburn, E.H. (1987). The telomere terminal transferase of *Tetrahymena* is a ribonucleoprotein enzyme with two kinds of primer specificity with permission from Elsevier.

We then faced our next challenge: we needed to have the sequence of the RNA component to determine if a template mechanism was, indeed, working. We tried a number of different methods to obtain the RNA sequence, including the very newly developed method termed PCR (polymerase chain reaction), which we had heard about but which had not yet even been published.

After trying to obtain a sequence for a number of months, I decided to take a more direct approach: I used direct RNA sequencing techniques to determine a partial sequence from those RNAs that emerged as good candidates. To do this, the RNAs of interest were cut out of a high-resolution gel and sequenced. This sequencing revealed that the very small RNAs that co-purified with telomerase activity were, in fact, tRNA contaminants that were present in the active fractions due to the high abundance of tRNA in the cell. For example, I sequenced one RNA that ran near a marker of 175 bases and found it was related to 7SL RNA, an RNA that had recently been associated with the signal recognition particle involved in ribosome function. Finding known RNAs was somewhat reassuring as it told us that the sequencing technique was working. However, it was somewhat of a disappointment as we were hoping to identify a new RNA that might provide the template for telomerase.

One RNA that I had my eye on after staring at many different purification experiments ran near the 154 base marker. No single experiment had pinpointed this RNA as the best candidate; my interest was really a hunch since I had seen this RNA repeatedly in many experiments. I decided to test thus hunch, however this "154 base RNA" as I referred to it proved harder to sequence. I was able to obtain only partial sequence from different regions of the RNA, but none of the partial sequences contained a telomere repeat as expected for a template and the full length RNA proved impossible to sequence.

All of the accumulated evidence indicated there must be a template in telomerase, but we did not have the final key piece of the puzzle. Since telomerase was an interesting enzyme and our experiments clearly suggested that an RNA was involved, we decided to write a paper about what we knew about telomerase, despite not having the actual RNA in hand. We wrote the paper on the biochemical characterization of telomerase and the fact that there was likely an RNA component; it was published in *Cell* in 1987 (Greider and Blackburn, 1987). Looking back it is now clear that many of the experiments in this paper that helped to characterize telomerase served as the basis for identifying telomerase in other organisms. After the identification of *Tetrahymena* telomerase, telomerase enzymes from other organisms such as *Euplotes, Oxytricha* and humans were characterized by other groups (Lingner and Cech, 1996; Lingner *et al.*, 1994; Morin, 1989; Shippen-Lentz and Blackburn, 1989)

A CHANGE IN VENUE: SEEING THE PUZZLE FROM A DIFFERENT PERSPECTIVE

I finished my Ph.D. at Berkeley in November 1987 and took a position as an independent fellow at Cold Spring Harbor Laboratory in January 1988. At Cold Spring Harbor I was still focused on identifying the RNA, but my exploration took a different approach. Rather than continuing with more RNA sequencing, I decided to see if I could clone the RNA gene directly from the partial sequence information that I had already obtained.

I designed several different oligonucleotide probes that were complementary to the regions of partial RNA sequence I had obtained, and made size- selected genomic libraries that were enriched for sequences that hybridized to these oligonucleotides. After a number of attempts, I obtained one clone that possessed both the correct partial sequence and the sequence CAACCCCAA. Seeing this sequence on the sequencing gel was exciting, as it mirrored what would be expected if a template mechanism for TTGGGG addition was indeed used by telomerase. This clone was clearly a central puzzle piece. I went on to verify that this RNA was expressed in *Tetrahymena* and was around 160 nucleotides in length. All the signs were that my earlier hunch about "154 base RNA" being a key player in telomerase activity had been right.

IS THIS THE RIGHT PUZZLE PIECE?

It was exhilarating to have the sequence of an RNA that might encode the RNA component of telomerase in hand. But I needed some evidence that this RNA was indeed the template. Again, just as when we were trying to first identify telomerase, there was no precedent for the experiments I was trying to do. It was fun to be creative and dream up ways to test whether this RNA was the right candidate. I talked to my friends at Cold Spring Harbor, listened to their advice and read about other enzymes and about how functional RNAs were identified.

My next step in characterizing telomerase activity was to test the function of my candidate RNA template. To do this, I decided to use the oligonucleotides I had used in the cloning experiments, which were complementary to telomerase. I thought that if I could inactivate the enzyme by specifically cleaving the candidate RNA, I would have strong evidence for the involvement of this specific RNA in telomerase activity. I think it was Adrian Krainer, a colleague at CSH, who suggested using RNase H in this experiment. RNase H is a specific RNase that will cleave the RNA of a DNA/RNA duplex. The thought was that, if the oligonucleotide hybridized specifically to regions of complementary RNA in telomerase, and subsequent RNase H cleavage inactivated the telomerase, we would have evidence for the involvement of the RNA to which the oligonucleotide had hybridized in telomerase-mediated telomere elongation (Figure 8).

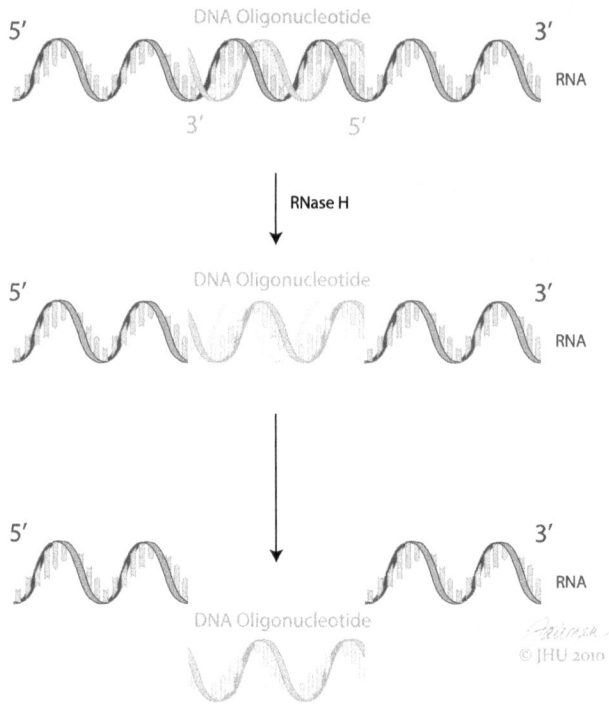

Figure 8. **RNase H cleaves the RNA of a DNA/RNA duplex**
Single stranded regions of RNA will hybridize to single stranded complementary oligo-nucleotides. When such hybrids are incubated with RNase H, the RNA part of the duplex will be degraded in only the regions that are complementary. In this example the RNA is cleaved in the middle leaving the two RNA halves intact.

As is often the case in science, unexpected results provided the most important puzzle pieces. The RNase H experiment relied on the oligonucleotide having access to the RNA to be able to hybridize with it, even when the RNA is bound by telomerase proteins. As this wasn't always the case, some of the oligonucleotides had no effect. However, two oligonucleotides did give completely unexpected results. Incubation of telomerase with *Oligo 3* inactivated telomerase even before RNase H was added, while, amazingly, incubation of telomerase with *Oligo 8* resulted in elongation of *Oligo* 8 itself (Figure 9). I had to sit and think about what this meant.

At first it was frustrating: if telomerase was already being inactivated by *Oligo3* and adding RNase H had no further effect, how could I do the experiment? This frustration soon faded when, having talked about this result with my friends and puzzling more, I realized there was a much more interesting explanation for these results. I had fortuitously found additional evidence of a role for the 159 nucleotide RNA in the telomerase reaction. *Oligo 3* was unique in that it hybridized to my 159 nt RNA in a region adjacent to the template region and also across it (Figure 9A). When anchored to the RNA by hybridization, this oligonucleotide would block binding of the TTGGGG oligonucleotide substrate, and thus block telomerase activity (Figure 9B).

Subsequent work showed that RNase H would cleave the 159 nt RNA when incubated with *Oligo* 3, showing that *Oligo* 3 does indeed hybridize to the 159 nt RNA as part of telomerase.

The other unusual puzzle piece was *Oligo8*. This oligonucleotide shared sequence similarity with *Oligo3* but its 3' end stopped just before the template region. This oligonucleotide also hybridized to the 159 nt RNA and the 3' end was positioned in exactly the right place for it to be elongated by telomerase (Figure 9A). This implied that the CAACCCCA sequence in the putative RNA did indeed serve as a template for TTGGGG repeat addition. Putting all of these pieces of the puzzle together, I felt there was good evidence to support the cloned RNA being the RNA component of telomerase.

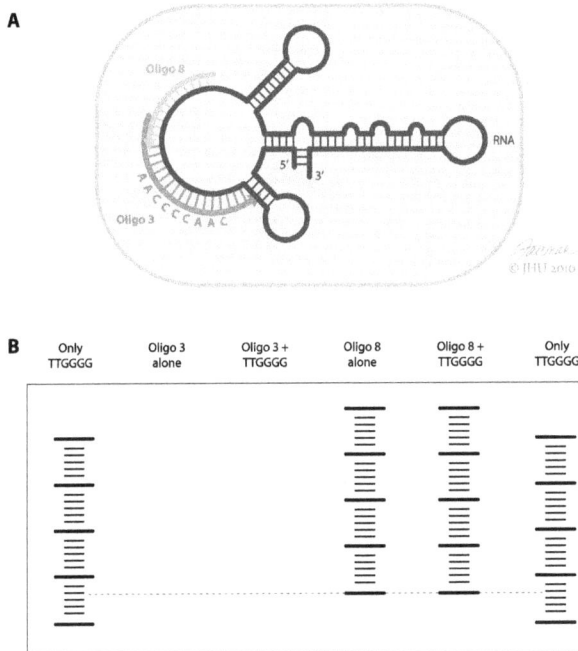

Figure 9. **Elongation of oligonucleotides complementary to the telomerase RNA**
A: the secondary structure of *Tetrahymena* telomerase. *Oligo 8* hybridizes adjacent to the template region with its 3' end positioned one base before the template. *Oligo 3* also hybridized adjacent to the template but also extends across the template region. B: Diagram of gel showing effects of *Oligo 3* and *Oligo 8* on telomerase activity. Lane 1, Telomerase will elongate $(TTGGGG)_4$ primer when it was added alone. Lane 2, *Oligo 3* alone was not elongated even though the 3' end contains TTGGGG sequence. Lane 3, *Oligo 3* and TTGGGG are added together there is no activity because *Oligo 3* binds to and blocks the active site of telomerase. This inhibition is the basis for the first telomerase inhibitor to be used in clinical trials against cancer (GRN163) (Asai *et al.*, 2003). Lane 4, *Oligo 8* was added alone and it was extended by telomerase generating a distinct banding pattern. This is because *Oligo 8* itself was being extended. Lane 6, When both *Oligo 8* and TTGGGG primer were added, the *Oligo 8* banding pattern was seen because it can out-compete the TTGGGG primer through its ability to hybridize to the telomerase RNA. (Adapted from Greider, C.W., and Blackburn, E.H. (1989), "A telomeric sequence in the RNA of Tetrahymena telomerase required for telomere repeat synthesis," *Nature* 337, 331–337.)

MODELS CAN SHOW THE SOLUTION TO THE PUZZLE

In writing the paper on the identification of the telomerase RNA (Greider and Blackburn, 1989), I was encouraged by Bruce Stillman to include a diagrammatic model for how I thought the enzyme might work.

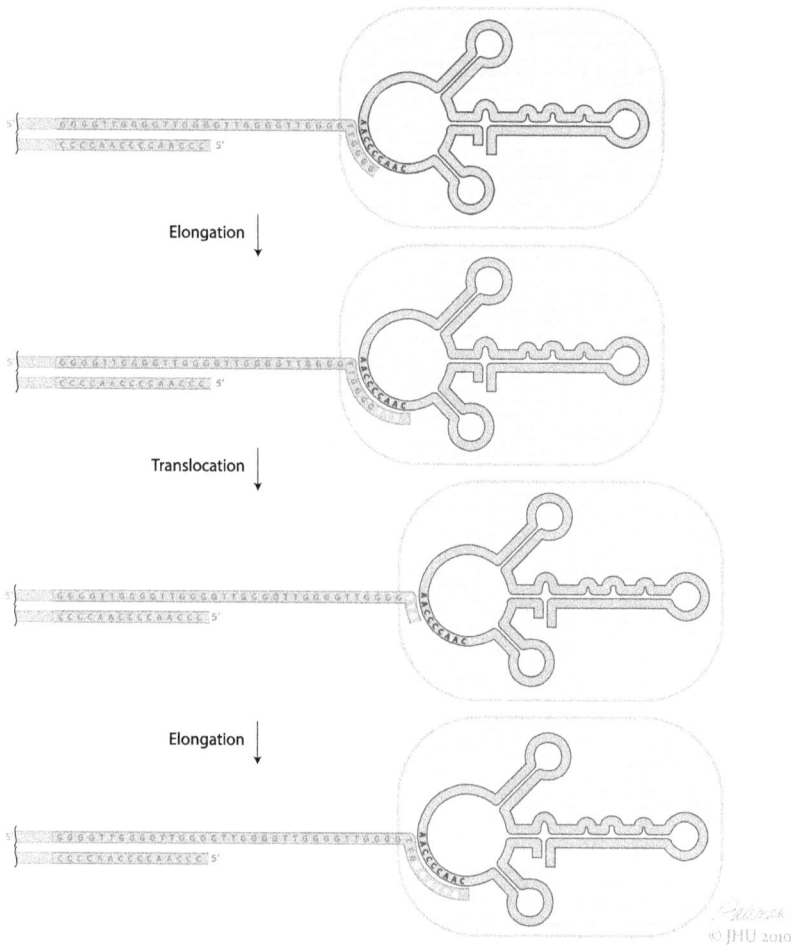

Figure 10. **Model for telomerase elongation of telomeres**
The sequence AACCCCAAC in the telomerase RNA serves as a template for extension of the telomere TTGGGG strand. The terminal TTGGGG on the telomere can base pair with the AACCCC in the RNA. This leaves three bases that can serve as a template for elongation. The activity of telomerase adds TTG and the end of the template is reached. Translocation can then reposition the 3' end such that base pairing between the terminal TTG and the telomerase RNA is maintained.

I was so caught up in the data that drawing a model was not foremost in my mind. However, I found that drawing out how I interpreted the results made everything even clearer. We knew from our early experiments that there must be a protein component to telomerase in addition to the RNA. I did not know how many protein components there might be but decided to draw just one for simplicity. The CAACCCCAA sequence in the RNA represented one and a half copies of the complementary strand of the TTGGGG repeat sequence. So the model I drew had the sequence GTTGGG base paired with the CAACCC, which left CAA free in the template sequence to be copied (Figure 10). With this in mind, I proposed that telomerase has an elongation phase during which the CAA is copied, followed by a translocation to reposition the growing sequence for another round of elongation. I wrote up the paper and, together with Liz Blackburn, in whose lab I had begun the sequencing, published a paper in *Nature* describing the RNA (Greider and Blackburn, 1989).

SOLUTIONS TO PUZZLES SHOW THE WAY TO MORE INTERESTING QUESTIONS

Drawing out the telomere elongation model helped to clarify my thinking about telomerase. Thinking about the model also immediately raised several new questions that I was curious to address. For example, does the proposed translocation step actually occur? That is, does telomerase hold on to the substrate it is elongating for a while, or does one enzyme only add one repeat, with a second repeat added during a second round of binding by a separate enzyme molecule? This question, which I had not thought of before I drew the model, was suddenly a burning one for me. I went on to tackle this next puzzle in a later paper (Greider, 1991). Many other questions arose as we continued our work on telomerase. The many different paths the research took and our later focus on telomerase in cancer and human disease is described in the Nobel Lecture presentation online at the Nobel Foundation website (www.nobelprize.org).

Putting together puzzle pieces is challenging, fun, and extremely gratifying, especially when they lead to new understanding in biology. This process of making a hypothesis and following leads is not a linear one: there are many twists and turns in the path. But the key is to keep the excitement and to follow the leads that are the most rewarding. I learned this during the first six years of working on telomerase, and it is the approach that I continue to follow. Many new questions often arise after one part of a puzzle is solved; the rewarding thing about curiosity-driven science is being able to pick from these new questions those that seem the most interesting to me. The pleasure of figuring out the puzzle and finding out things not known before is a great reward. Sharing that experience with friends and colleagues makes the reward even greater.

ACKNOWLEDGMENTS

I would like to thank all of the talented scientists who have worked with me over the years for their energy and ideas that have made solving the puzzles fun, and opened up new puzzles. I would also like to thank Jonathan Crowe and Mary Armanios for editing help with this manuscript and David Robertson for helpful advice on the accompanying autobiography. I would like to thank Jennifer Fairman for the figures in this manuscript and Bang Wong for the figures used in the Nobel Lecture PowerPoint presentation.

REFERENCES

1. Asai, A., Oshima, Y., Yamamoto, Y., Uochi, T.A., Kusaka, H., Akinaga, S., Yamashita, Y., Pongracz, K., Pruzan, R., Wunder, E., *et al.* (2003), "A novel telomerase template antagonist (GRN163) as a potential anticancer agent," *Cancer Res* **63**, 3931–3939.
2. Bernards, A., Michels, P.A.M., Lincke, C.R., and Borst, P. (1983), "Growth of chromosomal ends in multiplying trypanosomes," *Nature* **303**, 592–597.
3. Blackburn, E.H., and Gall, J.G. (1978), "A Tandemly Repeated Sequence at the Termini of the Extrachromosomal Ribosomal RNA Genes in *Tetrahymena*," *J Mol Biol* **120**, 33–53.
4. Boswell, R.E., Klobutcher, L.A., and Prescott, D.M. (1982), "Inverted terminal repeats are added to genes during macronuclear development in *Oxytricha nova*," *Proc Natl Acad Sci USA* **79**, 3255–3259.
5. Greider, C.W. (1991), "Telomerase is processive," *Mol Cell Biol* **11**, 4572–4580.
6. Greider, C.W. (1998), "Telomeres and senescence: the history, the experiment, the future," *Curr Biol* **8**, R178–181.
7. Greider, C.W., and Blackburn, E.H. (1985), "Identification of a specific telomere terminal transferase activity in Tetrahymena extracts," *Cell* **43**, 405–413.
8. Greider, C.W., and Blackburn, E.H. (1987), "The telomere terminal transferase of Tetrahymena is a ribonucleoprotein enzyme with two kinds of primer specificity," *Cell* **51**, 887–898.
9. Greider, C.W., and Blackburn, E.H. (1989a), "A telomeric sequence in the RNA of *Tetrahymena* telomerase required for telomere repeat synthesis," *Nature* **337**, 331–337.
10. Johnson, E.M. (1980),"A family of inverted repeat sequences and specific single-strand gaps at the termini of the Physarum rDNA palindrome," *Cell* **22**, 875–886.
11. Larson, D.D., Spangler, E.A., and Blackburn, E.H. (1987), "Dynamics of telomere length variation in *Tetrahymena thermophila*," *Cell* **50**, 477–483.
12. Lingner, J., and Cech, T.R. (1996), "Purification of telomerase from Euplotes aediculatus: requirement of a primer 3' overhang," *Proc Natl Acad Sci USA* **93**, 10712–10717.
13. Lingner, J., Hendrick, L.L., and Cech, T.R. (1994), "Telomerase RNAs of different ciliates have a common secondary structure and a permuted template," *Genes & Dev* **8**, 1984–1998.
14. McClintock, B. (1939), "The behavior in successive nuclear divisions of a chromosome broken at meiosis," *Proc Natl Acad Sci USA* **25**, 405–416.
15. McClintock, B. (1941), "The stability of broken ends of chromosomes in *Zea mays*," *Genetics* **26**, 234–282.
16. Morin, G.B. (1989), "The human telomere terminal transferase enzyme is a ribonucleoprotein that synthesizes TTAGGG repeats," *Cell* **59**, 521–529.
17. Muller, H.J. (1938), "The remaking of chromosomes," *Collecting Net* **13**, 181–198.
18. Olovnikov, A.M. (1973), "A theory of marginotomy," *J Theor Biol* **41**, 181–190.
19. Shampay, J., Szostak, J.W., and Blackburn, E.H. (1984), "DNA sequences of telomeres maintained in yeast," *Nature* **310**, 154–157.

20. Shippen-Lentz, D., and Blackburn, E.H. (1989), "Telomere terminal transferase activity from *Euplotes crassus* adds large numbers of TTTTGGGG repeats onto telomeric primers," *Mol Cell Biol* **9**, 2761–2764.

21. Szostak, J.W., and Blackburn, E.H. (1982), "Cloning yeast telomeres on linear plasmid vectors," *Cell* **29**, 245–255.

22. Watson, J.D. (1972), "Origin of concatameric T7 DNA," *Nature New Biol* **239**, 197–201.

Portrait photo of Professor Greider by photographer Ulla Montan.

JACK W. SZOSTAK

I greatly enjoy reading the biographies of scientists, and when doing so I always hope to learn the secrets of their success. Alas, those secrets generally remain elusive. Now that I find myself in the surprising situation of having to write my own biography, and thus to reflect on my career, I find the same mystery. I do not know why I have always been fascinated by science, or why I have been driven by the intense desire to make some original contribution. And although I have had some degree of success as a scientist, it is hard to say precisely why. Nevertheless, I have attempted to identify some of the incidents and decisions that helped or hindered me at various times, in the hope that these anecdotes might be helpful to those embarking on a scientific career.

I have generally sought to work on questions that I thought were both interesting and approachable, yet not too widely appreciated. To struggle to make discoveries that would be made by others a short time later seems futile to me. This, coupled with a distaste for direct competition, attracts me to areas of science that are less densely populated. On multiple occasions, I have been led into these new areas by talking to people working in fields quite different from mine. The confluence of ideas from distinct fields seems to create a kind of intellectual turbulence that is both exciting and productive.

My knowledge of the details of my family history is rather sketchy. My paternal great-grandfather was born near Cracow, and emigrated to New York City in the late 19th century, but ultimately settled in a small farming town in Saskatchewan, Canada where my father was born. Eager to escape the small town isolation, my father left as soon as he could by joining the Royal Canadian Airforce (RCAF) towards the end of World War II. He was trained as a pilot but fortunately the war ended before he could serve in combat, and he was then posted to Ottawa. My mother's family came from England but settled in Ottawa, where my mother was raised and met my father after the war. Shortly after they married my parents moved to England for my father's continued training in aeronautical engineering at Imperial College, London. I was born in London, England during the great fog of 1952, but survived the coal-fueled air pollution with no ill effects and after less than a year in England was carried to Canada by my parents. My father continued to work as an aeronautical engineer for the RCAF for the next twenty years, and our house was always decorated with models of the airplanes he worked on. After he retired my father joined the civil service, and for a time studied issues of Arctic transportation; I remember him telling me about the complex properties of Arctic sea ice. Some of my work has an engineering flavor, in that we build structures and test their properties, and it's possible that it may reflect some influence of my early home life. But a more direct influence stems from the fact that my father was often unhappy with his job, chafing at both his superiors and his subordinates. This I am sure made me seek out the academic life for its more egalitarian aspects. I have never felt like I worked for a boss or had employees who worked for me, just colleagues who like me were interested in learning more about the world around us.

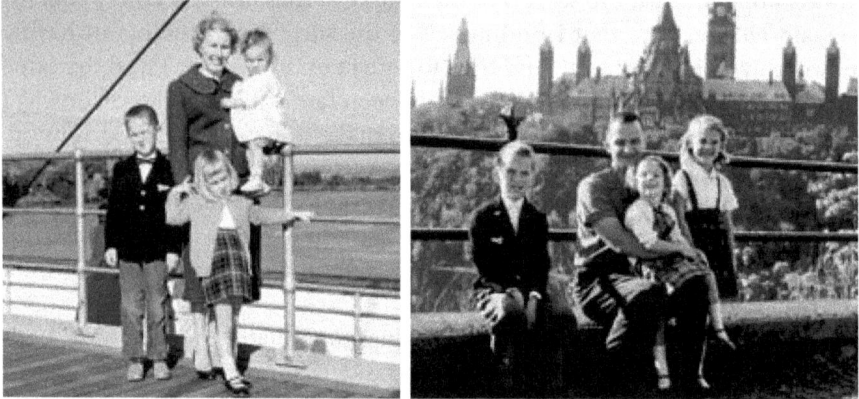

Figure 1. Family photos. 1A: Photo taken by my mother (Vi Szostak) on the HMS Homeric, as we returned to Canada in 1960 after three years on the RCAF base in Zweibrücken, Germany. Left to right, me, my mother and sisters Kathy and Carolyn. 1B: Photo taken by my mother, in Ottawa, 1963, showing me, my father (Bill Szostak), and sisters Carolyn and Kathy, with the Canadian Parliament Buildings in the background.

My childhood was punctuated by frequent moves, as my father was transferred to different Air Force postings in Germany, Montreal, and Ottawa (Figure 1). At the time many school systems encouraged students to advance as rapidly as possible; as a consequence I was often the youngest in my class. Although socially difficult, this was more than compensated for by making my classes more interesting than they would otherwise have been. Some of my earliest recollections involve grade school math. Learning about fractions was for some reason surprising enough to have stuck with me for the rest of my life; similarly, my discovery of quadratic equations in grade 5 was a revelation. Later, at Riverdale High School in suburban Montreal I was fortunate to have some exceptional teachers. Don Hall struggled to answer my strange science questions, and Irene Brun (now Winston) inspired a life-long love of biology. At the same time, my interest in science was encouraged at home. My father built a basement chemistry lab for me, and the experiments I conducted there often made use of remarkably dangerous chemicals that my mother was able to bring home from the company where she worked. My mother also helped me to get my first summer job, in a chemical testing laboratory at the same company. This was a good window into the importance of quantitative analysis, but the repetitive nature of the work was not at all interesting. Some of the experiments carried out in my basement lab were much more dramatic. For example, with my father's assistance, shortly after the tragic Apollo 1 fire, we prepared and collected a jar of pure oxygen. We then carefully lowered a small quantity of methanol into the oxygen reservoir. The transformation of the barely visible pale blue flame in air into an intense jet of fire in oxygen was amazing, but also horrifying in the context of the recent Apollo fire. Less carefully supervised experiments frequently led to explosions, which made chemistry seem much more dramatic than one would guess from the textbooks. My failure to carefully separate the hydrogen evolved during electrolysis from ambient air led to an impressive explosion which resulted in a glass tube being embedded in a wooden ceiling rafter. I also participated in more biologically oriented projects with my high school friend Joachim Sparkuhl. In the basement of his house we constructed a small hydroponics garden, inspired, I believe, by the idea that astronauts living on some future space station would need or want to grow their own fresh food.

In 1968 I began my undergraduate studies at McGill, at the age of 15. My first laboratory work at McGill involved helping a chemistry graduate student to purify cholesterol, the starting material for the synthesis of sterols. We started with large sacks of gallstones, which we would dissolve in hot solvent, and then recover the iridescent crystals of pure cholesterol after the solution cooled. While this was a useful experience, it did not inspire me to remain in chemistry, and the pull of biology increased as new opportunities opened up. To my surprise I was accepted into a summer research program for undergraduates at the Jackson Laboratories, a renowned mouse genetics institute on Mt. Desert Island off the coast of Maine. The environment was idyllic, and the program combined intense scientific education and hands-on experimental work with outdoor activities such as hiking up Cadillac Mountain and observing the beautiful organisms that populated the nearby tidal pools. The Jackson labs are a mouse genetics research facility, and this influenced my future scientific career in an unexpected way. My project, carried out under the guidance of Dr. Chen K. Chai, involved the analysis of thyroid hormones in various mutant strains. This required the careful dissection of the thyroid gland from many mice. Although I was, after much practice, able to remove the thyroid without (at least most of the time) severing any of the many nearby major blood vessels, I strongly disliked the process of killing and dissecting the animals, and by the end of the summer had vowed never again to work on animal models.

Back at McGill the next fall, this time as a resident student (my parents having moved back to Ottawa), I started spending less time in the lectures and more time in the library, and also searching out new labs in which to gain additional experience. I was always surprised when seemingly intimidating Professors welcomed me into their labs and invited me to join in ongoing research projects. During this year and the next I did work in several labs in the Biology and Biochemistry departments, generally on plant biology systems. Field trips with Kurt Meier, a specialist in bryophyte biology, inculcated an enduring affection for the simple mosses and liverworts. I apparently did well enough in a physiology course run by Ron Poole to land a summer job prototyping and testing new lab experiments for the following year's lab course. Although most of my lecture courses were uninspiring, John Southin's course in Molecular Biology was an incredible exception. I'll never forget entering the first class and being handed a thick book of printouts, which I assumed were a set of papers we were supposed to read. In fact the whole book was simply a list of references, which we were expected to read and absorb in the library. These readings from the frontiers of molecular biology were very impressive. We read and discussed the beautiful Meselson-Stahl experiment, which was just over a decade old at the time, and learned how the genetic code had been unraveled only a few years previously. The fact that one could deduce, from measurements of the radioactivity in fractions from a centrifuge tube, the molecular details of DNA replication, transcription and translation was astonishing to me. One of the intellectual highlights of my time at McGill was the open-book, open-discussion final exam in this class, in which the questions were so challenging that the intense collaboration of groups of students was required to reach the answers.

In my senior year, I began a project in Mel Goldstein's lab, together with my friend Joachim Sparkhul. The subject of our study was the beautiful colonial green flagellate *Eudorina elegans*, a smaller version of the more common *Volvox*. Over the school year and the following summer, we obtained evidence that these algae secreted a peptide hormone that induced spermatogenesis under favorable environmental conditions. This work led to our first scientific publication, which appeared the following year (1).

In the fall of 1972 I started my graduate studies at Cornell University in Ithaca, New York. I decided to attend Cornell in part because the A.D. White Fellowship would fully support me, but also because I would be able to pursue my work on *Eudorina* in the Department of Plant Physiology. At the time, I was enamored with a grandiose plan to develop *Eudorina* as a simple model system for studies in developmental genetics. This plan did not work out, for several reasons, not least the fact that this sort of ambitious program cannot be developed in isolation by an inexperienced student. Lacking the necessary genetic expertise, and because I was either unable or unwilling to seek out the necessary help, my project became mired in frustrating technical difficulties.

However, the periods spent waiting for my *Eudorina* cultures to grow allowed for plenty of time for conversations with my fellow graduate student John Stiles. John was approaching graduation and was thinking about what to do after the completion of his Ph.D., while I was gradually shifting from thinking about *Eudorina* to dreaming up some more productive project. We talked a lot about the emerging methods in molecular biology, which were clearly heading towards the ability to explore the structure and activity of individual genes at the molecular level; cloning and sequencing technologies were just beginning to emerge. John and I eventually came up with a specific proposal for a collaborative experiment. Our idea was to chemically synthesize a DNA oligonucleotide of sufficient length that it would hybridize to a single sequence within the yeast genome, and then to use it as an mRNA and gene specific probe. While conceptually simple, our idea was technically challenging. At the time, there was only one short segment of the yeast genome for which the DNA sequence was known, the region coding for the N-terminus of the iso-1 cytochrome c protein, which had been intensively studied by Fred Sherman for many years. The Sherman lab, in a tour de force of genetics and protein chemistry, had isolated double-frameshift mutants in which the N-terminal region of the protein was translated from out-of-frame codons. Protein sequencing of the wild type and frameshifted mutants allowed them to deduce 44 nucleotides of DNA sequence. John and I thought that if we could prepare a synthetic oligonucleotide that was complementary to the coding sequence, we could use it to detect the cytochrome-c mRNA and gene. At the time, essentially all experiments on mRNA were done on total cellular mRNA, rendering efforts to monitor the expression of individual genes almost impossible.

John and I were sufficiently confident of our ideas to begin contacting labs where we might pursue the work, with me doing the chemistry, and John working on the yeast biology. At Cornell, there was one laboratory that was the obvious place for such an experiment, and that was the lab of Ray Wu in the Department of Biochemistry. Ray was already well known for determining the sequence of the sticky ends of phage lambda, the first ever DNA to be sequenced, and his lab was deeply involved in the study of enzymes that could be used to manipulate and sequence DNA more effectively. John and I approached Professor Wu, who listened to our proposal and allowed that it was an interesting idea worth exploring. However he was reluctant to appear to be 'poaching' a graduate student from another lab and department; another complication was that the work would require collaboration with Fred Sherman's lab. John applied to Fred's lab in nearby Rochester, New York, for a postdoctoral position, and was accepted. At Cornell, I persisted and eventually Ray allowed me to transfer into his lab and begin the project.

The interlude between wrapping up my work in the Department of Plant Physiology and starting as a transfer student in the Department of Biochemistry provided me with the opportunity for an extended vacation and my first trip to Europe on my own. I began with a visit to Cambridge, England where I was very kindly hosted by Professor Poole (for whom I had worked at McGill), who was on sabbatical at the University of Cambridge. I explored the town and was incredibly impressed by the Chapel of King's College and the ethereal music therein. Even more impressive was the famous Laboratory of Molecular Biology at the MRC, where I talked with one of the iconic figures of molecular biology, Sydney Brenner. I was asked to wait for Sydney in his office, which I was surprised to notice held two large desks, both piled to the ceiling with papers. When Sydney arrived he told me about his remarkable new project involving the use of the nematode *Caenorhabditis elegans* as a model system for developmental genetics – this was an impressive if somewhat painful lesson on the right way to carry forward such an ambitious project.

I also learned why two desks crammed that small office – it turned out that Sydney shared an office with a fellow molecular biologist, one Francis Crick!

After a memorable month of art, architecture and music in Paris, I returned to Ithaca to start afresh in a new lab with a new project. My goal was clear – the chemical synthesis of the oligonucleotide needed for our gene detection scheme. At the time, this was still a challenging endeavor for a student such as myself with minimal synthetic skills. Ray had an ongoing collaboration with Saran Narang, who was developing the solution phase phosphotriester approach to oligonucleotide synthesis. Our plan was to use this approach to prepare large quantities of the five trimers needed to make a 15-mer, then link the trimers together to form 6-mers, a 9-mer and finally a 15-mer. I began the work under the tutelage of Chander Bahl, a postdoc who had some experience with this technology. Unfortunately our lab was better equipped for enzymology than synthesis, and we lacked a critical mass of experienced chemists. After a year of work, I was still far from my goal and becoming increasingly frustrated. Fortunately Ray Wu realized that I needed help, and arranged for me to visit Saran Narang's lab in Ottawa. There I was fortunate to receive training from Keichi Itakura, who later became famous for synthesizing the gene for insulin. After two weeks of intense training, I returned to Ithaca, and attacked my synthesis with fresh energy. A few months later, I was rewarded with several milligrams of our long sought 15-mer. In collaboration with John Stiles and Fred Sherman, who sent us RNA and DNA samples from appropriate yeast strains, we were able to show that we could use the labeled 15-mer as a probe to detect the *cyc1* mRNA, and later the gene itself. This was quite exciting, and seeing our work published in *Nature* (2) was a great boost to my confidence after years of work with little to show. It was also an important lesson in effective research strategy, imprinting on me the value of seeking help from knowledgeable people when faced with difficulties. One of the delights of the world of science is that it is filled with people of good will who are more than happy to assist a student or colleague by teaching a technique or discussing a problem.

The completion of my Ph.D. in 1977 marked the beginning of a major scientific transition for me. Against all commonsense advice, I decided to remain in Ray's lab for postdoctoral work, but in a very different scientific area. The decision was triggered by the arrival in Ray's lab of a new postdoc, Rodney Rothstein, from Fred Sherman's lab in Rochester. Rod was already a seasoned yeast geneticist, but had little experience with molecular biology; in contrast my graduate work was in molecular biology but I had no practical experience with genetics. We hit it off and essentially trained each other through our collaborative work on yeast transformation. Our frequent discussions were long and often loud, sometimes triggering mild protests from Ray who would emerge from his office and ask us to turn it down a notch when he needed a quieter atmosphere in which to work. The combination of the molecular biology I learned in Ray's lab and the genetics I learned there from Rod prepared me well for the next decade of my work on yeast, first in recombination studies, and later in telomere studies and other aspects of yeast biology. Ray was a wonderful advisor (3), and in addition to his scientific advice I absorbed much of his way of running a lab, which in essence was to be there when advice was needed but otherwise to let creative students and postdocs run with their ideas (Figure 2).

Figure 2. Ray Wu's lab, circa 1978. Top left, my graduate and postdoctoral advisor Ray Wu; next to him is Rodney Rothstein, who introduced me to yeast genetics. I am seated at the lower right.

My postdoctoral studies of recombination in yeast were enabled by the discovery, in Gerry Fink's lab at Cornell, of a way to introduce foreign DNA into yeast (4). These pioneering studies of yeast transformation showed that circular plasmid DNA molecules could on occasion become integrated into yeast chromosomal DNA by homologous recombination. Rod and I began to search for ways of increasing the frequency with which transformants were recovered. Increasing the target size for recombination seemed like a good possibility, and indeed when I transformed yeast with plasmids containing fragments of rDNA, I did recover more transformants, and these contained plasmid DNA integrated at the rDNA locus. These strains allowed me to initiate studies of unequal sister chromatid exchange in rDNA locus, resulting in my first publication in the field of recombination (5). Towards the end of my stay in Ray Wu's lab, Rod and I came upon the first hints of double-strand break stimulated recombination in yeast. Our preliminary experiments suggested that cutting plasmid DNA within a region of homology to yeast chromosomal DNA led to an increase in the recovery of transformants, presumably reflecting increased recombination of the input DNA with the homologous chromosomal locus. The idea that you could increase transformation frequency by cutting the input DNA was pleasingly counter-intuitive and led us to continue our exploration of this phenomenon.

My first independent position was at the Sidney Farber Cancer Institute (now the Dana-Farber Cancer Institute). I owe a great debt to Professor Ruth

Sager, who was the main force behind hiring me. She established a terrific group of young investigators in her division, including Richard Kolodner and Gerry Rubin, creating a superb intellectual atmosphere. Ironically, I heard many years later that Ruth was only able to hire me over the objections of some of the senior clinical faculty, who did not believe that studies of yeast had any place in a cancer institute. Times have changed, and fortunately model systems are now much more widely appreciated. My graduate students came from the graduate program at Harvard Medical School, where I had an academic appointment in the Department of Biological Chemistry. These students were wonderful, and together we made rapid progress in setting up a productive yeast genetics lab.

Our initial focus was the study of double-strand breaks in DNA and their repair by recombination. This work was spearheaded by my first graduate student, Terry Orr-Weaver, who is now a Professor at the Whitehead Institute and MIT. Terry's work, and our continuing interactions with Rod Rothstein, led us to think intensively about the kinds of reactions engaged in by DNA ends (6). There was considerable debate about different models for recombination within the wider DNA repair and recombination community, and seminars and conferences were important means for the exchange of the latest information. For many years, the major international recombination meeting was held in Aviemore, Scotland, which afforded the opportunity to sample diverse single-malts while discussing the intricacies of genetic exchange. I do recall that excessive sampling at one Aviemore meeting did make it difficult for me to present my work the next morning.

I also enjoyed attending Gordon Conferences and Cold Spring Harbor meetings, which were small and highly interactive meetings that provided wonderful opportunities for young scientists to present their work and meet and talk to people doing the best and most important current work. In the summer of 1980, I attended the Nucleic Acids Gordon Conference, expecting to hear the latest advances in DNA synthesis, sequencing and repair. However, for me the high point of the meeting was hearing Liz Blackburn talk about her work on telomeres in *Tetrahymena*. Our subsequent discussion led to the initiation of a collaboration in which we decided to test the ability of *Tetrahymena* telomeres to function in yeast. Those experiments are described in my Nobel Lecture; here I will just say that it was an incredibly exciting time for me. I performed the experiments myself, and experienced the thrill of being the first to know that our wild idea had worked. It was clear from that point on that a door had been opened and that we were going to be able to learn a lot about telomere function from studies in yeast. Within a short time I was able to clone bona fide yeast telomeres, and in a continuation of the collaboration with Liz Blackburn's lab we soon obtained the critical sequence information that led us to propose the existence of the key enzyme, telomerase.

With the success of the recombination and telomere projects, my lab began to grow. My second graduate student, Andrew Murray, now a Professor at Harvard, began to work on building artificial chromosomes. Andrew was a brilliant and energetic student who was fun to talk with about any conceivable experiment; his colorful personality (and dress) enlivened the lab. My collaboration with Rod and Terry grew to include Frank Stahl, the world's leading expert on the genetics of meiotic recombination, with whom we had many detailed discussions of the genetic implications of specific physical models. I particularly remember an afternoon I spent at Frank and Mary Stahl's house in Eugene, Oregon, going back and forth with Frank about different versions of the double-strand break repair model as we worked on our manuscript (7). It was an intense and stimulating experience that I still treasure.

After five very productive years at the Farber, a remarkable opportunity induced me to move to the fledgling Department of Molecular Biology at the Massachusetts General Hospital (MGH). Howard Goodman, the founder of the Department and a major figure in the emerging field of bio-technology, had arranged an extremely interesting and innovative academia/industry collaborative venture. In this deal, the pharmaceutical giant Hoechst AG agreed to fully support all research in the MGH Department of Molecular Biology for a period of about ten years, in return for limited intellectual property rights. This was extremely attractive to me, as it promised to allow me to pursue research in any direction that I found to be of interest, without having to worry about obtaining traditional grant support for novel and hence untried ideas. Thus, in the summer of 1984 I moved my lab from the Farber to our new home in the downtown Boston campus of MGH (humorously referred to by colleagues at MIT's Whitehead Institute as "one of the finest research institutes in down-town Boston").

At that time, I was actively exploring the possibility of moving into other fields. By 1984, I had a growing feeling that my work in yeast was becoming less significant, in the sense that other people would inevitably end up doing the same experiments we were doing in a few months or years at the most. To learn more about other fields and to prepare myself to work in a new area I audited several courses at Harvard. A delightful course by Steve Kosslyn on cognitive psychology explored the fascinating correlations between localized brain lesions and cognitive deficits, and highlighted the emerging neuroim-aging technologies that were promising to revolutionize studies of brain function. I also audited an applied math course to brush up on the skills I would need should I decide to seriously enter into structural biology. Finally an outstanding course on enzymology and catalytic mechanisms by the late Jeremy Knowles stimulated my interest in catalysis. Later, when Jeremy left science to become Dean of the Faculty of Arts and Sciences at Harvard, I had the good fortune to "inherit" one of his graduate students, Jon Lorsch, who migrated to my lab and did outstanding work on ribozyme selections and mechanistic enzymology.

The combination of Jeremy's enzymology course and the recent discovery of ribozymes by Tom Cech and Sid Altman (who shared the 1989 Nobel Prize in Chemistry for their work), ultimately led me to begin a transition to work on ribozymes. This seemed like a reasonably conservative way to switch fields, since the methods used to study ribozymes were largely a combination of molecular biology and chemistry. I was surprised that so few people were entering the field, since I thought that there were major questions to be addressed in terms of understanding the origins of biological catalysis in the hypothetical RNA world that preceded the evolution of protein synthesis.

I began to work with RNA myself, playing around with Cech's *Tetrahymena* ribozyme, which I obtained from the same piece of DNA that contained the *Tetrahymena* telomeres I had worked on just a few years earlier. The first student to join me in this new area was Jennifer Doudna. Jennifer had actually come to my lab to work on yeast genetics, but I was fortunate to persuade her that the future lay in RNA. Jennifer's energy and determination drove our efforts to convert self-splicing introns into an RNA replicase. We were soon joined by Rachel Green and several other dedicated students, techs, postdocs, and a memorable sabbatical visitor, François Michel, who impressed everyone with his work ethic, his uncanny ability to intuit structure from phylogeny, and his parallel career in butterfly evolution.

Even as I pushed our gradual transition to a focus on RNA, I maintained a substantial effort in yeast genetics for several years during the mid to late 1980s. My interest in recombination and telomeres had not disappeared, and I wanted to bring our earlier advances to a satisfying conclusion. Recombination remained a large part of the lab, with Doug Treco, Alain Nicolas, Neil Schultes and Hong Sun maintaining a focus on the role of double-strand breaks in meiotic recombination. Most important for the telomere story was Vicki Lundblad's ground-breaking work on telomere genetics in yeast, which provided a link between telomere maintenance and senescence and aging (8). Barbara Dunn linked the telomere and recombination realms by study the transfer of sub-telomeric repeats between chromosomes by recombination.

Figure 3. The Szostak lab, circa 1985. Top row, from left: Neil Schultes, Andrew Murray, Dean Dawson, Neil Sugawara, Hong Sun. Bottom row, from left: Vicki Lundblad, Barbara Dunn, Jack Szostak, Stephanie Ruby.

By end of the 80s, our yeast work was almost done, and the lab was increasingly focused on RNA. The RNA floodgates really opened with the work of Andy Ellington on *in vitro* selection (9), which ushered in a new era of work on the *in vitro* directed evolution of new functional molecules. Over time we came to feel that we could evolve a binding site for virtually any target molecule, using any kind of nucleic acid. This confidence led us to try to evolve new catalysts, and, returning to the RNA world hypothesis for inspiration, we aimed for the chemistry of nucleic acid polymerization (10). This was the basis of Dave Bartel's ground-breaking work on the selection of ribozyme ligases, which he subsequently (in his own lab at the Whitehead Institute and MIT) evolved into an RNA molecule with bona fide RNA polymerase activity. Our advances fueled my interest in the role of RNA in early evolution and seemed to bring the resurrection of the RNA world almost within reach. Our ability to evolve new aptamers and ribozymes was so intoxicating that my lab spent most of the 90s exploring the range of possibilities and the limitations of what RNA could do. Our advances began to attract attention, leading to my election to the National Academy of Sciences and appointment as a Howard Hughes Investigator in 1998. At the same time, the Hoechst funding of my department was winding down, making my HHMI appointment particularly welcome as a means of enabling ventures into new scientific areas.

As other labs also started to evolve new and interesting ribozymes, the difficulty of evolving *de novo* proteins began to seem the greater challenge. We entered the field of protein and peptide evolution when Richard W. Roberts, a postdoc in my lab, learned how to trick the translation apparatus into covalently linking a newly translated protein to its own mRNA through the action of the antibiotic puromycin (11). Galvanized by this advance, I encouraged several new lab members to develop and use this mRNA-display technology to address fundamental questions about the origin of protein structure. Most significantly, Tony Keefe used this method to evolve a novel ATP-binding protein from a large library of random sequence polypeptides (12). Remarkably, this non-biological protein looks indistinguishable from any normal biologically derived small protein domain. Postdoctoral fellows John Chaput and Sheref Mansy continued to evolve this protein and study its structure over the following years.

The development of this protein evolution technology led me to co-found a startup biotechnology company, together with Rich and my colleague Brian Seed. Although the company was not a business success, it was a very interesting and educational experience. The collaborative efforts of a team of scientists ranging from protein biophysicists to people with clinical drug development experience allowed us to evolve a small protein domain with therapeutic potential; this artificially evolved protein is now in clinical trials. While I have continued to maintain a focus on fundamental questions in my laboratory, I firmly believe that small startup companies are the best way to develop more applied research to the point that it can eventually be therapeutically useful.

By the year 2000, I started to pay more attention to fundamental questions related to the origin of life. My interest in the role of compartmentalization and cellular structure in the origin of life was stimulated by discussions with Pier Luigi Luisi and David Bartel. A year of debate led to a synthesis of our views on the roles of genetics, compartmentalization and evolution, which we expressed in our 2001 *Nature* paper Synthesizing Life (13). This paper catalyzed my entry into the field of membrane biophysics, for I felt that having proposed a model for early cells in which bilayer membranes played a crucial role, it was incumbent on us to show that such models were physically plausible. I have to admit that I was somewhat surprised to find myself working with lipids and membranes, which are remarkably squishy and ill-defined by comparison with nucleic acids. However, in at least one way, the study of membranes composed of prebiotic building blocks such as fatty acids was perfect for me, since this field was filled with important yet technically addressable questions. When postdoctoral fellow Marty Hanczyc and graduate student Shelly Fujikawa joined the project, we were able to make rapid progress, and within a few years had demonstrated a proof-of-principle path for vesicle growth and division based solely on physical processes. I began to grow more confident that it might ultimately be possible to deduce plausible explanations for at least some of the mysterious steps in the origin of life. My enthusiasm grew when Irene Chen, a brilliant biophysics graduate student, made further progress by demonstrating a pathway for competition between protocells. We worried that our model protocells would not be able to take up nutrients, such as the nucleotides needed for the replication of their genetic material, but Sheref Mansy showed that this was not a problem. Most recently, another graduate student, Ting Zhu has come up with a very

Figure 4. Szostak lab group photo taken shortly after the Nobel Prize announcement.

attractive pathway for spontaneous, coupled growth and division, so it is beginning to seem that the assembly and replication of protocell membranes is not as difficult as we once thought.

The dramatic progress in the identification of pathways for the self-replication of protocell membranes has encouraged us to focus on the hardest remaining problem, the replication of the genetic material. Here the big question is whether RNA was in fact the first genetic polymer, or whether RNA was preceded by some simpler, easier to make or more robust genetic material. This question has driven the most recent transformation of my lab (Figure 4), into a well equipped synthetic organic chemistry lab. We are synthesizing amino-nucleotides, the building blocks for phosphoramidate polymers, due to their greater reactivity than normal nucleotides. Alonso Ricardo, a postdoc, and Jason Schrum, a graduate student, have recently made very significant progress in the template-directed synthesis of 2'-5' linked phophoramidate DNA (14), and we are now exploring a series of related polymers in a search for even better self-replicating genetic materials. The complexity and fragility of RNA long made it seem an unlikely candidate for the first genetic material, but this prospect has been revived by the brilliant recent work from John Sutherland's lab in Manchester (15). With John's former graduate student Matt Powner now in my lab as a postdoc, we are eagerly exploring new avenues to the chemical replication of RNA. It is thrilling to me to see people in my lab developing new approaches to the synthesis of modified nucleic acids, but the suspense is almost unbearable as we await the results of template-directed polymerization experiments.

From our current vantage point, it is not clear whether there will be many solutions to the problem of chemically replicating genetic polymers, or just one, or none, but in any case it is an exciting quest. Encouraged by our small advances on the way, we are continuing to feel our way towards the tantalizing goal of building replicating, evolving chemical systems.

REFERENCES

1. Szostak, J.W., Sparkuhl, J., Goldstein M.E., "Sexual induction in Eudorina: effects of light, nutrients and conditioned medium," *J Phycol.* 1973; 9:215–218.
2. Szostak, J.W., Stiles, J.I., Bahl, C.P., Wu, R., "Specific binding of a synthetic oligonucleotide to yeast cytochrome c mRNA," *Nature* 1977; 265:61–63.
3. Szostak, J.W., "Ray Wu, as remembered by a former student,". *Sci China C Life Sci.* 2009; 52:108–110.
4. Hinnen, A., Hicks, J.B., Fink, G.R., "Transformation of yeast," *Proc. Natl. Acad. Sci. USA* 75, 1929–1933 (1978).
5. Szostak, J.W., Wu, R., "Unequal crossing over in the ribosomal DNA of Saccharomyces cerevisiae," *Nature* 1980; 284:426–430.
6. Orr-Weaver, T.L., Szostak, J.W., Rothstein, R.J., "Yeast transformation: a model system for the study of recombination," *Proc. Natl. Acad. Sci. USA* 1981; 78:6354–6358.
7. Szostak, J.W., Orr-Weaver, T.L., Rothstein R.J., Stahl, F., "The double-strand-break repair model for recombination," *Cell* 1983; 33:25–35.
8. Lundblad, V. and Szostak, J.W., "A mutant with a defect in telomere elongation leads to senescence in yeast," Cell 57, 633–643 (1989).
9. Ellington, A.E. and Szostak J.W., "In vitro selection of RNA molecules that bind specific ligands," *Nature* 346, 818–822 (1990).
10. Bartel, D.P. and Szostak, J.W., "Isolation of new ribozymes from a large pool of random sequences," *Science* 261, 1411–1418 (1993).
11. Roberts, R.W. and Szostak, J.W., "RNA-peptide fusions for the in vitro selection of peptides and proteins," *Proc. Natl. Acad. Sci. USA* 94, 12297–12302 (1997).
12. Keefe, A.D. and Szostak, J.W., "Functional proteins from a random sequence library," Nature 410, 715–718 (2001).
13. Szostak, J.W., Bartel, D.P., Luisi, P.L., "Synthesizing life," *Nature,* 2001; 409:387–390.
14. Schrum, J., Ricardo, A., Krishnamurthy, K., Blain, J.C. and Szostak, J.W., "Efficient and rapid template-directed nucleic acid copying using 2′-amino-2′, 3′-dideoxyribonucleoside-5′-phosphorimidazolide monomers," *J. Am. Chem. Soc.* 31, 14560–14570 (2009).
15. Powner, M.W., Gerland, B., Sutherland, J.D. 2009, "Synthesis of activated pyrimidine ribonucleotides in prebiotically plausible conditions," *Nature* 459:239.

DNA ENDS: JUST THE BEGINNING

Nobel Lecture, December 7, 2009

by

JACK W. SZOSTAK

Harvard Medical School; Howard Hughes Medical Institute; Massachusetts General Hospital, Boston, MA, U.S.A.

The contributions of my laboratory to our understanding of telomere function and maintenance by telomerase were made over a limited period of time early in the development of this story, from 1980 to 1989. What I would like to discuss here are some of the problems that we had to overcome, especially the preconceptions we had about models for telomere function and how hard it was to let go of those models. Fortunately the evidence we uncovered was strong enough to bring us to the right conclusions! Then, since I left the telomere field fairly early on, I would like to take this opportunity to briefly review some of the work that we've done since, primarily to show students who are just entering science that it is not only possible but really fun to address very different questions in different fields during one's career.

There were two well-known and long-standing puzzles associated with the nature of eukaryotic chromosome ends, or telomeres: the problem of the stability of the ends of chromosomes, and the problem of complete replication. My first introduction to these issues came when I was an undergraduate student at McGill University in Montreal. The first of those two problems, the reactivity of chromosome ends, had been a puzzle for many decades, ever since the pioneering work of Hermann Muller[1] and Barbara McClintock[2] in the 1930s. Muller used X-rays to create breaks in DNA, while McClintock used cytogenetic tricks to break chromosomes. But both came to the same conclusion, which is that the ends of broken chromosomes are very reactive and do things that normal chromosome ends never do. This is dramatically illustrated by the famous breakage-fusion-bridge cycle explored by McClintock (Figure 1). The basic observation is that the replication of a chromosome with a broken end results in two ends that can join together, generating a chromosome with two centromeres. When those centromeres are pulled towards opposite poles of the spindle during cell division, the chromosome is broken again, regenerating chromosomes with broken ends. This results in continuing cycles of fusion and breakage, a consequence of which is the formation of cells that have lost important parts of chromosomes. Not surprisingly many dead cells are generated in this process.

The contributions of my laboratory to our understanding of telomere function and maintenance by telomerase were made over a limited period of time early in the development of this story, from 1980 to 1989. What I would like to discuss here are some of the problems that we had to overcome, especially the preconceptions we had about models for telomere function and how hard it was to let go of those models. Fortunately the evidence we uncovered was strong enough to bring us to the right conclusions! Then, since I left the telomere field fairly early on, I would like to take this opportunity to briefly review some of the work that we've done since, primarily to show students who are just entering science that it is not only possible but really fun to address very different questions in different fields during one's career.

There were two well-known and long-standing puzzles associated with the nature of eukaryotic chromosome ends, or telomeres: the problem of the stability of the ends of chromosomes, and the problem of complete replication. My first introduction to these issues came when I was an undergraduate student at McGill University in Montreal. The first of those two problems, the reactivity of chromosome ends, had been a puzzle for many decades, ever since the pioneering work of Herman Müller[1] and Barbara McClintock[2] in the 1930s. Müller used X-rays to create breaks in DNA, while McClintock used cytogenetic tricks to break chromosomes. But both came to the same conclusion, which is that the ends of broken chromosomes are very reactive and do things that normal chromosome ends never do. This is dramatically illustrated by the famous breakage-fusion-bridge cycle explored by McClintock (Figure 1). The basic observation is that the replication of a chromosome with a broken end results in two ends that can join together, generating a chromosome with two centromeres. When those centromeres are pulled towards opposite poles of the spindle during cell division, the chromosome is broken again, regenerating chromosomes with broken ends. This results in continuing cycles of fusion and breakage, a consequence of which is the formation of cells that have lost important parts of chromosomes. Not surprisingly many dead cells are generated in this process. Normal chromosomes never do this, so it was clear that there was something very special and different going on at the ends of normal chromosomes that prevents end-to-end joining. But at the time of this work it wasn't even known that DNA was the genetic material, so they had no way to think in molecular terms about what was going on.

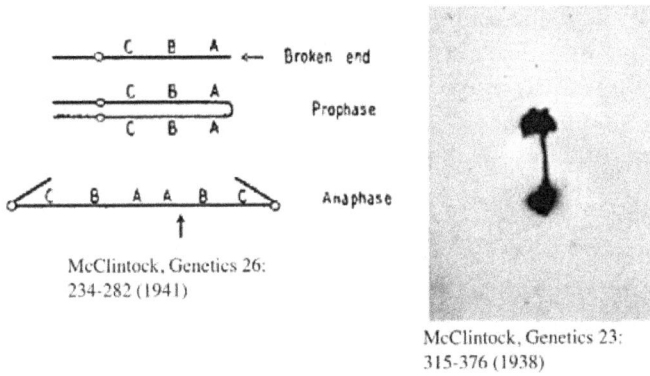

McClintock, Genetics 26:
234-282 (1941)

McClintock, Genetics 23:
315-376 (1938)

Figure 1. The chromosomal breakage-fusion-bridge cycle explored by Barbara McClintock. Left: After the replication of a broken chromosome, the two broken ends join together, creating a dicentric chromosome. When the two centromeres are pulled to opposite poles of the dividing cell, the chromosome breaks, and the new broken chromosomes continue the cycle. Right: micrograph of a dicentric chromosome bridging the two poles of a mitotic spindle.

Much later on, long after it was recognized that DNA was the genetic material in chromosomes, an additional problem was discussed by Watson[3] and by Olovnikov[4], who recognized that the replication of the very ends of DNA molecules posed a special problem (Figure 2). When a replication fork heads towards the end of the chromosome, the leading strand can go all the way to the end, but the lagging strand cannot since it is generated by the extension of an RNA primer by DNA polymerase. If this RNA primer is generated at an internal site, any distal DNA will remain unreplicated; even if the RNA primer was made at the very end, after the RNA primer is degraded, a short region of unreplicated DNA would remain. In the absence of some compensatory mechanism, the ends should get shorter and shorter, and since that doesn't happen, there must be some unknown process to counterbalance the necessarily incomplete replication.

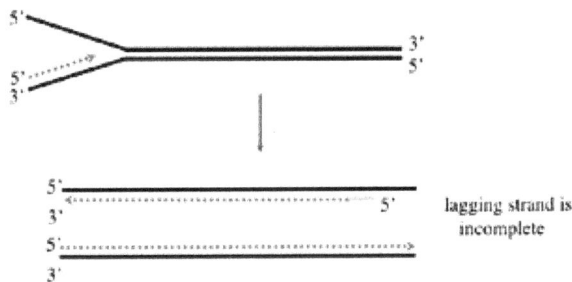

Figure 2. The end-replication problem as posed by Watson ([3]) and by Olovnikov ([4]). When a replication fork reaches the end of a chromosome, the lagging strand will necessarily be incomplete as a result of the removal and potentially internal location of the last primer generated by primase.

Although I learned about these problems as a student, I can't say that they made a very big impression on me and I didn't really think about them very much until years later, when I began working on the molecular reactions engaged in by broken pieces of DNA. This was work that I started as a post-doc at Cornell with Ray Wu, working in collaboration with my friend and colleague, Dr. Rodney Rothstein. Means for introducing DNA molecules into yeast cells, a process referred to as yeast transformation, had just been discovered down the road from our lab at Cornell in Gerry Fink's lab[5]. The ability to do this opened up a huge number of interesting experiments. Rod and I started to examine some variations on the initial procedure, such as cutting the circular DNA molecules before putting them into yeast. Shortly thereafter, when I moved to Boston and was setting up my lab at the Sidney Farber Cancer Institute, we continued this collaboration with the additional participation of my first graduate student, Terry Orr-Weaver[6,7].

In the course of our experiments on transformation and recombination, we observed a process that is analogous to the fusion events studied by McClintock in maize decades earlier (Figure 3). We began with a circular DNA molecule that was able to replicate as a circular DNA plasmid in yeast because it contained a yeast origin of DNA replication[8,9]. Intact circular DNA of that plasmid yielded a high frequency of yeast transformants, because chromosomal integration was not required for plasmid maintenance. When we made a cut in the DNA with a restriction enzyme, in a region of the DNA that is not found in any yeast chromosome, we recovered many fewer trans-formants. When we analyzed the few transformants that we did recover, the cut DNA ends had been joined back together, presumably by the action of the enzyme DNA ligase[7]. In many cases some DNA was lost as the ends were chewed back by exonucleases before being joined together by ligase. As with McClintock's much earlier results, these DNA reactions are very different from anything that would happen at the ends of natural chromosomes.

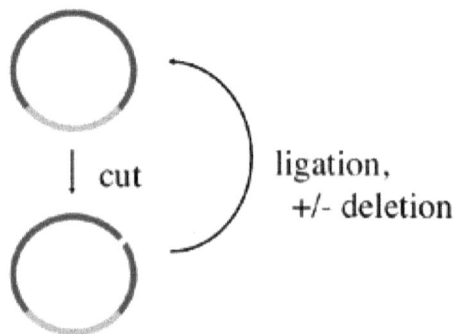

Figure 3. Non-homologous end-joining in yeast. A circular plasmid, cut with a restriction enzyme in a region of DNA that is not homologous to any yeast chromosomal DNA adjacent to the cut site may be degraded prior to ligation of the ends.

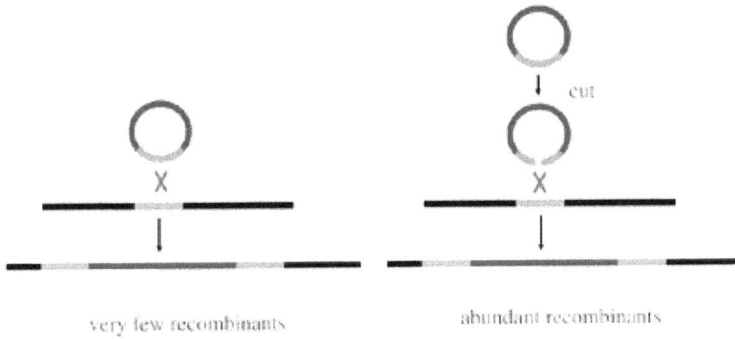

Figure 4. Double-strand breaks in DNA stimulate recombination. Intact circular DNA lacking a replication origin yields few transformants, because recombination events leading to chromosomal integration are rare. The same plasmid, when cut within a region of homology to a yeast chromosome, yields many more integrated transformants.

Terry, Rod and I actually spent most of our effort looking at what happened when we made cuts in regions of plasmid DNA that were homologous to a segment of a yeast chromosome (Figure 4). When a circular DNA molecule containing a region of homology with a chromosome is used to transform yeast, the occasional recombination event will occur, resulting in the plasmid becoming integrated into the yeast chromosome. This was the pathway found in the Fink lab in their early studies of transformation. What Terry, Rod and I found was that cutting the DNA in this region of homology led to a greatly increased frequency of such recombination events[6]. We continued to follow this up by studying the reactions that broken DNA ends undergo (Figure 5). If a DNA molecule is broken by cutting with a restriction enzyme, then in the cell the ends can be chewed back by nucleases, and exonucleases can generate single-stranded ends that can invade a homologous sequence. Strand invasion allows repair synthesis to begin using DNA polymerases, and Holliday junctions can be formed which can branch migrate. After repair synthesis, the Holliday junctions can be resolved by special enzymes called resolvases, to yield crossover or non-crossover configurations. This work eventually led us to propose, along with Frank Stahl, that cells entering meiosis engage in the programmed breakage of their chromosomal DNA as a means of initiating meiotic recombination by double-strand-break repair[10]. So broken DNA ends do a lot of things, but they are all things that don't happen with normal chromosome ends. I mention them here because these are the reactions I was thinking of before I entered the telomere field.

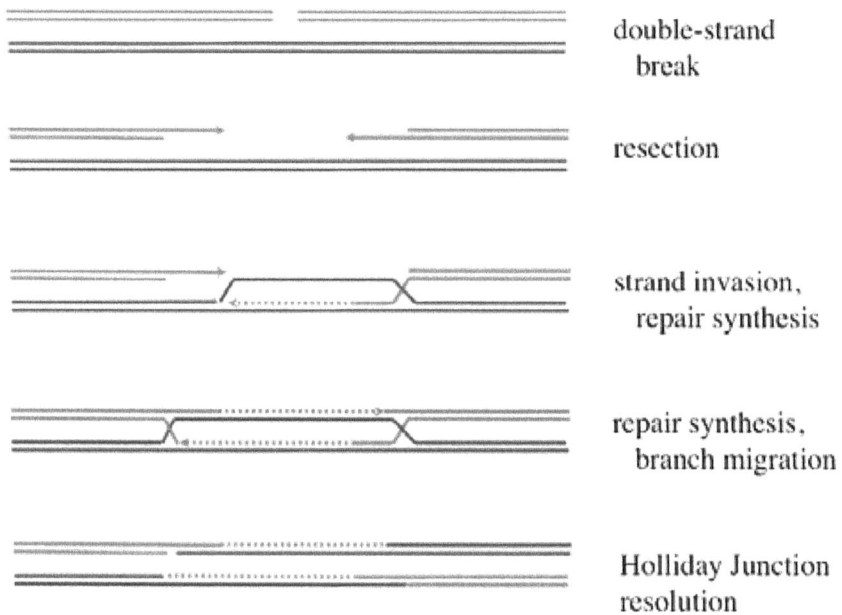

Figure 5. The double-strand break repair model for recombination. Two homologous chromosomes (red and blue) recombine when one is broken. The initial cut is further processed by nucleases, exposing single-stranded DNA, which invades the homologous duplex. Repair synthesis and branch migration generate Holliday junctions, the resolution of which generates recombinant DNA products.

In the summer of 1980, I attended the Nucleic Acids Gordon Conference and heard, for the first time, Elizabeth Blackburn talk about her amazing work on the stable DNA ends from *Tetrahymena thermophila*[11]. This unicellular organism is very divergent from metazoans, and has an unusual cell biology characterized by the presence of both a micronucleus with normal chromosomes and a macronucleus in which the chromosomal DNA has been chopped into thousands of small fragments, many of which become highly amplified. Liz talked about the very simple repetitive sequences, just stretches of a GGGGTT repeats, that she had found at the ends of these very abundant short DNA molecules in the large macronucleus of *Tetrahymena* (Figure 6). It was incredibly striking that these little pieces of DNA were stable ends, and were apparently fully replicable, i.e. they seemed to behave just like normal chromosomal telomeres. They clearly behaved completely differently from the DNA ends that we were studying in my lab, in yeast cells. After Liz's talk I sought her out to discuss these experiments, and we realized that there was a really simple and potentially very interesting experiment that we could do to see if the telomeric ends from *Tetrahymena* would work as stable telomeric ends in yeast cells. Neither of us thought that the experiment was very likely to work, because *Tetrahymena* and yeast are so very distantly related. On the other hand, we had all the necessary bits and pieces and technically the experiment was quite trivial, so we decided to go ahead. Liz sent me some DNA that she had painstakingly purified from *Tetrahymena*, and I took this little restriction fragment from the end of the ribosomal DNA of *Tetrahymena* and put it into yeast to see how it would function.

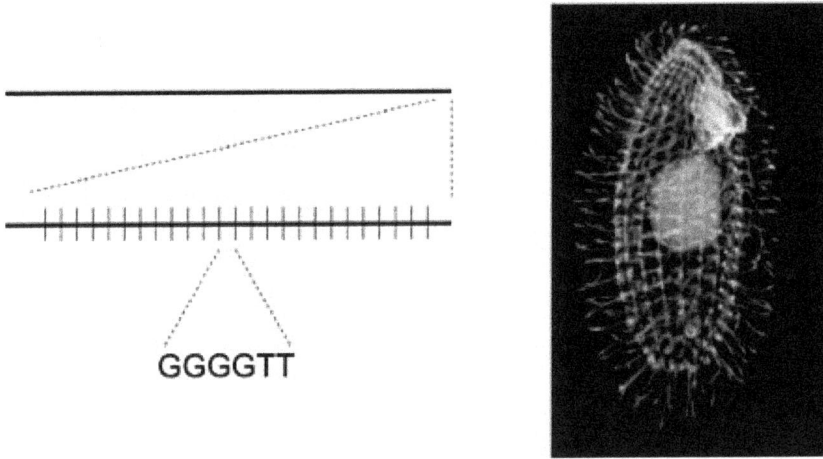

Figure 6. Telomeres from *Tetrahymena*. Left: DNA from the macronuclear fragments end in a series of tandem repeats of the hexanucleotide GGGGTT. These DNA ends are stable and fully replicated. Right: Image of *Tetrahymena*, showing the large macronucleus (blue).

There is an amazing aspect of this piece of DNA from *Tetrahymena* that I would like to comment on before describing the yeast experiment (Figure 7). Right next door to the telomere sequence, just a couple of kilobases in, is the primary ribosomal RNA transcript of *Tetrahymena*. In that transcript there is a little intron, just over 400 bases long, and that intron is the first self-splicing intron ever discovered[12], in the work for which Tom Cech was awarded the Nobel Prize in chemistry in 1989. A very nice piece of DNA indeed!

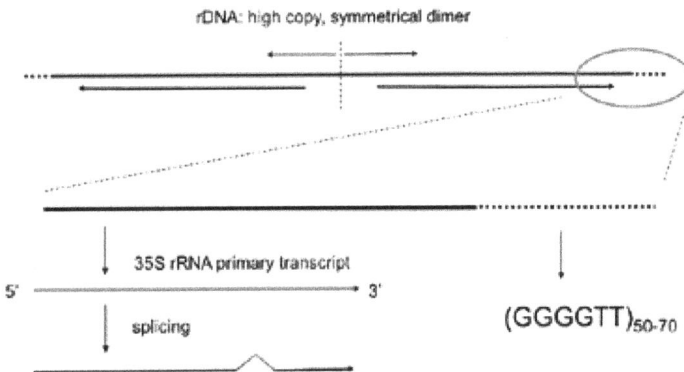

Figure 7. A very special piece of DNA. The *Tetrahymena* ribosomal DNA fragment from the macronucleus is a symmetrical dimer. The ends are telomeres and consist of GGGGTT repeats. Close to the ends is a region of the rRNA genes coding for a self-splicing intron.

Returning to the experimental test of *Tetrahymena* telomere function in yeast, what we really wanted to do was to test the idea that the biochemical machinery underlying telomere function might have been very highly conserved. If that turned out to be true, then the mechanisms that were being learned about in *Tetrahymena* might apply broadly to eukaryotic organisms, which would make the whole process much more significant. This was the motivation for the experiment that Liz Blackburn and I collaborated on. What we had available at that time, in my lab, were circular DNA plasmids containing yeast genes[13,14] so that we could select for yeast transformants, i.e. cells that had taken up the DNA. These plasmids also contained origins of replication (known then as autonomous replication sequences or ARS elements[8,9]) so that they could replicate independently of integration into the chromosome. When intact circular plasmid DNA of this type is used to transform yeast cells, many transformants are recovered and they almost all contain replicating circular DNA molecules. As I explained above, if the plasmid DNA is cut with a restriction enzyme (in a region that is not homologous to yeast genomic DNA) so as to generate linear DNA with 'broken ends', those ends do not function as stable telomeric ends and as a result very few transformants are recovered.

The critical experiment was to take the little pieces of telomeric *Tetrahymena* DNA ending in G_4T_2 repeats, and ligate them onto each end of the linearized plasmid DNA (Figure 8). I carefully purified the ligated DNA, put that into yeast, and recovered transformants. I was then able to ask whether the plasmid DNA was replicating as a linear molecule, which would mean the telomeres were working, or whether I had only recovered standard replicating circular plasmids. I distinguished between linear and circular DNA forms by preparing DNA from a dozen or so transformants, and analyzing the DNA by gel electrophoresis. When DNA molecules are separated by gel electrophoresis, circles generate a series of bands corresponding to monomers and multimers, and relaxed and supercoiled forms, leading to a complicated pattern. Linear DNA molecules don't have any of those alternative forms, so they migrate as a single band. The two possible results of the DNA analysis were therefore quite distinct. When I analyzed the DNA from the transformants that I had recovered, about half of them contained plasmid DNA that migrated as a single band on the gel. This was perhaps the most clear-cut experiment I have ever done. It was immediately obvious that the experiment had worked, and that the *Tetrahymena* ends were able to act as functional telomeres in yeast[15]. We therefore knew immediately that the underlying biochemical machinery must be very broadly conserved because these two organisms were so distantly related to each other. It also meant that we could now use all of the tools of yeast genetics and molecular biology to study telomeres in yeast.

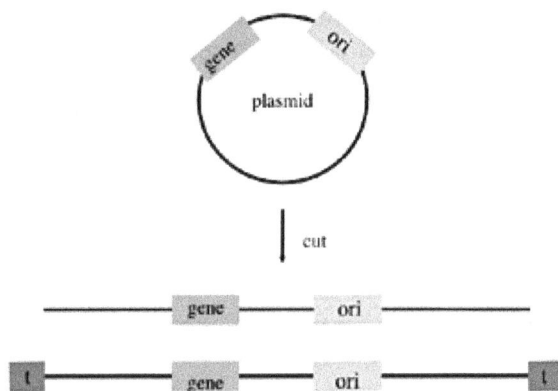

Figure 8. Moving *Tetrahymena* telomeres into yeast. A yeast plasmid vector containing selectable markers and an origin of replication was linearized by digestion with a restriction enzyme. *Tetrahymena* telomeres were ligated onto both ends, and the ligated DNA was purified and used to transform yeast cells. The resulting transformants contained replicating linear plasmids.

One of the first things that I wanted to do with the new linear plasmid with two *Tetrahymena* ends was to use it as a vector for cloning natural telomeres from the ends of yeast chromosomes. That experiment was extremely simple conceptually (Figure 9). I began with yeast chromosomal DNA, and cut it up with restriction enzymes into lots of pieces. Most of them were internal fragments, but the occasional fragment from the end of a chromosome would have one restriction cut end and one end derived from a yeast telomere. I then took our vector DNA, the linear plasmid with two *Tetrahymena* ends, cut off one end, and carefully purified the resulting DNA. This DNA molecule, which had one functional telomeric end and one non-functional 'broken' end, could not be maintained in yeast cells. The yeast telomere cloning experiment then simply involved joining the yeast DNA fragments and the purified vector DNA together using DNA ligase. Every now and then, this would result in a molecule with a *Tetrahymena* telomere at one end and a normal yeast telomere at the other end, and those rare molecules were expected to be able to replicate as linear molecules in yeast cells. I did recover some transformants with the expected linear structure[15], and I was able to confirm through a variety of tests that one end was indeed a yeast telomeric DNA fragment. This allowed us to start looking at the structures found in normal yeast telomeres, including the DNA sequences characteristic of yeast telomeres. We didn't expect the repeat sequences to be the same, since the *Tetrahymena* sequences didn't cross-hybridize with yeast DNA. Other hybridization experiments, done in collaboration with Tom Petes[16], showed that yeast telomeres contained stretches of alternating GT repeats. Still, when Janis Shampay, a graduate student in Liz's lab, sequenced the yeast telomeres I had cloned, we were all a bit surprised to see a somewhat irregular sequence, summarized as $G_{1-3}T$ repeats[17]. This was independently confirmed in the Tye and Petes laboratories based on the cloning of telomeric ends by hybridization with $(GT)_n$ probes[18]. While yeast did fit the general finding of a GT rich 3'-terminal strand, the absence of simple repeats was puzzling, and didn't seem to fit easily into the prevailing recombination-based models of telomere replication (19). It was the resolution of that puzzle that would eventually lead to us to telomerase.

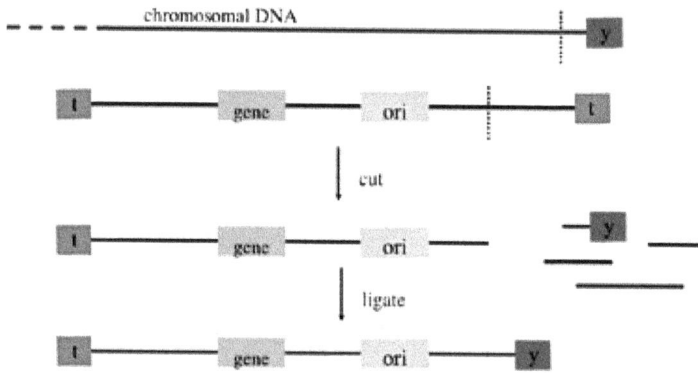

Figure 9. Cloning yeast telomeres. Yeast chromosomal DNA was digested with a restriction enzyme, as was the linear plasmid with two *Tetrahymena* telomeres. The purified vector fragment was ligated to the yeast DNA fragments, and the resulting mixture was used to transform yeast. A few linear plasmids were recovered, in which one end of the linear vector was replaced by a yeast telomere.

At this point, I would like to take a little digression to describe how we used these new telomeric DNA fragments as a tool to study the requirements for proper chromosome function in yeast. This work was done by Andrew Murray, my second graduate student. What we did was to take an engineering approach to seeing if we really understood the elements of chromosome structure. With telomeres in hand, we thought that we had all of the pieces that would be required to generate a fully functional chromosome. We had centromeric DNA, first cloned in John Carbon's lab[20]; we had various genes such as LEU2 and HIS3[13,14], and we had origins of replication, first cloned by Kevin Struhl and Dan Stinchcomb in Ron Davis's lab[8,9]. Those were all of the elements known at the time to be important in terms of chromosomal function. We thought that it would be interesting to put them all together and see if we could make something that behaved like a natural chromosome. To do this we constructed a circular plasmid that had all of the known chromosomal elements (Figure 10), linearized it so that it had two telomeric ends, and put it into yeast. Despite the fact that this DNA molecule had all the pieces (an origin of replication, a centromere, genes, and telomeres), when we put it into yeast it didn't behave at all like a proper chromosome. During mitosis it displayed a very high frequency of segregation errors, so that instead of being maintained over many cell cycles it was lost at a high frequency[21]. This was a very interesting result, because it said there was something going on that we didn't understand. What could be missing? What were the potential problems that prevented accurate inheritance of this mini-chromosome? We tested many possible explanations. Eventually, Andrew figured out that what was missing was just more DNA[21,22]. By simply adding enough non-yeast DNA from phage lambda to our small artificial chromosomes, he was able to make much bigger DNA molecules that now exhibited stable inheritance and behaved much more like natural yeast chromosomes (Figure 11). We considered various models for this, and based on the observation that the linear centromeric plasmid was much less mitotically stable than a similar circular centromeric plasmid, we proposed that the intertwining of DNA after the completion of DNA replication[23] played a role in holding sister chromatids together. This was long before the modern story of cohesin and separase[24] and the complex biochemistry that underlies the adherence and separation of sister-chromatids after replication. Our artificial chromosomes were also technically useful, at least for a little while, in the early days of genomic sequencing because it turns out that they are very nice vectors for cloning extremely large pieces of DNA, up to a megabase or two in length[25].

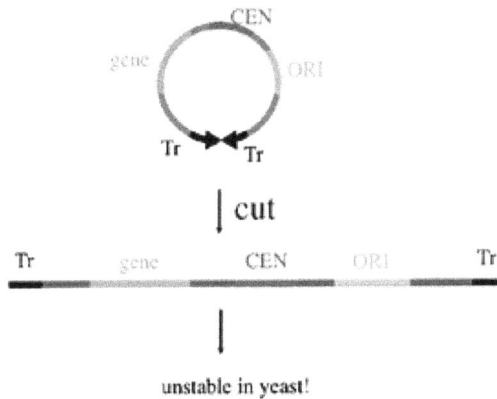

Figure 10. Our first attempt to make an artificial chromosome. We constructed a circular plasmid containing yeast genes, an origin of replication, a centromere, and telomeric DNA (Tr). This was linearized by cutting between the telomeric sequences, then introduced into yeast, where the DNA was maintained as a linear plasmid. Unexpectedly, this DNA molecule did not behave like a normal chromosome – it was mitotically unstable due to a high frequency of segregation errors.

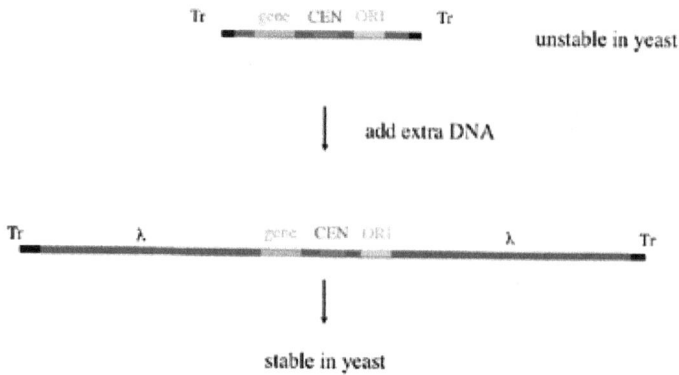

Figure 11. Successful construction of a yeast artificial chromosome. The addition of 50 to 150 kb of non-yeast DNA from phage λ greatly improved the mitotic stability of the DNA molecule, conferring improved chromosome-like behavior.

Returning once more to the story of telomeres and how they are fully replicated, all of our early models for thinking about this problem were based on recombination and the various kinds of reactions known to be engaged in by DNA ends. A very simple model that seemed quite attractive after Liz Blackburn's discovery of the short repetitive sequences of *Tetrahymena* telomeres was that recombination between different ends, perhaps biased in some way, could generate ends that were longer than either of the input DNA ends (Figure 12)[26]. Alternatively, strand-invasion by the 3' end of one telomere into the repeats of another telomere could lead to repair synthesis which would result in elongation of that end (Figure 12). Another model that we considered invoked Holliday junction resolution. This model was based on idea that the very end of telomeric DNA was actually a hairpin, i.e. the strand loops around at the end. That was attractive because it meant that there was no actual DNA end, and a hairpin could act as a relatively inert DNA terminus. Replication would generate an inverted repeat structure, which could isomerize into a central Holliday junction, resolution of which by the corresponding recombination enzyme would generate two new hairpin terminated telomeres (Figure 13). A more complex variant of this model that originated in Piet Borst's lab[27] was that internal nicks within the repeats were sites of unpairing followed by gap-filling synthesis, leading to synthesis of new repeat units. These were the kinds of recombination based models that we discussed in the early years of thinking about telomere replication. How did we finally let go of these models and come to the correct explanation? Remarkably, we were driven to the answer by analyzing the sequences of *Tetrahymena* telomeres after their replication in yeast.

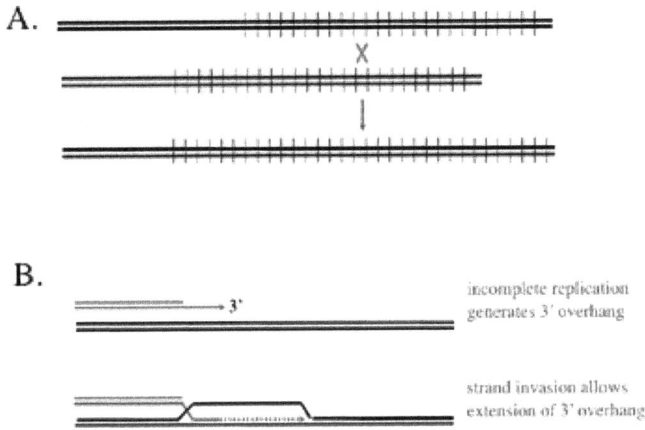

Figure 12. A. Telomere lengthening by recombination. B. Telomere lengthening by repair synthesis.

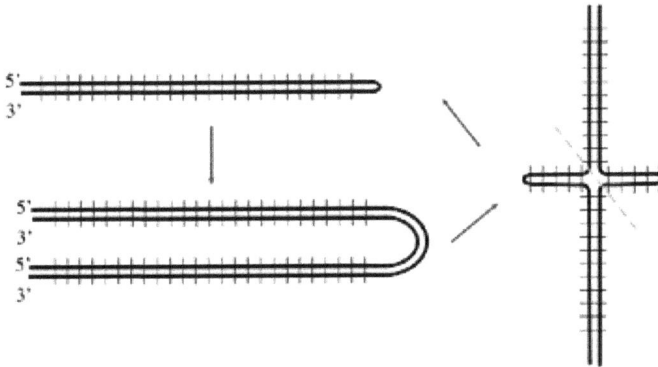

Figure 13. Telomere replication by Holliday junction resolution. Early models of telomeric DNA proposed a hairpin terminus. Replication would generate an inverted repeat, which could isomerize to form a Holliday junction, resolution of which would regenerate the original structures.

To understand why the replication of *Tetrahymena* telomeres in yeast was so important, consider again the linear plasmid with *Tetrahymena* ends. Those telomeric ends began as a restriction fragment of a certain size, but we noticed that after their maintenance in yeast that they had grown longer, by as much as a few hundred base-pairs, as well as becoming heterogeneous in size. We didn't know where this extra DNA had come from, but there were several possible explanations. It could have been, for example, a result of recombination between *Tetrahymena* ends on different molecules, or a result of strand-invasion and repair synthesis. Eventually, we cloned some of these lengthened *Tetrahymena* ends and, in a continuation of the collaboration with Liz, sent those DNA samples to Liz's lab where once again Janice Shampay did the actual sequencing. To our complete shock, we found that the actual structure consisted of G_4T_2 repeats from the *Tetrahymena* ends joined directly to the irregular $G_{1-3}T$ repeats that were characteristic of yeast telomeres (Figure 14)[17]. Thus the reason the DNA had become longer was that the yeast-specific sequence had become appended to the *Tetrahymena* ends. This new DNA seemed to have just dropped out of the sky. Such a different and irregular sequence couldn't possibly have been generated by any recombinational process, so we immediately knew that all of our early models were wrong. The new sequencing data led directly to the idea that there must be a specific new enzyme that adds extra DNA to chromosomal ends. Shortly after these results and our prediction of this new enzyme, of course, Carol Greider went on to identify the predicted enzyme activity biochemically[29]. Characterization of the purified enzyme, later named telomerase, showed that it is a ribonucleoprotein enzyme that contains an RNA template that specifies the telomeric repeat sequences, which are synthesized by a reverse transcriptase component of the enzyme[30]. We now know that the different telomeric repeats found in different organisms are specified by the RNA templates of their particular telomerase enzymes. A great deal of work has been done to characterize telomerase in many organisms, including *Tetrahymena*, yeast and humans, by Elizabeth Blackburn, Carol Greider, Tom Cech and many other people.

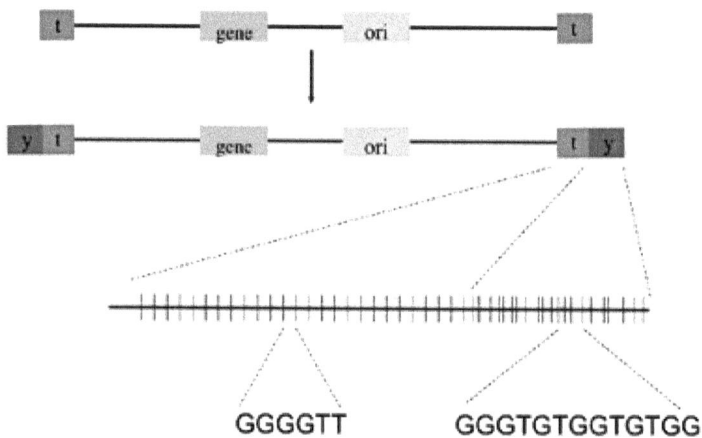

Figure 14. Yeast adds new DNA to *Tetrahymena* telomeres. Cloning and sequence analysis of *Tetrahymena* telomeres after replication in yeast (as the telomeres of a linear plasmid) revealed the addition of yeast telomeric sequences.

It is interesting to revisit the end replication problem in light of the activity of telomerase. As mentioned above, one of our early models was that the actual end was a hairpin structure. Of course that also turned out to be wrong, and the proper structure is a 3'-end overhang consisting of GT-rich repeats (Figure 15). This was originally worked out in a different ciliated protozoan, *Oxytricha*, in the lab of David Prescott[28], and then found to be a universally conserved aspect of telomere structure. If we consider the replication of DNA with a 3' overhang, the end-replication problem is actually a little bit different from that noted earlier by Watson[3] and by Olovnikov[4]. A replication fork heading towards this kind of end retains the previously noted problem of incomplete replication of the 3' end strand, but a much worse problem in that the leading strand can go to the end, but can't regenerate a 3' overhang. The 3' overhang will therefore be lost in every cycle of replication, unless there is a compensatory process. This, of course, is the role of the telomerase enzyme, which adds extra repeats to telomeric ends and thereby on average maintains the proper telomeric length and structure. The regulation of proper telomere length and structure has turned out to be quite elaborate, and the biochemistry of the corresponding protein-DNA interactions is remarkably complex and interesting [31].

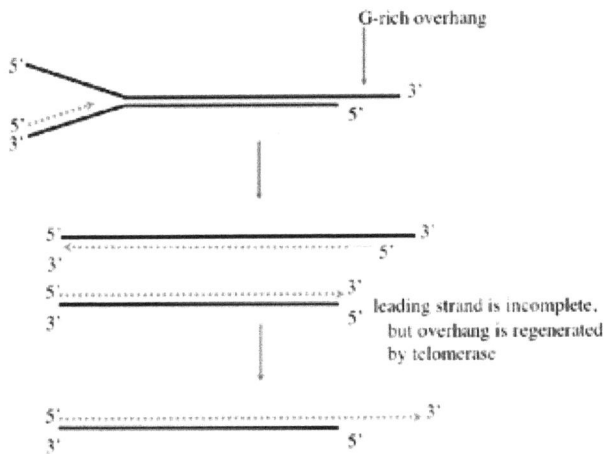

Figure 15. New model for telomere shortening, and the role of telomerase in telomere maintenance. When a replication fork reaches the end of a DNA duplex, the leading strand cannot regenerate the 3'-overhang. This is done by telomerase.

The activity of telomerase and its associated regulatory machinery in controlling telomere length turns out to have important biological consequences. Cells with high levels of telomerase activity can divide without limit, because they maintain functional telomeres. In contrast, cells with insufficient telomerase activity cannot maintain telomere length, and as a result have limited division potential. This prediction was initially verified by Vicki Lundblad, who came to my lab as a postdoc and decided to address this issue genetically in yeast[32]. What Vicki did was to set up a large and actually quite difficult screen for mutants that would be unable to maintain telomeres at their proper average length. She was able to recover mutants that had the property we were looking for, namely that telomeres would get shorter and shorter over an increasing number of cell divisions (Figure 16, part A). The first mutation with that property was named *est-1*, for '*ever shorter telomeres*'. The most interesting property of this mutation (and similar mutations recovered later) is that it confers a delayed senescence phenotype, just as predicted (Figure 16, part B). This phenotype is visually apparent in colonies of the mutant strain of yeast that have been grown for different numbers of generations. After 25 generations, the mutant colonies look just like wild-type colonies. After 46 generations the colonies are a little more irregular, and there are some small colonies; by 60 to 70 generations they are quite small and irregular, and after 80 to 90 generations the mutant strain can hardly grow at all. There are many dead cells in the small colonies, and there is a very high level of chromosome loss. Because the telomeres are getting shorter and shorter, eventually proper telomeric structure isn't maintained. As a consequence, ends are getting joined together leading to chromosome breakage and loss, so that cells are generated that are missing big chunks of their DNA. This was the first experimental demonstration that an inability to maintain normal telomere length would lead to a senescence phenotype, and therefore this inability to maintain telomeres might have an important role in problems of cellular senescence in higher organisms. At about the same time very similar experiments were done in Liz's lab, using *Tetrahymena*, and led to the same conclusion[33]. We thought this was a potential explanation for the senescence seen during repeated passage of primary cells in tissue culture, and by extension perhaps to problems of ageing related to a gradual decline in tissue renewal, perhaps due to limited cell division potential. The shortening of telomeres during passage of fibroblasts was soon demonstrated by Carol Greider[34], and the causal role of this shortening in cellular senescence was later proven[35]. Of course, this has turned out to be a very important aspect of our growing understanding of ageing and age-related diseases[36]. The complementary aspect of this has turned out to be very important for our understanding of cancer. In the vast majority of cancer cells, which have unlimited division potential, the telomerase gene has been up-regulated and functional telomeres are maintained indefinitely[36,37].

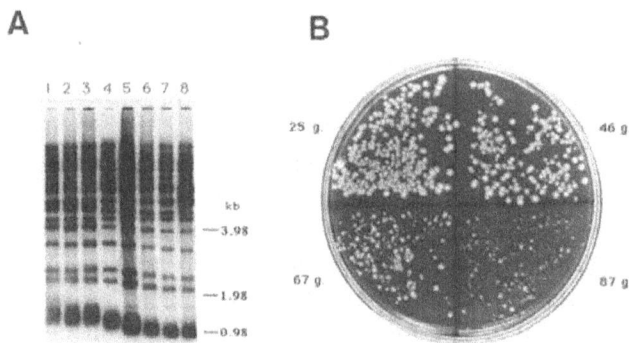

Figure 16. Senescence of yeast EST-1 cells. A: Telomeric yeast DNA fragments from an EST-1 mutant strain are visualized by Southern blotting. Lanes 1 through 8 represent increasing numbers of generations of growth. B: A mutant EST-1 strain streaked out on an agar plate after 25, 46, 67 and 87 generations of prior growth.

At that point in my career it became clear that many people would soon be exploring the roles of telomeres and telomerase in cancer and aging. I felt that the main questions were clear, and that they would be addressed whether or not I remained active in the field of telomere biology. I therefore began to look for other interesting questions that could be addressed experimentally, but where there were not too many people trying to look at the same issues.

Even as Vicki was doing her genetic work on telomere maintenance in yeast, I was already becoming interested in ribozymes, because Tom Cech's discovery of the self-splicing introns was very new and exciting[12]. I thought there were many interesting questions, and I was surprised that more people weren't entering that field. In particular, I was attracted by the RNA world hypothesis[38] and the idea that RNA might be able to catalyze its own replication without protein enzymes. Since the experiments were largely molecular biology in nature, I thought that we might be able to make some contributions to that nascent field. For several years we studied the group I introns and tried to use various molecular techniques to force them to catalyze RNA replication reactions. Several of my students including Jennifer Doudna[39] and Rachel Green[40] worked on that problem, with some success. But eventually we came to the conclusion that the ribozymes available from nature were not good enough. Those ribozymes were doing jobs that they had evolved to do in modern organisms, and what we were primarily interested in were questions about what RNA could have done much earlier.

In the late 1980s we started to think about ways of evolving new RNA molecules that would do things that we were interested in. The basic idea was simple: prepare huge collections of random sequences, and then isolate the rare functional molecules that did what we wanted. The technology for doing this *in vitro* selection, or directed evolution, was worked out by Andy Ellington when he was a postdoc in my lab[41], and independently by Craig Tuerk in Larry Gold's lab[42]. We spent most of the 90s applying this kind of selection technology to the laboratory evolution of RNA and DNA molecules that could do all kinds of interesting things. For example, an RNA molecule isolated by Mandana Sassanfar when she was a postdoc in the lab folds up into a three-dimensional shape that contains a binding site for ATP (Figure 17)[43]. Subsequently, we and others were able to show that it is possible to evolve, in the laboratory, RNA and DNA sequences that will fold into defined shapes that can bind almost any target molecule of interest. Ongoing studies in several different labs and companies are aimed at exploring potential therapeutic uses of these target binding RNA molecules, known as aptamers, perhaps doing some of the things that we use antibodies to do today.

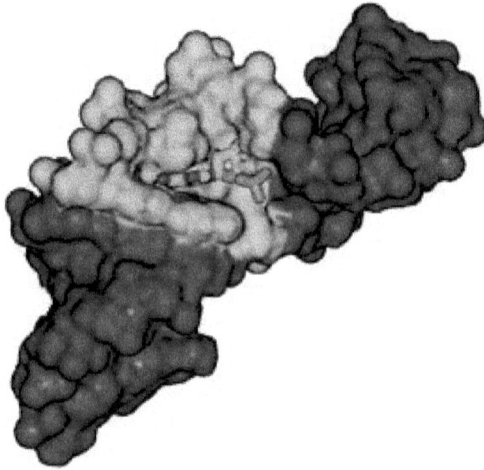

Figure 17. An ATP binding RNA molecule. This RNA was evolved from an initially random population of sequences. Dark blue: double-helical regions; light blue: folded recognition loop; stick figure presents bound AMP.

Once we were able to evolve aptamers routinely we turned our attention to evolving RNA molecules that could catalyze interesting reactions. Dave Bartel, when he was a graduate student in the lab, isolated a surprisingly intricate RNA molecule that catalyzes a joining reaction between two adjacent RNAs aligned on a template (Figure 18)[44]. It uses the same chemistry that RNA and DNA polymerases use, i.e. the 3'-prime hydroxyl of one RNA substrate attacks the α-phosphate of the triphosphate of the other RNA substrate, generating a new phosphodiester bond. The ribozyme has an intricate folded secondary[45] and three-dimensional structure[46]. This was a very exciting demonstration that RNA could catalyze the chemistry of RNA replication. Subsequently, in his own lab at the Whitehead Institute at MIT, Dave Bartel further evolved this ribozyme into an actual RNA polymerase that can copy RNA templates using nucleoside triphosphates as substrates[47]. This is a marvelous 'proof-of-principle' of the plausibility of the RNA world hypothesis. Unfortunately the current versions of this RNA polymerase are not yet good enough to copy themselves and exhibit full cycles of replication, so there is plenty of scope for additional evolutionary optimization.

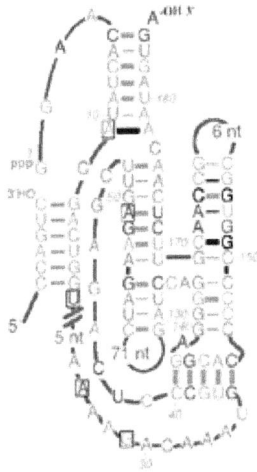

Figure 18. Secondary structure of the class I ribozyme ligase. This ribozyme catalyzes template-directed RNA-RNA ligation. It was evolved from an initially random population of RNA sequences.

More recently we have applied RNA *in vitro* selection to the analysis of human genomic sequences, in work done by Kouresh Salahi-Ashtiani and Andrej Luptak when they were postdocs in my lab (Figure 19)[48]. Kourosh began this project by generating a large library of pieces of human DNA. He then transcribed them into RNA, and selected for molecules that could cut themselves at a unique site. He recovered four distinct self-cleaving RNAs or ribozymes. One of these is found in the CPEB3 gene, which has been implicated in memory[49], possibly through a role in controlling localized protein translation at synapses. There are two interesting things about this self-cleaving human genomic ribozyme. One is that it turns out to have exactly the same structure as a well known viral ribozyme, the HDV ribozyme of the hepatitis delta virus. The fact that there is a version of this ribozyme in the human genome suggests that the viral ribozyme may be derived from the genomic copy. Another potentially very interesting observation is that there is a polymorphism in the human population at a position within this ribozyme that affects its activity. A recent genetic study done by a group in Switzerland[50] has found an association between this polymorphism and performance on a word-recall memory test. A lot more work needs to be done on this, but the possibility that a self-cleaving catalytic RNA may play a role in human memory is fascinating.

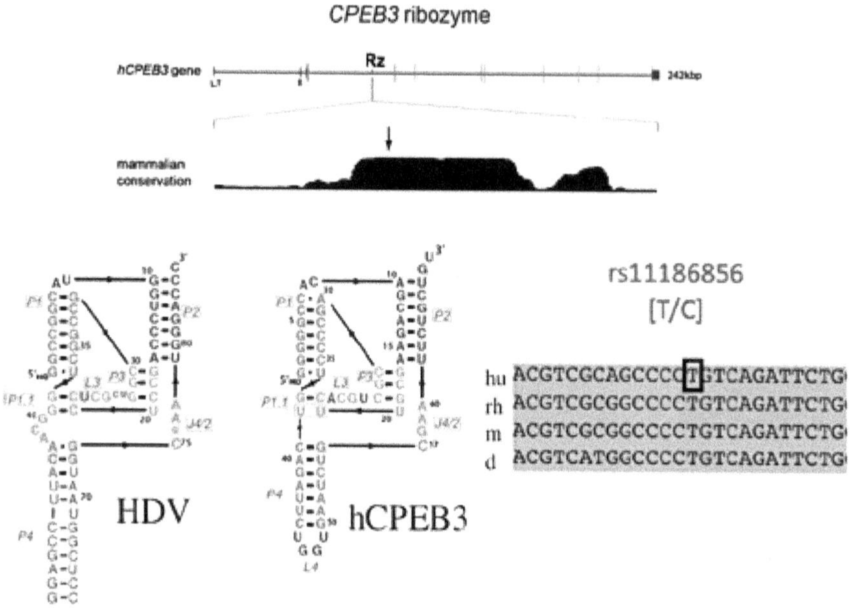

Figure 19. An HDV ribozyme in the human genome. Top: The self-cleaving ribozyme is located within the second intron of the CPEB3 gene. The ribozyme sequence is highly conserved relative to flanking intron sequences. Bottom left: The secondary structures of the human genomic ribozyme and the HDV ribozyme are virtually identical. Bottom right: A polymorphism with the ribozyme sequence affects ribozyme activity, and may affect human memory.

In the 1990s we extended our work on RNA and DNA directed evolution by developing methods for evolving proteins. Rich Roberts developed a clever means of tricking the ribosome into chemically linking a nascent peptide or protein chain to its own mRNA[51], so that selection for a functional protein would also enrich the corresponding coding mRNA. This approach was used in later work done by Tony Keefe, who isolated a small ATP-binding protein from a library of completely random protein sequences[52]. This little protein domain looks indistinguishable from any natural biological protein domain. These kinds of laboratory evolution experiments showed that it is relatively easy to evolve functional RNAs, DNAs, and even proteins out of completely random collections of sequences.

The above experiments showed very directly that Darwinian evolution, applied to populations of molecules, is a powerful means of generating functional sequences. That led us to deeper questions: how did evolution get started? How did the transition from chemistry to Darwinian evolution first happen on the early earth? These are the central questions concerning the origin of life, and addressing these questions has become the main focus of my laboratory. The approach that we are taking is essentially a synthetic or engineering approach. We have a simple model for what we think early cells might have looked like (Figure 20)[53]. This is not by any means a universally accepted model, but it is our view of what a very primitive cell might have looked like, and we are trying to construct such systems in order to define possible pathways from chemistry to biology. We think that a primitive cell would have two critical components, the first of which is a cell membrane. In our experiments we make these membranes out of simple molecules that might have been around on the early earth, such as fatty acids. The cell membrane has to be able to grow spontaneously and divide to make daughter cells. The other important component of a primitive cell would be a polymer that could mediate the inheritance of genetic information. Here the big question is whether this could be RNA itself, or is it more likely to be some simpler progenitor material that was subsequently replaced by RNA? In either case, this material has to be able to replicate spontaneously without any of the highly sophisticated evolved machinery that is used by modern biology. The key question is therefore: how could both cell membranes and early genetic materials replicate prior to the evolution of complex biological machinery? The approach that we are taking is to try to divide this big problem up into simpler pieces that can be addressed separately. I will briefly describe a few of the experiments that we have done in the last six or seven years.

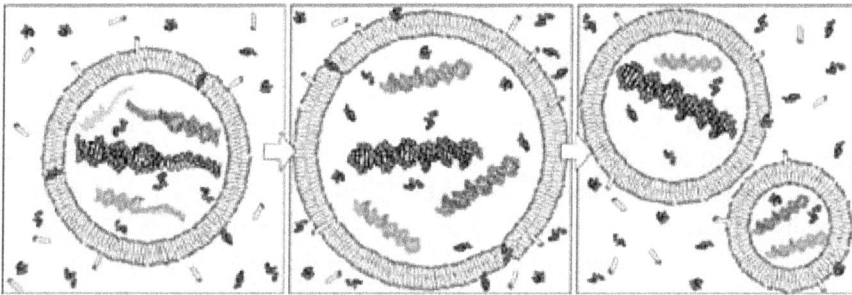

Figure 20. Schematic model of a protocell. A simple cell might be based on a replicating vesicle for compartmentalization, and a replicating genome to encode heritable information. A complex environment provides nucleotides, lipids and various sources of energy. Mechanical energy (for division), chemical energy (for nucleotide activation), phase transfer and osmotic gradient energy (for growth) may be used by the system.

About six years ago Marty Hancyzc, a postdoc, and Shelly Fujikawa, a graduate student in the lab, became interested in how protocell-like assemblies could be formed. They found that a common clay mineral, formed from volcanic ash and seawater, can facilitate this assembly process in a surprising way[54]. This clay mineral is well known in the prebiotic chemistry community because it had been shown several years previously by Jim Ferris and Leslie Orgel to catalyze the assembly of RNA from activated nucleotides[55]. Marty and Shelly showed that the same mineral could catalyze the assembly of membranes. Moreover, it can bring genetic polymers, such as RNA, into the vesicles it helps to assemble (Figure 21). Thus a common mineral can help to make genetic materials, help to assemble membranes, and bring them together[54], all of which is very attractive in terms of the assembly of early cellular structures.

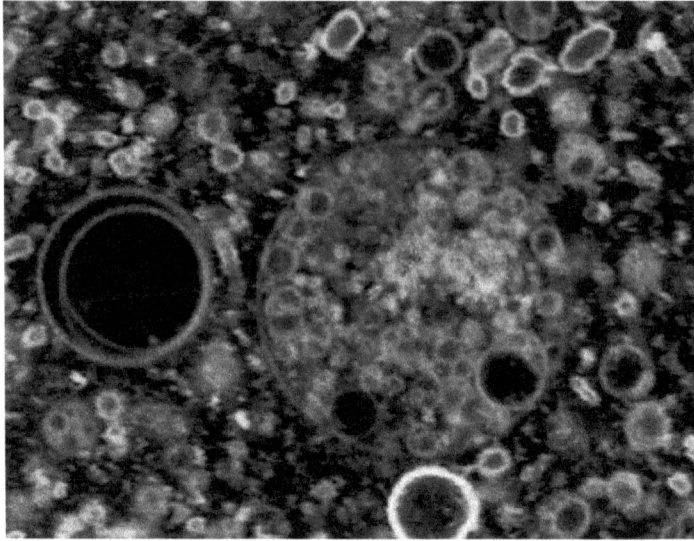

Figure 21. Montmorillonite can bring RNA into vesicles. Fluorescently labeled RNA (orange) on the surface of a clay particle is trapped inside a large vesicle (green) along with numerous small vesicles, all assembled as a result of the catalytic activity of the clay particle.

The replication of protocell-like structures is much more difficult than their assembly. However, the growth and division of the protocell membrane, which looked like an almost impossible problem just a few years ago, has actually turned out to be relatively simple. Our current model for what an early cell cycle might have looked like with respect to the cell membrane is based on the work of Ting Zhu, a graduate student in the lab[56]. We prepare large multilamellar vesicles, and feed them with new fatty acids. Remarkably, they grow into long filaments, which are quite fragile; in response to gentle agitation, such as might result from waves on a pond, they break up into daughter cells (Figure 22). That generates a robust cycle that can be carried out indefinitely. Thus, the spontaneous growth and division of membrane compartments appears to be a relatively straightforward process.

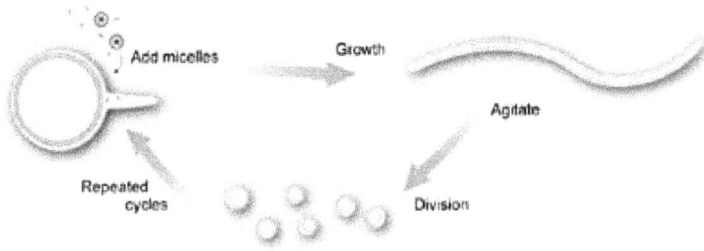

Figure 22. Cycles of growth and division of a model protocell membrane. Large multilamellar vesicles grow into long hollow vesicles following the addition of excess fatty acids. The filamentous vesicles are fragile and fragment in response to mild shear stress. The smaller daughter vesicles can grow and repeat the cycle.

What about the replication of genetic information? At the moment, this still seems to be difficult, because we don't understand how to accomplish this step. The RNA world hypothesis is based on the idea of RNA catalyzing its own replication[38], but that has turned out to be a harder problem than we thought. Could genetic replication have begun as a chemical, i.e. non-enzymatic, process? Almost twenty years ago, Leslie Orgel, one of the giants of prebiotic chemistry, proposed that chemical means of replicating genetic polymers should be found fairly easily by chemists, and that the solution to that problem would be relevant to the origin of life[57]. That hasn't happened, perhaps because it's a harder problem than anybody thought, but also perhaps because there are not that many people working on this problem. I think that makes it a perfect problem to tackle because it is important, interesting and there are many reasonable experimental approaches. What we are doing is making synthetic nucleotides that are modified so as to become more reactive (Figure 23). For example, changing the hydroxyl nucleophile to an amine results in nucleotides that spontaneously extend a primer in a template-directed manner, without any enzyme[58]. We do not yet have a robust and general replication system, but that is our goal.

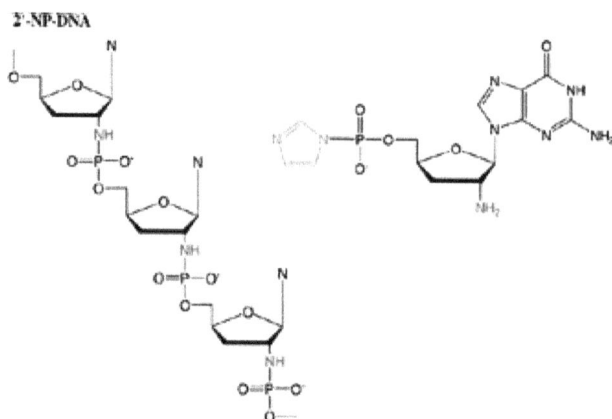

Figure 23. Typical monomer for spontaneous nucleic acid synthesis and the corresponding polymer. Left: 2'-5' linked phosphoramidate DNA. Right: The activated 2'-amino monomer. Note the 2'-amino nucleophile (blue) and the imidazole leaving group (green) on the 5'-phosphate. The combination of a good nucleophile with a good leaving group allows for rapid non-enzymatic polymerization of this class of monomer when aligned on an appropriate template.

There is an interesting aspect of the problem of chemical replication that we have just recently started to think about, namely, how can the very ends of our sequences be copied in the absence of telomerase? This turns out to be very interesting. Chemical replication results in spontaneous template-directed primer-extension, but once the end of the template is reached, the reaction slows down but often doesn't stop entirely. Depending on the conditions, we sometimes see chemical extension beyond the end of the template, generating a 3' overhang (Figure 24)[58]. Thus complete replication of a template does not seem to be a problem, and in fact this process generates new sequences. It is interesting to speculate that this spontaneous chemical reaction might have something to do with eventual emergence of genetically encoded catalysts that would control and exploit this process, eventually leading to the telomerase enzyme that has been the main subject of my lecture.

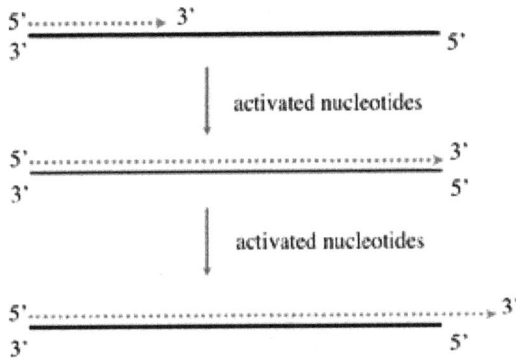

Figure 24. Origin of telomerase in spontaneous copying chemistry? Under certain conditions non-enzymatic primer-extension proceeds past the end of the template, generating a 3' overhang. Enzymatic control and elaboration of this chemical process could provide an evolutionary path towards telomerase.

ACKNOWLEDGEMENTS

I would like to thank all of the many brilliant students, postdocs, friends, and collaborators who contributed to this work.

REFERENCES

1. Muller, H.J., "The remaking of chromosomes," *Collecting Net* 13, 181–198 (1938).
2. McClintock, B., "Cytological observations of deficiencies involving known genes, translocations and an inversion in Zea mays," *Missouri Agr. Exp. Sta. Res. Bull.* 163, 4 (1931).
3. Watson, J.D., "Origin of concatameric T7 DNA," *Nature New Biol.* 239, 197–201 (1972).
4. Olovnikov, A.M., "A theory of marginotomy," J. Theor. Biol. 41, 181–190 (1973).
5. Hinnen, A., Hicks, J.B. and Fink, G.R., "Transformation of yeast," *Proc. Natl. Acad. Sci. USA* 75, 1929–1933 (1978).
6. Orr-Weaver, T.L., Szostak, J.W. and Rothstein, R.J., "Yeast transformation: A model system for the study of recombination," *Proc. Natl. Acad. Sci. USA* 78, 6354–6358 (1981).
7. Orr-Weaver, T.L. and Szostak, J.W., "Yeast recombination: The association between double-strand-gap repair and crossing-over," *Proc. Natl. Acad. Sci. USA* 80, 4417–4421 (1983).
8. Struhl, K., Stinchcomb, D.T., Scherer, S. and Davis, R.W., "High-frequency transformation of yeast: autonomous replication of hybrid DNA molecules," *Proc. Natl. Acad. Sci. USA* 76, 1035–1039 (1979).
9. Stinchcomb, D.T., Struhl, K. and Davis, R.W., "Isolation and characterisation of a yeast chromosomal replicator," *Nature* 282, 39–43 (1979).
10. Szostak, J.W., Orr-Weaver, T.L., Rothstein, R.J. and Stahl, F., "The double-strand-break repair model for recombination," *Cell* 33, 25–35 (1983).
11. Blackburn, E.H. and Gall, J.G., "A tandemly repeated sequence at the termini of the extrachromosomal ribosomal RNA renes in Tetrahymena," *J. Mol. Biol.* 120, 33–53 (1978).
12. Kruger, K., Grabowski, P.J., Zaug, A.J., Sands, J., Gottschling, D.E. and Cech, T.R., "Self-splicing RNA: autoexcision and autocyclization of the ribosomal RNA intervening sequence of Tetrahymena," *Cell* 31, 147–157 (1982).

13. Ratzkin, B. and Carbon, J., "Functional expression of cloned yeast DNA in Escherichia coli," *Proc. Natl. Acad. Sci. USA* 74, 487–491 (1977).

14. Struhl, K. and Davis, R.W., "Production of a functional eukaryotic enzyme in Escherichia coli: cloning and expression of the yeast structural gene for imidazole-glycerolphosphate dehydratase (his3)," *Proc. Natl. Acad. Sci. USA* 74, 5255–5259 (1977).

15. Szostak, J.W. and Blackburn, E.H., "Cloning yeast telomeres on linear plasmid vectors," *Cell* 29, 245–255 (1982).

16. Walmsely, R., Petes, T.D. and Szostak, J.W., "Is there left-handed DNA at the ends of yeast chromosomes?" *Nature* 302, 84–86 (1983).

17. Shampay, J., Szostak, J.W. and Blackburn, E.H., "DNA sequences of telomeres maintained in yeast," *Nature* 310, 154–157 (1984).

18. Walmsley, R.W., Chan, C.S.M., Tye, B.-K. and Petes, T.D., "Unusual DNA sequences associated with the ends of yeast chromosomes," *Nature* 310, 157–160 (1984).

19. Szostak, J.W., "Replication and Resolution of Telomeres in Yeast," Cold Spring Harbor Symp. Quant. Biol. 47, 1187–1194 (1983).

20. Clarke, L. and Carbon, J., "Isolation of a yeast centromere and construction of functional small circular chromosomes," *Nature* 287, 504–509 (1980).

21. Murray, A.W. and Szostak, J.W., "Construction of artificial chromosomes in yeast," *Nature* 305, 189–193 (1983).

22. Murray, A.W., Schultes, N.P. and Szostak, J.W., "Chromosome length controls mitotic chromosome segregation in yeast," *Cell* 45, 529–536 (1986).

23. Sundin, O. and Varshavsky, A., "Arrest of segregation leads to accumulation of highly intertwined catenated dimers: dissection of the final stages of SV40 DNA replication," *Cell* 25, 659–669 (1981).

24. Nasmyth, K., "Segregating sister genomes: the molecular biology of chromosome separation," *Science* 297, 559–565 (2002).

25. Burke, D.T., Carle, G.F. and Olson, M.V., "Cloning of large segments of exogenous DNA into yeast by means of artificial chromosome vectors," *Science* 236, 806–812 (1987).

26. Blackburn, E.H. and Szostak, J.W., "The molecular structure of centromeres and telomeres," *Ann. Rev. Biochem.* 53, 163–194 (1984).

27. Bernards, A., Michels, P.A., Lincke, C.R. and Borst, P., "Growth of chromosome ends in multiplying trypanosomes," *Nature* 303, 592–597 (1983).

28. Klobutcher, L.A., Swanton, M.T., Donini, P. and Prescott, D.M., "All gene-sized DNA molecules in four species of hypotrichs have the same terminal sequence and an unusual 3' terminus," *Proc. Natl. Acad. Sci. USA* 78, 3015–3019 (1981).

29. Greider, C.W. and Blackburn, E.H., "Identification of a specific telomere terminal transferase activity in Tetrahymena extracts," *Cell 43*, 405–413 (1985).

30. Greider, C.W. and Blackburn, E.H., "The telomere terminal transferase of Tetrahymena is a ribonucleoprotein enzyme with two kinds of primer specificity," *Cell 51*, 887–898 (1987).

31. de Lange, T., "How telomeres solve the end-protection problem," *Science* 326, 948–952 (2009).

32. Lundblad, V. and Szostak, J.W., "A mutant with a defect in telomere elongation leads to senescence in yeast," *Cell 57*, 633–643 (1989).

33. Yu, G.L., Bradley, J.D., Attardi, L.D. and Blackburn E.H., "*In vivo* alteration of telomere sequences and senescence caused by mutated *Tetrahymena* telomerase RNAs," *Nature* 344, 126–132 (1990).

34. Harley, C.B., Futcher, A.B. and Greider, C.W., "Telomeres shorten during ageing of human fibroblasts," *Nature* 345, 458–460 (1990).

35. Bodnar, A.G., Ouellette, M., Frolkis, M., Holt, S.E., Chiu, C.P., Morin, G.B., Harley, C.B., Shay, J.W., Lichtsteiner, S. and Wright, W.E., "Extension of life-span by introduction of telomerase into normal human cells," Science 279, 349–352 (1998).

36. Blackburn E.H., Greider C.W. and Szostak J.W., "Telomeres and telomerase: the path from maize, *Tetrahymena* and yeast to human cancer and aging," *Nat. Med. 12,* 1133–1138 (2006).

37. Greider, C.W., "Telomerase activation. One step on the road to cancer?" *Trends Genet.* 15, 109–112 (1999).

38. Gilbert, W., "The RNA World," *Nature* 319, 618 (1986).

39. Doudna, J.A. and Szostak, J.W., "RNA catalyzed synthesis of complementary strand RNA," *Nature* 339, 519–522 (1989).

40. Green, R. and Szostak, J.W., "Selection of a ribozyme that functions as a superior template in a self-copying reaction," *Science* 258, 1910–1915 (1992).

41. Ellington, A.E. and Szostak J.W., "In vitro selection of RNA molecules that bind specific ligands," *Nature* 346, 818–822 (1990).

42. Tuerk, C. and Gold, L., "Systematic evolution of ligands by exponential enrichment: RNA ligands to bacteriophage T4 DNA polymerase," *Science* 249, 505–510 (1990).

43. Sassanfar, M. and Szostak, J.W., "An RNA motif that binds ATP," *Nature* 364, 550–553 (1993).

44. Bartel, D.P. and Szostak, J.W., "Isolation of new ribozymes from a large pool of random sequences," *Science* 261, 1411–1418 (1993).

45. Ekland, E.H., Szostak, J.W. and Bartel, D.P., "Structurally complex and highly active RNA ligases derived from random RNA sequences," *Science* 269, 364–370 (1995).

46. Shechner, D.M., Grant, R.A., Bagby, S.C., Koldobskaya, Y., Piccirilli, J.A. and Bartel D.P., "Crystal structure of the catalytic core of an RNA-polymerase ribozyme," *Science* 326, 1271–1275 (2009).

47. Ekland, E.H. and Bartel, D.P., "RNA-catalysed RNA polymerization using nucleoside triphosphates," *Nature* 383, 192–198 (1996).

48. Salehi-Ashtiani, K., Luptak, A., Litovchick, S. and Szostak, J.W., "A genome wide search for ribozymes reveals an HDV-like sequence in the human CPEB3 gene," *Science* 313, 1788–1792 (2006).

49. Theis, M., Si, K. and Kandel, E.R., "Two previously undescribed members of the mouse CPEB family of genes and their inducible expression in the principal cell layers of the hippocampus," *Proc. Natl. Acad. Sci. USA* 100, 9602–9607 (2003).

50. Vogler, C., Spalek, K., Aerni, A., Demougin, P., Müller, A., Huynh, K.-D., Papassotiropoulos, A. and de Quervain, D.J.-F., "CPEB3 is associated with human episodic memory," *Front. Behav. Neurosci.* 3, 1–5 (2009).

51. Roberts, R.W. and Szostak, J.W., "RNA-peptide fusions for the in vitro selection of peptides and proteins," *Proc. Natl. Acad. Sci. USA* 94, 12297–12302 (1997).

52. Keefe, A.D. and Szostak, J.W., "Functional proteins from a random sequence library," *Nature* 410, 715–718 (2001).

53. Mansy, S.S., Schrum, J.P., Krishnamurthy, M., Tobé, S, Treco, D. and Szostak, J.W., "Template-directed synthesis of a genetic polymer in a model protocell," *Nature* 454, 122–125 (2008).

54. Hanczyc, M.M, Fujikawa, S.M. and Szostak, J.W., "Experimental models of primitive cellular compartments: Encapsulation, growth and division," *Science* 2003; 302, 618–622 (2003).

55. Ferris, J.P., Hill, A.R., Jr, Liu, R. and Orgel, L.E., "Synthesis of long prebiotic oligomers on mineral surfaces," *Nature* 381, 59–61 (1996).

56. Zhu, T.F. and Szostak, J.W., "Coupled growth and division of model protocell membranes," *J. Am. Chem. Soc.* 131, 5705–5713 (2009).

57. Orgel, L.E., "Molecular replication," *Nature* 358, 203–209 (1992).

58. Schrum, J., Ricardo, A., Krishnamurthy, K., Blain, J.C. and Szostak, J.W., "Efficient and rapid template-directed nucleic acid copying using 2′-amino-2′, 3′-dideoxyribonucleoside-5′-phosphorimidazolide monomers," *J. Am. Chem. Soc.* 31, 14560–14570 (2009).

Portrait photo of Professor Szostak by photographer Ulla Montan.

Physiology or Medicine 2010

Robert G. Edwards

"for the development of in vitro fertilization"

THE NOBEL PRIZE IN PHYSIOLOGY OR MEDICINE

Speech by Professor Christer Höög of the Nobel Assembly at Karolinska Institutet. Translation of the Swedish text.

Your Majesties, Your Royal Highnesses, Ladies and Gentlemen,

The 2010 Nobel Prize in Physiology or Medicine rewards one of the great medical advances of our age, in vitro fertilisation – also called test tube fertilisation. Like a miracle, this method has enabled many involuntarily childless couples to have babies.

Behind this scientific breakthrough was British researcher Robert Edwards. Starting with his basic scientific research on reproductive biology, he saw the potential for treating infertility, a medical condition that afflicts more than 10 per cent of humanity.

Infertility is often caused by the failure of sperm and eggs to meet in the natural way. In the early 1960s, Edwards decided to try to develop a method for fertilising human eggs outside the body and only then returning them to the woman. This visionary research project ran into resistance from the establishment, since it raised questions about the beginnings of life and the natural limitations of humans. At an early stage, Edwards thus initiated an ethical debate about in vitro fertilisation – a debate that helped this method to become generally accepted over time.

Many scientific issues needed to be resolved in order for in vitro fertilisation to become a reality – issues that Edwards addressed systematically. During the 1960s he clarified the maturation process of human eggs outside the body and in what way various hormones affect this process. In 1969, Edwards and his colleagues showed that human eggs could be fertilised outside the body, a revolutionary discovery.

Working with gynaecological experts, Edwards then began the research work that was required in order to transform basic scientific insights into clinical treatment. They succeeded in showing that a human egg fertilised outside the body could generate an early embryo. But the most important question remained: Could an egg fertilised outside the body lead to a pregnancy and the birth of a baby?

After many years of experiments, this question was finally answered. On July 25, 1978, Louise Joy Brown was born – the first baby conceived through in vitro fertilisation. Today in vitro fertilisation is an established method for treating infertility, and with the help of this method more than four million babies have been born. These babies are as healthy as those born in the natural way, and many of them have now reached adulthood and have had babies of their own.

By charting a new territory in medicine, Robert Edwards has given new hope to millions of people who were involuntarily childless. With his scientific vision and his personal courage, he has showed all of us an example of how a medical therapy, though highly controversial at first, can become established over time.

Dear Robert Edwards,

The development of human in vitro fertilisation has made it possible to treat infertility, a medical condition that afflicts a large proportion of humanity. Your pioneering work therefore represents a monumental achievement that truly can be said to confer the greatest benefit to mankind. The result of your work has touched us all, giving millions of infertile couples a precious gift, a child.

On behalf of the Nobel Assembly at Karolinska Institutet, it is my great privilege to convey to you our warmest congratulations and our deepest admiration.

In the absence of this year's Nobel Laureate in Physiology or Medicine, I ask Professor Edwards' wife and long-term scientific companion, Dr Ruth Fowler Edwards, to come forward and receive his Prize from the hands of His Majesty the King.

ROBERT EDWARDS: NOBEL LAUREATE IN PHYSIOLOGY OR MEDICINE

Nobel Lecture/Nobel Prize Symposium in Honour of
Robert G. Edwards, December 7, 2010

by

MARTIN H. JOHNSON*

Cambridge University, Cambridge, UK.

Today we are here to celebrate the achievements of Robert Edwards. It is however a celebration tinged with sadness. Sadness that Bob himself is not well enough to be here in person, so in preparing this lecture, I have tried to weave into it some of Bob's written and spoken words. Sadness also that neither Patrick Steptoe (1913–1988; Fig. 1) [1] nor Jean Purdy (1946–1985; Fig. 2) [2], two of his key collaborators, are alive to celebrate with him.

TO BEGIN AT THE BEGINNING

Robert Geoffrey Edwards was born on the 27th of September 1925 in the small Yorkshire mill town of Batley. He arrived into a working-class family, the second of three brothers – an older brother, Sammy and a younger, Harry. These brothers Bob describes as competitive, "all determined to win or, if not to win, to go down fighting" [3]. Bob's mother, Margaret, was a machinist in a local mill. She came originally from Manchester, to where the family relocated when Bob was about 5, and where he was educated. In those days, bright working class kids could take a scholarship exam at age 10 or 11 in competition for the few coveted places at a grammar school: the potential pathway out of poverty and even to University. All three brothers passed the exam, but Sammy decided against Grammar School, preferring to leave education as soon as he could to start earning. His mother was furious at this wasted opportunity, and so when her two younger sons passed the exam, there was no question but that they would continue in education. So it was that Bob progressed in 1937 to Manchester Central Boy's High School, which, incidentally, also claims Sir James Chadwick, FRS (1891–1974), another Cambridge professor and Nobel Laureate (in Physics in 1935 for discovery of the neutron [4]), as a former pupil. Bob's summers were spent in the Yorkshire Dales, where their mother took her sons to be closer to their father's place of work. There Bob laboured on the farms and developed an enduring love for the place.

---

* At the Nobel Prize Symposium in Honour of Robert G. Edwards, Martin H. Johnson delivered this lecture.

Figure 1. Patrick Steptoe (1913–1988) (courtesy Andrew Steptoe).

Figure 2. Jean Purdy (1946–1985) (courtesy Barbara Rankin).

These early experiences were formative for Bob. He became a life–long egalitarian, for five years a Labour Party councillor [5], willing to listen to and to talk with all and sundry, regardless of class, education, status and background. Second, he developed an enduring love for and curiosity about natural history and especially the reproductive patterns that he observed among the farm's sheep, pigs and cattle in the Dales. Finally, he took great pride in being a 'Yorkshire man' – with traditional attributes of affability and generosity of spirit combined with no-nonsense blunt-speaking. Indeed, following his only meeting with Gregory Pincus (1903–1967) [6] at a conference in Venice in May 1966, at which Bob, the young pretender, clashed with the 'father of the pill' over the timing of egg maturation in humans, Bob paid Pincus the biggest compliment he could imagine, saying "He would have made a fine Yorkshireman!" [7].

Figure 3. Bob on National Service,1940s (courtesy Ruth Edwards).

The aftermath of war was to provide an extended interruption to Bob's education: when he left school in 1943, he was conscripted into the British Army for almost four years. To his surprise as someone who was from a working class family, he was identified as potential officer material and sent on an officer-training course, before being commissioned in 1946 (Fig. 3). However, the alien life-style of the officers' mess was not to his taste and served to reinforce his socialist ideals. The years in the army were broken by 9 months compassionate leave back in the Dales, to which he was released to help out when his farmer friend there became ill. So engaged did he become in farming life that, after discharge from the army in 1948, he returned home to Manchester, from where he applied to read agricultural sciences at the University College of North Wales at Bangor.

Figure 4. John Slee, 1963 (courtesy Ruth Edwards).

Having gained a place and a grant to fund it, the 6 or so months that intervened were occupied in a Government desk job in Salford, Greater Manchester, work experience that reinforced the attractions of agricultural science. So his disappointment in the course offered at Bangor was acute. By that time he was an experienced 23 year old, described by his impressionable 18 year old public-school educated and self-described "unlikely" friend John Slee (Fig. 4), as being "both ambitious and flexible, and unusually confident in his own judgement" [8]. And in Bob's confident judgement, the course on offer was not 'scientific', and he was bored through two tedious years of agricultural descriptions, after which he reported that his teachers were "glad to see the back of him" in Zoology for a year. The Zoology Department offered a course much more to his style and led by the more intellectually challenging Rogers Brambell, FRS (1901–1970) [9]. However, that year was not enough to salvage an honours degree, and in 1951, aged 26 he gained a simple pass. Unbeknown to him at the time, he was not alone in this undistinguished academic embarrassment, as neither "Tibby" Marshall, FRS (1878–1949), the founder of the Reproductive Sciences [10], nor Sir Alan Parkes, FRS (1900–1990), the first Professor of Reproductive Sciences at Cambridge [11], and who was later to recruit Bob there, distinguished themselves as undergraduates. In 1951, however, Bob "was disconsolate. It was a disaster. My grants were spent and I was in debt. Unlike some of the students I had no rich parents… I could not write home, 'Dear Dad, please send me £100 as I did badly in the exams.'" [12].

Characteristically, however, Bob's low spirits did not last long. He learned that John Slee had been accepted on a Diploma course in Animal Genetics at Edinburgh University under Conrad Waddington, FRS (1905–1975) [13]. Bob applied, and, despite his pass degree and to his amazement, he was accepted. That summer, he worked in various labouring jobs to earn enough to pay his way in Edinburgh [14]. It is tempting to see in these experiences of the youthful Bob Edwards consequences for his later approach to life: he learned that the accepted hierarchies of organisation and ideas were there to be challenged, not simply accepted, and that recovery was possible after what might seem the severest of knock-backs. These were lessons that Bob was to draw on later in his career.

Figure 5. Ruth Fowler in laboratory, Edinburgh, 1950s (courtesy Ruth Edwards).

In Edinburgh, Bob not only started to map out his scientific career, but importantly also met Ruth Fowler (Fig. 5), who was to become his life-long scientific collaborator, and whom he was to marry in 1954, their five daughters following between 1959 and 1964: Caroline, Sarah, Jenny, and twins, Anna and Meg. Bob initially found himself somewhat overwhelmed, even "intimidated" by Ruth's august family background. Her father, Sir Ralph Fowler, FRS (1889–1944) [15], and her maternal grandfather, Lord Ernest Rutherford, FRS (1871–1937) [16], were not only both 'titled', but both also had the most impressive academic credentials imaginable. Ralph Fowler was Plummer Professor of Mathematical Physics in Cambridge from 1932 to 1944, whilst Rutherford was the first Nobel Laureate in Ruth's family, having been awarded the 1908 Nobel Prize in Chemistry 'for his investigations into the disintegration of the elements, and the chemistry of radioactive substances'.

BOB EDWARDS, THE RESEARCH SCIENTIST

The intellectual spirit of scientific enquiry that Bob experienced in Edinburgh obviously fitted his aptitudes perfectly, for Waddington rewarded his Diploma year with a three year PhD place and funded it with the princely sum of £240.00 per year [17]. Bob's chosen field of research was the developmental biology of the mouse. Bob saw that to understand development involved engaging in an interdisciplinary mix, not just of embryology and reproduction, the conventional view at the time, but also of genetics. Given the scientific and social emphasis on genetics over the last 40 or so years, it is difficult now to realise how advanced a view this was in the 1950s, when genetic knowledge was still rudimentary and largely alien to the established developmental and reproductive biologists of the day, as Bob himself was later to comment [18]. For example, it was in the 1950s that DNA was established as the molecular carrier of genetic information [19–22], that it was first demonstrated that each cell of the body carried a full set of DNA/genes [23–25], and that genes were selectively expressed as mRNA to generate different cell phenotypes [26]. Perhaps of greater importance for Bob at that time, it was only by the late 1950s that cytogenetic studies led to the accepted human karyotype as 46 chromosomes [27–28], that agreement was reached on the Denver system of classification of human chromosomes [29], and that the chromosomal aneuploidies underlying developmental anomalies such as Down, Turner and Klinefelter Syndromes were described [30–32].

Figure 6. Alan Gates at a meeting in Cambridge in the late 1950s (possibly 1957).

Bob worked under his supervisor, Alan Beatty, to generate haploid, triploid and aneuploid mouse embryos and studied their potential for normal development. In order to undertake what were, in effect, early attempts at 'genetic engineering' in mammals, he needed to be able to manipulate the chromosomal composition of eggs, sperm and embryos. Whilst in mice, sperm were abundant, eggs were not, and overcoming this deficit led him to two major discoveries that proved to be of later significance. First, with Ruth, they worked to devise ways of increasing the numbers of synchronised eggs recoverable from adult female mice through a series of papers on the control of ovulation induced by use of exogenous hormones [33]. In doing so, they overturned the conventional wisdom that super-ovulation of adult females was not possible. Second, working with an American post-doc, Alan Gates (Fig. 6), Bob described the remarkable timed sequence of egg chromosomal maturation events that led up to ovulation after injection of the ovulatory hormone (human chorionic gonadotrophin; hCG) [34]. His six years, between 1951 and 1957, in Edinburgh give an early taste of his prodigious energy, resulting in 38 papers. Indeed so productive was this period that the last of the papers resulting from his Edinburgh work did not appear in print until 1963.

It was also in Edinburgh that Bob's interest in ethics was first sparked by the interdisciplinary debates among scientists and theologians that Waddington organised, and, as a result, Bob went on what he describes as a "church crawl", trying the ten of so variants of Christianity on offer in 1950s Edinburgh. He did not emerge from his consumer testing "God-intoxicated" [35], but convinced that man held his own future in his own hands. Bob's humanist ethical sympathies were to be developed further in all his later encounters.

AN AMERICAN DIVERSION

These early 1950s studies in science and ethics were to form the platform on which Bob's later IVF work was to be based, but before that his interests and life took a diversion to the California Institute of Technology for the year 1957–8. Bob describes his year at CalTech as being "a bit of a holiday", but it was a holiday which, with hindsight, had distracting consequences. He went there to work with Albert Tyler (1906–1968) [36], an influential elder statesman of American reproductive science, working on sperm–egg interactions. CalTech was then a hotbed of developmental biology, and Tyler had clustered around him an exciting group of young scientists, which included that year a visit by the English doyen of fertilisation, Lord Victor Rothschild, FRS (1910–1990) [37], who was later to clash scientifically with Edwards over his IVF work [38]. In this clash, needless to say, the younger man triumphed [39], just as he had with Pincus. Tyler was exploring the molecular specificity of egg–sperm interactions and had turned as a model to immunology. Immunology was then at a very exciting phase in its development, with the engaging Sir Peter Medawar, FRS (1915–1987, Nobel Laureate in Physiology or Medicine, 1960) [40], influentially for Bob, extending his ideas on immunological tolerance to the paradox of the 'fetus as an allograft': a semi-paternal graft nonetheless somehow protected from maternal immune attack inside the mother's uterus. This confluence of reproduction and immunology excited Bob's restless curiosity and hence the choice of Tyler. The subject also offered funding possibilities via the Ford and Rockefeller Foundations and the Population Council, which were increasingly concerned about world population growth and the need for better methods to control fertility. Immuno-contraception then seemed to offer tantalisingly specific possibilities.

So when Bob returned to the UK from CalTech in 1958 at Alan Parkes' invitation to join him at the Medical Research Council (MRC) National Institute for Medical Research (NIMR) at Mill Hill in north London, it was to work on the science of immuno-contraception [5]. This period in the USA initiated a series of 24 papers on the immunology of reproduction between 1960 and 1976. It also prompted Bob's first involvement in founding an international society in 1967 in Varna, Bulgaria when the International Coordinating Committee for the Immunology of Reproduction was created [41]. It was, in retrospect, to prove a distracting diversion from what was to become Bob's main work, albeit one that continued to enthuse Bob for many years, witnessed not least by my own recruitment to enter this field of study with him as a graduate student in 1966. Nonetheless, the period at Mill Hill, between 1958 and 1962, seems to have been a period of increasing intellectual conflict for Bob. Whilst enthusiastically working on the science underlying immuno-contraception, his old interests in eggs, fertilisation and, in particular, the genetics of development were gradually reasserting themselves. His day job was therefore increasingly supplemented by evening and weekend flirtations with egg maturation.

THE CRUCIAL EGG MATURATION STUDIES

Bob claims that the stimulus reawakening his interests in eggs was provided by the then recent consensus about the number of human chromosomes and, more particularly the descriptions in 1959 of the pathologies in man that resulted from chromosomal anomalies [42]. Might these anomalies result from errors in the complex chromosomal dance that he and Alan Gates had observed in maturing mouse eggs? The possible clinical relevance of his work on egg maturation and aneuploidy in the mouse was becoming significant.

So Bob resumed his experimenting with mice, trying to mimic *in vitro* the *in vivo* maturation of eggs. He tried releasing the immature eggs from their ovarian follicles into culture medium containing the ovulatory hormone hCG, to see whether he could simulate their *in vivo* development. Amazingly he found it worked the first time: but it did so whether or not the hormone had been added. It seemed that the eggs were maturing spontaneously when released from their follicles. And the same happened in rats and hamsters.

If this also happened in humans, then the study of the chromosomal dance during human egg maturation was a realistic practical possibility, as was *in vitro* fertilisation and thereby studies on the genetics of early human development. However, Bob's excitement at seeing eggs spontaneously maturing was temporarily blunted by his discovery that Gregory Pincus in the 1930s [43–44] and M.C. Chang (1908–1991) [45–46] in the 1950s had been there before him, using both rabbit and, Pincus claimed, human eggs.

In order to pursue his basic science studies on maturation, he needed a reliable supply of human ovarian eggs. This requirement posed difficulties for a scientist with no medical qualifications, given the elitist attitudes and scientific illiteracy then prevalent amongst most of the UK's gynaecologists. His break-through came initially with Molly Rose, at whose door in the nearby Edgeware General Hospital he arrived after a recommendation from a fellow kindred spirit in John Humphrey, FRS (1915–1997) [47], ten years Bob's senior and the medically qualified Head of Immunology at Mill Hill. Notwithstanding his more privileged social background, Humphrey shared Bob's passion for science, its social application and utility, and his left wing politics – indeed he had been a Marxist until 1940. Bob asked John if he knew anyone who might be helpful, and John suggested Molly Rose and offered to arrange an introduction. So off Bob went, and Molly Rose provided human ovarian biopsy samples intermittently for the next ten years.

Between 1960 and 1962, Bob tried to repeat and extend Pincus' observations, using not only human but also dog, monkey and baboon eggs, but with such limited success compared with smaller rodents that in a 1962 *Nature* paper [48], he carefully interprets the few maturing human and baboon eggs that he observed as artefacts. But by this time, Bob's quest for human eggs, and his dreams of IVF and studying early aneuploidies in human embryos, had reached hostile ears, most notably those of the then Director of the Institute, Sir Charles Harington, FRS (1897–1972), who banned any work on human IVF at NIMR [49]. Alan Parkes was no longer able to defend Bob, having left in 1961 to take up his chair in Cambridge and, although he had asked Bob to join him, there was no post until 1963. By the time Bob left Mill Hill in 1962 for a year in Glasgow, he had encountered just a taste of the opposition to come.

THE MOVE TO CAMBRIDGE

Bob had been invited to Glasgow University's Biochemistry Department by John Paul, then the acknowledged master of tissue culture in the UK, who had heard of Bob's attempts to generate stem cells from rabbit embryos [18]. The invitation was to result in a paper [50] remarkable for its prescience – the *first of eight landmark papers* (Table 1) that I identify in this contribution. It describes the production of embryonic stem cells from rabbit embryos – capable of proliferating through over 100 generations and of differentiating into various cell types. This report was published some 18 years before Evans and Kaufman described the derivation of ES cells from mouse embryos [51].

That this work has largely been ignored by those in the stem cell field is probably mainly attributable to its being too far ahead of its time. Thus, reliable molecular markers for different types of cells were not available then, nor were appropriate techniques with which to critically test the developmental potential of the cultured cells.

Table 1. The eight landmark papers

1. Cole R.J., *Edwards R.G.*, Paul J. (1965) Cytodifferentiation in cell colonies and cell strains derived from cleaving ova and blastocysts of the rabbit. *Exp. Cell Res.* 37: 501–4.
2. *Edwards R.G.* (1965) Maturation in vitro of human ovarian oocytes. *Lancet* 286: 926–9.
3. Gardner R.L., *Edwards R.G.* (1968) Control of the sex ratio at full term in the rabbit by transferring sexed blastocysts. *Nature* 218: 346–9.
4. *Edwards R.G.*, Bavister B.D., Steptoe P.C. (1969) Early stages of fertilization in vitro of human oocytes matured in vitro. *Nature* 221: 632–5.
5. Steptoe P.C., *Edwards R.G.* (1970) Laparoscopic recovery of preovulatory human oocytes after priming of ovaries with gonadotrophins. *Lancet* 295: 683–9.
6. Steptoe P.C., *Edwards R.G.*, Purdy JM. (1971) Human blastocysts grown in culture. *Nature* 229: 132–3.
7. *Edwards R.G.*, Sharpe DJ. (1971) Social values and research in human embryology. *Nature* 231: 87–91.
8. Steptoe P.C., *Edwards R.G.* (1978) Birth after the reimplantation of a human embryo. *Lancet* 312: 366.

Bob arrived in Cambridge from Glasgow in 1963. He describes how he immediately reacted against the then extant "misogynist public-school traditions; the exclusivity...; the privileges given to the already privileged". But he set against that the "sheer beauty of the place... the concern with the truth and high seriousness... the ambience of scientific excellence... I was surrounded by so many talented young men and women." [52] He continued to pursue both the immunology of reproduction and egg maturation, working furiously on the latter to collect pig, cow, sheep, the odd monkey and some human eggs. Eventually, he was able to show that eggs of all these species would indeed mature *in vitro*, but that the eggs of larger animals simply needed longer than those of smaller ones, human eggs taking some 36 hours rather than the 12 or less hours erroneously reported by Pincus [44]. These cytogenetic studies were reported in two seminal papers in 1965 [53–54], both of which are primarily concerned with understanding the kinetics of the meiotic chromosomal events. As the *second landmark paper*, I have selected the one in *The Lancet*, in which Bob's breath-taking clarity of vision is evident as he sets out a programme of research that predicted the events of the next 20 years and beyond (Table 2). You will notice the heavy focus on the early study and detection of genetic disease compared with the slight emphasis on infertility alleviation, unsurprising given Bob's research interests. Indeed, within three years he had, with my fellow graduate student Richard Gardner, provided proof of principle for preimplantation genetic diagnosis (PGD), in a paper on rabbit embryo sexing published in 1968 [55] and my *third landmark paper*. Later, in the 1980s, Bob was to play a key role in promoting the development of PGD clinically [56], and PGD was to prove a powerful political tool in convincing the UK Parliament to permit research on human embryos. And then by 1969 he had reported the first step towards PGD in humans by describing IVF [57] – a *fourth landmark paper* in as many years.

Table 2. Key points in the programme of research laid out in the Discussion to Edwards' 1965 *Lancet* paper (*landmark paper 2*) [54].
1. Studies on non-disjunction of meiotic chromosomes as a cause of aneuploidy in humans.
2. Studies on the effect of maternal age on non-disjunction in relation to the origins of trisomy 21.
3. Use of human eggs in IVF.
4. Culture of fertilised human eggs in vitro.
5. Use of priming hormones to increase the number of eggs per woman available for study/use.
6. Study of early IVF embryos for evidence of (ab)normality – especially aneuploidies arising prior to or at fertilisation.
7. Control of some of the genetic diseases in man.
8. Control of sex-linked disorders by sex detection at blastocyst stage and transfer of only female embryos.
9. Para-cervical transfer of IVF embryos into the uterus.
10. Use of IVF embryos to circumvent blocked tubes.
11. Avoidance of a multiple pregnancy (as observed after hormonal priming and *in vivo* insemination) by transfer of a single IVF embryo.

THE PROBLEM OF FERTILISATION RESOLVED

Underlying both the 1965 and 1969 papers are two scientific struggles: the first being simply but critically the continuing difficulty in obtaining a regular supply of ovarian tissue. Local Cambridge sources proved unreliable, and Molly Rose was now 2–3 hours' drive away in north-west London, so during the summer of 1965, Bob turned to the USA for help and initiated his now famous contacts with Howard and Georgeanna Jones [58], then at the Johns Hopkins Medical School in Baltimore. This supply of American eggs allowed Bob to confirm the maturation timings published in 1965 [54]. However, it was the second struggle that was by then occupying most of his attention, namely that in order to fertilise these *in vitro* matured eggs, he had to 'capacitate' the sperm. 'Capacitation' is a final maturation process, which sperm will usually undergo physiologically in the uterus, and that is essential for the acquisition of fertilising competence. Failing to achieve this convincingly at Johns Hopkins, he made a second transatlantic summer journey in 1966 to visit Luther Talbot and his colleagues at Chapel Hill. Bob applied his usual ingenuity to try a variety of ways to overcome the problem of sperm capacitation, but no reliable evidence for success was forthcoming [59, 60]. Then in 1968 both struggles began to resolve.

Figure 7. Bob with Bunny Austin, 1960s (courtesy Ruth Edwards).

Resolving the problem of sperm capacitation was the initial attraction to Bob of Patrick Steptoe's laparoscopic technique, Bob seeing it as a way of recovering capacitated sperm from the oviduct [61]. However, the actual solution to this problem lay nearer home. Parkes had retired as Professor in 1967, to be replaced by Colin 'Bunny' Austin (1914–2004) [62] (Fig. 7). In the early 1950s, Bunny, and independently M.C. Chang [45], had discovered the requirement for sperm capacitation [63–64], and so Bunny set his graduate student, Barry Bavister (1943–), to work to try and resolve how to reliably capacitate hamster sperm *in vitro*. Bavister demonstrated a key role for pH in a short paper published in 1969 that showed how higher rates of fertilisation could be obtained by simply increasing the alkalinity of the medium [65]. Bob seized on this observation and co-opted Barry to his project of capacitating human sperm. That proved to do the trick, leading to the 1969 paper [57].

The problem of the intermittent egg supply in the UK was also resolving. Bob continued to rely on surgeons to provide him with ovarian biopsies from which to mature eggs *in vitro*, indeed four are thanked in the 1969 *Nature* paper. According to Bob [66], Molly Rose provided the first group of eggs to be fertilised, and although invited to be a co-author, declined for reasons unknown. Also thanked are Norman Morris [67], Janet Bottomley and Sanford Markham, and although it is not known whether they provided any of the ovarian eggs for *in vitro* maturation described in the paper, Patrick clearly did so [66], and at last provided for Bob a potentially more stable clinical partnership.

The 1969 *Nature* paper describing IVF in humans [57] makes modest claims, only two of 56 eggs reaching the two-pronuclear stage. But, like Bob's other papers, it is a model of clarity, describing well-controlled experiments, cautiously interpreted. This paper convinced where previous claims [68–73] had failed, precisely because the skilled hands and creative intellect that lay behind it are so evident from its text. With its publication, announced to the media on St Valentine's Day [74], all hell was let loose.

THE BATTLES BEGIN

So 1969 seemed to Bob to be a good year. Not only did IVF succeed at long last, and his partnership with Patrick seemed set to flourish, but also so impressed were the Ford Foundation with Bob's work that they paid for him to be awarded a Ford Foundation Readership (a half way step to a professorship) in the University. Elated by his promotion and their achievement, Bob and Patrick pressed on, the latter's laparoscopic skills coming to the fore, first in 1970 with the collection of *in vivo* matured eggs from follicles after mild hormonal stimulation [75], and then achieving regular fertilisation of these eggs and their early development through cleavage to the blastocyst stage [76] – my *fifth and sixth landmark papers*. So well was the work going that in February 1971 they confidently applied to the UK Medical Research Council for funding to bring Patrick to Cambridge from Oldham General Hospital in Greater Manchester, where Patrick worked [77].

However, any illusions that Bob may have had that their achievements would prove a turning point in his fortunes were soon shattered, and just 2 months later on April Fool's Day 1971 the MRC decided to reject the grant application [77]. The practical consequences of this rejection were profound – both psychologically and physically – not least that for the next 7 years, Bob shuttled on the 12 hour round trip between Cambridge and Oldham, leaving Ruth and his five daughters in Cambridge.

The professional attacks on Bob and his work took a number of forms [77], and one must try to make a mental time trip back to the 1960s/70s to understand their basis. Despite the nature of the political and religious battles to come, his scientific and medical colleagues did not focus on the special status of the human embryo as an ethical issue. However, ethical issues were raised professionally, but took quite a different form. Thus, it is difficult now to comprehend the complete absence of infertility from the consciousness of most gynaecologists in the UK at the time, to which Patrick Steptoe was a remarkable exception [78]. Indeed, Bob's strong commitment to treating infertility came to the fore only after he teamed up with Patrick, his previous priority being the study and prevention of genetic and chromosomal disorders. In the several reports from the Royal College of Obstetricians and Gynaecologists and the MRC during the 1960s examining the areas of gynaecological ignorance that needed academic attention, infertility simply did not feature [77]. Overpopulation and family planning were seen as dominant concerns and the infertile were ignored as at best a tiny and irrelevant minority and at worst as a positive contribution to population control. This was a values system that Bob simply could not accept, and the many encouraging letters Bob was to receive from infertile couples provided a major stimulus to his continued work later, despite so much professional and press antagonism and so many set-backs. For his medico-scientific colleagues, however, the fact that infertility was not seen as a clinical issue, meant that any research designed to alleviate it was not viewed as experimental treatment, but as using humans for experiments. Given the sensitivity to Nazi 'medical experiments', and the public reaction and disquiet surrounding the recent publication of '*The Human Guinea-pig*' [79], this distinction was critical. The MRC, in rejecting the grant application, took the position that what was being proposed was human experimentation, and so were very cautious, emphasising risks rather than benefits, of which they saw few if any [77].

Bob and Patrick were also attacked for their willingness to talk with the media. It is even more difficult nowadays, when the public communication of science is so embedded institutionally, to understand how damaging to them this was. The massive press interest of the late 1960s was unabated in the ensuing years, and so Bob was faced with a choice: either he could keep his head down and allow press fantasies and speculations to go unanswered and unchallenged, or he could engage, educate and debate. For Bob this was no choice, regardless of the consequences for him professionally. His egalitarian spirit demanded that he trust to common people's common sense. His radical political views demanded that he fought the corner of the infertile, the underdog with no voice. The Yorkshireman in him relished engagement in the debate and argument. In the *seventh landmark paper* selected, published in *Nature* in 1971 with Dave Sharpe [80], he acknowledges the risk to his own interests of so doing. Risky it clearly was, one of the scientific referees on their MRC grant application starting his referee's report declaring his strong distaste for all the media exposure [77]. Bob was a pioneer in the public communication of science, and paid a heavy price for being so.

The Edwards and Sharpe paper [80] is a *tour de force* in its survey of the scientific benefits and risks of the science of IVF, in the legal and ethical issues raised by IVF, and in the pros and cons of the various regulatory responses to them. It sets out the issues succinctly and anticipates social responses that were some 13–19 years into the future. In subsequent years, Bob built on his strong commitment to social justice based on a social ethic, as he engaged at every opportunity with ethicists, lawyers and theologians, arguing, playing 'devil's advocate' (literally, in the eyes of some), and engaging in what we would now call practical ethics, as he hammered out his position and felt able to fully justify his instincts intellectually.

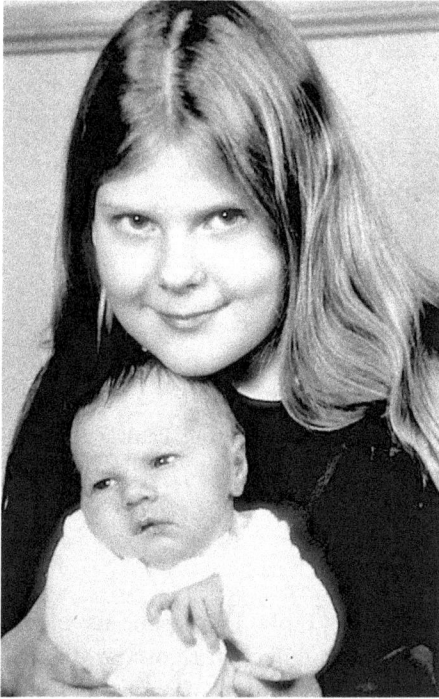

Figure 8. Louise Brown holding the 1000th Bourn Hall baby, 1987 (courtesy Bourn Hall Clinic).

But all this was to little avail. Indeed, Bob was continually frustrated at the unwillingness of most of the establishment to engage seriously in ethical debates in advance of the final validation of IVF that was to come in 1978 with the birth of Louise Brown (Fig. 8), and my *eighth and final landmark* paper [81]. Only then did most UK social hierarchies, such as the MRC, the British Medical Association, the Royal Society and Government move gradually from their almost visceral reactions against IVF and its possibilities to serious engagement with the issues [56]. Then, to their credit, both the MRC and the Thatcher Government of the time came on board, but it was not until 1989, 24 years after Bob's 1965 landmark paper in the *Lancet*, that the UK Parliament finally gave its stamp of approval to his visionary work, and then only after a fierce battle lasting some 11 years [82]. Eleven years since the eighth and final landmark paper was published announcing the birth of Louise Brown. And of course, it has taken 45 years since that 1965 paper for us to be celebrating the award of the Nobel Prize to this remarkable man.

Figure 9. Bob, Jean and Patrick at Bourn Hall, 1981 (courtesy Bourn Hall Clinic).

CONCLUSION

The eight landmark papers I have selected (Table 1) present us with a man of vision and foresight, imagination and intellectual rigour, and extraordinary energy and drive: witness his prodigious output of papers between 1954–2008 [83]. A man who could inspire colleagues to tread with him on a difficult scientific path besieged by public and professional animosity, of whom two in particular, Patrick and Jean, we miss today (Fig. 9). And these papers do not even begin to touch Bob's other academic and personal qualities, so evident in the way he has founded and generously nurtured journals and international societies [84, 85], transforming the intellectual landscape not just of gynaecology, but also of ethics and social anthropology. Truly the 'father of Assisted Reproductive Technology' in its widest interdisciplinary sense: the only sense in which Bob knows it. For Bob, it is truly the Nobel Prize for Physiology AND Medicine: there can be no OR about it.

ACKNOWLEDGEMENTS

I thank the Edwards family for their help in writing this account, for which however I take full responsibility. I also thank Kay Elder and Sarah Franklin for their unfailing wisdom and helpful advice. I thank Andrew Steptoe for permission to reproduce Figure 1, Barbara Rankin for permission to reproduce Figure 2, Ruth Edwards for permission to reproduce Figures 3–7, and Bourn Hall Clinic for permission to reproduce Figures 8 and 9. The research underpinning this account was supported by a grant from The Wellcome Trust [088708], which otherwise had no involvement in the research or its publication. This chapter is adapted from "Robert Edwards: the path to IVF" by M.H. Johnson published in *Reproductive BioMedicine Online* (Elsevier) at doi:10.1016/j.rbmo.2011.04.010 and is included in this volume with permission.

REFERENCES

1. Edwards, R.G., "Patrick Christopher Steptoe, C. B. E. 9 June 1913–22 March 1988," *Biog Mems Fell R Soc*, 1996. 42: p. 435–452.

2. Edwards, R.G. and P.C. Steptoe, *"Preface,"* In: Edwards, R.G., Purdy, J.M., Steptoe, P.C. (eds). Implantation of the Human Embryo, 1985. Academic Press: London, UK, p. vii–viii.

3. Edwards, R.G. and P. Steptoe, *A Matter of Life: The Story of a Medical Breakthrough.* 1980, Hutchinson: London, UK. p. 25.

4. Massey, H. and N. Feather "James Chadwick. 20 October 1891 – 24 July 1974," *Biog Mems Fell R Soc*, 1976. 22: p. 10–70.

5. Ashwood-Smith, M.J., "Robert Edwards at 55," *Reprod BioMed Online*, 2002. 4 (Suppl.1): p. 2–3.

6. Ingle, D.J., "Gregory Goodwin Pincus. April 9,1903–August 22,196," *Biog Mems Natl Acad Sci*, 1971: p. 229–270.

7. Edwards, R.G. and P. Steptoe, *A Matter of Life: The Story of a Medical Breakthrough.* 1980, Hutchinson: London, UK. p. 43.

8. Slee, J., "RGE at 25 – personal reminiscences," *Reprod BioMed Online*, 2002. 4 (Suppl.1): p. 1.

9. Oakley, C.L., "Francis William Rogers Brambell. 1901–1970," *Biog Mems Fell R Soc*, 1973. 19: p. 129–171.

10. Parkes, A.S., "Francis Hugh Adam Marshall. 1878–1949," *Biog Mems Fell R Soc*, 1950. 7: p. 238–251.

11. Polge, C., "Sir Alan Sterling Parkes. 10 September 1900 – 17 July 1990," *Biog Mems Fell R Soc*, 2006. 52: p. 263–283.

12. Edwards, R.G. and P. Steptoe, *A Matter of Life: The Story of a Medical Breakthrough.* 1980, Hutchinson: London, UK. p. 7.

13. Robertson, A., "Conrad Hal Waddington. 8 November 1905 – 26 September 1975," *Biog Mems Fell R Soc,* 1977. 23: p. 575–622.

14. Edwards, R.G. and P. Steptoe, *A Matter of Life: The Story of a Medical Breakthrough.* 1980, Hutchinson: London, UK. p. 18.

15. Milne, E.A., "Ralph Howard Fowler. 1889–1944," *Biog Mems Fell R Soc*, 1945. 5: p. 60–78.

16. Eve, A.S. and Chadwick, J., "Lord Rutherford. 1871–1937," *Biog Mems Fell R Soc*, 1938. 2: p. 394–423.

17. Edwards, R.G. and P. Steptoe, *A Matter of Life: The Story of a Medical Breakthrough.* 1980, Hutchinson: London, UK. p. 20.

18. Edwards, R.G., "An astonishing journey into reproductive genetics since the 1950's," *Reprod Nutr Dev*, 2005. 45: p. 299–306.

19. Watson, J.D. and F.H. Crick, "Genetical implications of the structure of deoxyribonucleic acid," *Nature*, 1953. 171: p. 964–967.

20. Watson, J.D. and F.H. Crick, "Molecular structure of nucleic acids: a structure for deoxyribose nucleic acid," *Nature*, 1953. 171: p. 737–738.

21. Franklin, R. and R. Gosling, "Molecular configuration in sodium thymonucleate," *Nature*, 1953. 171: p. 740–741.

22. Wilkins, M.H.F., A.R. Stokes and H.R. Wilson, "Molecular structure of deoxypentose nucleic acids," *Nature*, 1953.171: p. 738–740.

23. Gurdon, J.B., "Adult frogs derived from the nuclei of single somatic cells," *Dev Biol*, 1962. 4: p. 256–273.

24. Gurdon, J.B., "The developmental capacity of nuclei taken from intestinal epithelium cells of feeding tadpoles," *Development*, 1962. 10: p. 622–640.

25. Gurdon, J.B., T.R. Elsdale and M. Fischberg, "Sexually mature individuals of Xenopus laevis from the transplantation of single somatic nuclei," *Nature*, 1958. 182, p. 64–65.

26. Weinberg, A.M., 2001. "Messenger RNA: origins of a discovery," *Nature*, 2001. 414: p. 485.

27. Tjio, J.H. and A. Levan, "The chromosome number of man," *Hereditas*, 1956. 42: p.1–6.
28. Ford, C.E. and J.L. Hamerton, "The chromosomes of man," *Nature*, 1956. 178: p. 1020–1023.
29. Denver Conference, "A proposed standard system of nomenclature of human mitotic chromosomes," *Lancet*, 1960. 275: p. 1063–1065.
30. Ford, C.E., P.E. Polani, J.H. Briggs and P.M. Bishop, "A presumptive human XXY/XX mosaic," *Nature*, 1959. 183: p. 1030–1032.
31. Ford, C.E., K.W. Jones, P.E. Polani, J.C. De Almeida and J.H. Briggs, "A sex-chromosome anomaly in a case of gonadal dysgenesis (Turner's syndrome)," *Lancet*, 1959. 273: p. 711–713.
32. Lejeune, J., M. Gautier and R. Turpin, "Etude des chromosomes somatiques de neuf enfants mongoliens," *Comptes Rendus Hebd Seances Acad Sci*, 1959. 248: p. 1721–1722.
33. Fowler, R.E. and R.G. Edwards, "Induction of superovulation and pregnancy in mature mice by gonadotrophins," *J Endocr*, 1957. 15: p. 374–384.
34. Edwards, R.G. and A.H. Gates, "Timing of the stages of the maturation divisions, ovulation, fertilization and the first cleavage of eggs of adult mice treated with gonadotrophins," *J Endocr*, 1959. 18: p. 292–304.
35. Edwards, R.G. and P. Steptoe, *A Matter of Life: The Story of a Medical Breakthrough*. 1980, Hutchinson: London, UK. p. 23–4.
36. Horowitz, N.H., C.B. Metz, J. Piatigorsky, L. Piko, J.D. Spikes, and M. Ycas, "Albert Tyler," *Science*, 1969. 163: p. 424.
37. Reeve, S., "Nathaniel Mayer Victor Rothschild, G.B.E., G.M. Third Baron Rothschild. 31 October 1910 – 20 March 1990," *Biog Mems Fell R Soc*, 1994. 39: p. 364–380.
38. Rothschild, "Did fertilization occur?' *Nature*, 1969. 221: p. 981.
39. Edwards, R.G., B.D. Bavister and P.C. Steptoe, "Did fertilization occur?" *Nature*, 1969. 221: p. 981–982.
40. Mitchison, N.A., "Peter Brian Medawar. February 1915 – 2 October 1987," *Biog Mems Fell R Soc*, 1990. 35: p. 282–301.
41. Rukavina, D., "The history of reproductive immunology: my personal view," *Am J Reprod Immunol*, 2008. 59: p. 446–450.
42. Edwards, R.G. and P. Steptoe, *A Matter of Life: The Story of a Medical Breakthrough*. 1980, Hutchinson: London, UK. p. 38.
43. Pincus, G. and E.V. Enzmann, "The comparative behavior of mammalian eggs in vivo and in vitro i. the activation of ovarian eggs," *J Exp Med*, 1935. 62: p. 665–675.
44. Pincus, G. and B. Saunders, "The comparative behavior of mammalian eggs *in vivo* and *in vitro*. VI. The maturation of human ovarian ova," *Anat Record*, 1939. 75: p. 537–545.
45. Greep, R.O., "Min Chueh Chang. *October 10, 1908 — June 5, 1991*," *Biog Mems Natl Acad Sci*. http://www.nap.edu/readingroom.php?book=biomems&page=mchang.html
46. Chang, M.C., "The maturation of rabbit oocytes in culture and their maturation, activation, fertilization and subsequent development in the Fallopian tubes," *J Exp Zool*, 1955. 128: p. 379–405.
47. Askonas, B.A., "John Herbert Humphrey. 16 December 1915–25 December 1987," *Biog Mems Fell R Soc*, 1990. 36: p. 274–300.
48. Edwards, R.G., "Meiosis in ovarian oocytes of adult mammals" *Nature*, 1962. 196: p. 446–450.
49. Edwards, R.G. and P. Steptoe, *A Matter of Life: The Story of a Medical Breakthrough*. 1980, Hutchinson: London, UK. p. 48.
50. Cole, R.J., R.G. Edwards and J. Paul, "Cytodifferentiation in cell colonies and cell strains derived from cleaving ova and blastocysts of the rabbit," *Exp Cell Res*, 1965. 37: p. 501–504.
51. Evans, M.J. and M.H. Kaufman, "Establishment in culture of pluripotential cells from mouse embryos," *Nature*, 1981. 292: p. 154–156.

52. Edwards, R.G. and P. Steptoe, *A Matter of Life: The Story of a Medical Breakthrough.* 1980, Hutchinson: London, UK. p. 51.
53. Edwards, R.G., "Maturation *in vitro* of mouse, sheep, cow, pig, rhesus monkey and human ovarian oocytes," *Nature,* 1965. 208: p. 349–351.
54. Edwards, R.G., "Maturation in vitro of human ovarian oocytes," *Lancet,* 1965. 286: p. 926–929.
55. Gardner, R.L. and R.G. Edwards, "Control of the sex ratio at full term in the rabbit by transferring sexed blastocysts," *Nature,* 1968. 218: p. 346–349.
56. Theodosiou, A.A. and M.H. Johnson, "The politics of human embryo research and the motivation to achieve PGD," *Reprod BioMed Online,* 2011. 22: p. 457–471.
57. Edwards, R.G., B.D. Bavister and P.C. Steptoe, "Early stages of fertilization in vitro of human oocytes matured in vitro," *Nature,* 1969. 221: p. 632–635.
58. Jones Jr., H.W., "From reproductive immunology to Louise Brown," *Reprod BioMed Online,* 2002. 4(Suppl.1): p. 6–7.
59. Edwards, R.G., R.P. Donahue, T.A. Baramki and H.W. Jones Jr., "Preliminary attempts to fertilize human oocytes matured in vitro," *Am J Obstet Gynec,* 1966: 96, p. 192–200.
60. Edwards, R.G., L. Talbert, D. Israelstam, H.N. Nino, and M.H. Johnson, "Diffusion chamber for exposing spermatozoa to human uterine secretions," *Am J Obstet Gynec,* 1968. 102: p. 388–396.
61. Edwards, R.G., interviewed in: *To Mrs. Brown a daughter,* 1980. Peter Williams TV: The Studio, Boughton, Faversham, UK.
62. Anon, "Colin Austin," *Austral Acad Sci Newsletter,* 2004. 60: p.
63. Chang, M.C., "Fertilizing capacity of spermatozoa deposited into the fallopian tubes," *Nature,* 1951. 168: p. 697–698.
64. Austin, C.R., (1951) "Observations of the penetration of sperm into the mammalian egg," *Austral J Sci Res Series B,* 1951. 4: p. 581–596.
65. Bavister, B.D., "Environmental factors important for *in vitro* fertilization in the hamster," *Reproduction,* 1969. 18: p. 544–545.
66. Edwards, R.G., and P. Steptoe, *A Matter of Life: The Story of a Medical Breakthrough.* 1980, Hutchinson: London, UK. p. 81–83.
67. "*Morris, Prof. Norman Frederick,*" In: Who Was Who, A & C Black, 1920–2008: Oxford University Press. http://www.ukwhoswho.com/view/article/oupww/whowaswho/U28190
68. Rock, J. and M. Menkin, "In vitro fertilization and cleavage of human ovarian eggs," *Science,* 1944. 100: p. 105–107.
69. Shettles, L.B., "A morula stage of human ovum developed in vitro," *Fert Steril,* 1955. 9: p. 287–289.
70. Petrov, G.N., "[Fertilization and first stages of cleavage of human egg in vitro]," *Arkhiv Anatomii, Gistologii i Embriologii,* 1958. 35: p. 88–91.
71. Yang, W.H., "[The nature of human follicular ova and fertilization in vitro]," *J Jap Obstet Gynec Soc,* 1963. 15: p. 121–130.
72. Petrucci, D., "Producing transplantable human tissue in the laboratory," *Discovery,* 1961. 22: p. 278–283.
73. Hayashi, M., "Fertilization in vitro using human ova," In: *Proceedings of the 7th International Planned Parenthood Federation, Singapore,* 1963. Excerpta Medica International Congress Series No. 72: Amsterdam, Netherlands. pp. 505–510.
74. "New step towards test-tube babies," Nature-Times News Service. *The Times, 1969.* Friday, Feb 14: p. 1.
75. Steptoe, P.C. and R.G. Edwards, "Laparoscopic recovery of preovulatory human oocytes after priming of ovaries with gonadotrophins," *Lancet,* 1970. 295: p. 683–689.
76. Steptoe P.C., R.G. Edwards and J.M. Purdy, "Human blastocysts grown in culture," Nature, 1971. 229: p. 132–133.

www.ingramcontent.com/pod-product-compliance
Lightning Source LLC
Chambersburg PA
CBHW061744210326
41599CB00034B/6783